Natural Experiments

American and Comparative Environmental Policy

Sheldon Kamieniecki and Michael E. Kraft, series editors

A complete list of books published in the American and Comparative Environmental Policy series can be found at the end of this book.

Natural Experiments

Ecosystem-Based Management and the Environment

Judith A. Layzer

The MIT Press
Cambridge, Massachusetts
London, England

For information about special quantity discounts, please email special_sales@ mitpress.mit.edu

This book was set in Sabon by SPi Publisher Services, Puducherry, India.
Printed on recycled paper and bound in the United States of America.

Library of Congress Cataloging-in-Publication Data

Layzer, Judith A.
Natural experiments : ecosystem–based management and the environment / Judith A. Layzer.
    p. cm. — (American and comparative environmental policy)
Includes bibliographical references.
ISBN 978-0-262-12298-6 (hardcover : alk. paper) — ISBN 978-0-262-62214-1 (pbk. : alk. paper) 1. Ecosystem management. 2. Ecology. I. Title.

QH75.L367   2008
333.95—dc22

2008017039

10 9 8 7 6 5 4 3 2 1

Contents

Series Foreword

Social scientists are taught early on in their careers about the importance of challenging conventional wisdom in their research. They are instructed to never assume anything until rigorous, balanced inquiry is conducted. Blindly adopting a particular orientation can restrict the use of competing theoretical and methodological approaches and can lead to false findings. Many if not most of the important breakthroughs that have been reported in the social sciences (and other fields) tend to debunk a major assumption about some aspect of society. Findings drawn from survey research and reported in early studies of voting behavior and public opinion, for example, challenged long-held beliefs about the kinds of citizens who identify with a political party as well as who consider themselves political independents. Likewise, contemporary policy research has questioned previously held notions concerning the role of public opinion and interest groups in American politics and policymaking and has significantly improved our knowledge of influence, power, and decisionmaking.

An assumption held by many environmental policy scholars is that collaboration among diverse interests and constituencies in conflicts over natural resources and pollution control is "a good thing" because it tends to facilitate bargaining, negotiation, and compromise. Thus, instead of allowing environmental conflicts to endure for a very long time, collaboration among contending parties at the local or regional level is assumed to shorten the time required to reach a settlement. The literature also implies that final decisions from collaborative processes are likely to generate effective environmental policies which will endure over time. Beginning in the 1990s, these assumptions have led to a widespread call among policy analysts, as well as government officials, to decentralize environmental policymaking and to pursue the development and implementation of collaborative strategies to resolve environmental and natural resource disputes at the local and regional level.

Judith Layzer's outstanding study presents an in-depth and comprehensive examination of whether the expectations underlying collaborative, place-based environmental problem solving are empirically valid. Her goal is to improve understanding of how ecosystem-based management (EBM) works in practice. EBM, according to her, "entails collaborative, landscape-scale planning and implementation that is flexible and adaptive." Professor Layzer analyzes seven prominent cases of such landscape-scale initiatives, four of which involve extensive use of EBM. The case studies possess a number of key similarities and differences, allowing her to draw insightful generalizations across a variety of terrestrial and aquatic collaborative efforts in different parts of the country. Her analytic approach makes it possible for her to determine whether EBM produces in practice the benefits promised in theory. Specifically, she investigates to what extent, how, and under what circumstances EBM leads to sustained, environmentally protective policies and practices that represent improvements on the status quo and are likely to conserve natural resources and improve environmental quality.

The findings reported in this book are quite surprising given the results of previous research, and they are likely to lead to intense debate over the value and efficacy of EBM. Professor Layzer finds that the seven case studies of EBM initiatives she examines have produced "land use or natural resource management plans that are more holistic and comprehensive than the piecemeal approaches they replaced." Moreover, each case produced concrete achievements that furthered environmental protection and natural resource conservation. However, her study finds that "the initiatives whose goals were set in collaboration with stakeholders have produced environmental policies and practices that are less likely to conserve and restore ecological health than those whose goals were set through conventional politics." Her book presents a number of explanations for these findings and discusses the implications of the results for future efforts to enhance environmental quality at the local and regional level.

The analyses presented in this book illustrate well our purpose in the MIT Press series in American and Comparative Environmental Policy. We encourage work that examines a broad range of environmental policy issues. We are particularly interested in volumes that incorporate interdisciplinary research and focus on the linkages between public policy and environmental problems and issues both within the United States and in cross-national settings. We welcome contributions that analyze the policy dimensions of relationships between humans and the environment from either a theoretical or empirical perspective. At a time when environmental policies are increasingly seen as controversial and new approaches are being implemented widely, we especially encourage studies that assess policy successes and failures, evaluate new institutional arrangements and policy tools, and clarify new directions for environmental politics and policy. The books in this series are written for a wide audience that includes academics, policymakers, environmental scientists and professionals, business and labor leaders, environmental activists, and students concerned with environmental issues. We hope they contribute to public understanding of environmental problems, issues, and policies of concern today and also suggest promising actions for the future.

Sheldon Kamieniecki, *University of California, Santa Cruz*
Michael Kraft, *University of Wisconsin-Green Bay*
American and Comparative Environmental Policy Series Editors

Preface and Acknowledgments

This book grew out of my impatience with the euphoria that accompanied the explosion of collaborative, place-based, environmental problem-solving in the 1990s. I was familiar with the myriad critiques of the regulatory framework erected in the 1970s, and like many others I found the idea that we could avert some of the existing system's worst failings appealing. The more I looked for credible evidence that newer approaches actually benefited the environment, however, the more frustrated I became. After working to furnish some of that evidence, I understand why so few people have undertaken such a project. Ecosystem-based management (EBM) initiatives, my particular focus, are extraordinarily complicated; they unfold over a very long period of time; and it is difficult to document their results. Because each is invented in its own place, comparison among them can be hazardous. Nevertheless, that's precisely what I have tried to do in this book. My goal is straightforward and pragmatic: I want to improve our understanding of how EBM works in practice, so that we can get on with restoring the ecological resilience of the landscapes we rely on for our sustenance, both physical and spiritual.

Although I cannot pretend to be neutral, I have done my best to be transparent in the criteria I used to select and interpret data. The fierce and admirable commitment of those who have been devising and implementing the programs I looked at made it difficult to render a dispassionate (and ultimately critical) assessment. What I hope to make clear is that participants in these "experiments" are working under real and potent political constraints, many of which are rooted in structural power differences. To this end, I have tried to acknowledge their many valuable achievements, even as I question the overall effectiveness of the EBM initiatives I studied.

A few additional caveats may be helpful. I have written for a broad audience that includes not just scholars but also planners, administrators, elected officials, and advocates. Therefore, in composing the cases I have tried to do two things at once: tell a compelling story and provide a systematic analysis of the evidence. To spare the reader excessive citation, I generally have not referenced sources of widely confirmed information. I have refrained from quoting or citing the people I interviewed unless they were my only source for a claim, to enable them to be as honest as possible. In my effort to maintain a practical focus, I have not asked readers to wade through extensive theoretical discussions about some topics—such as collaboration, deliberation, and consensus–building—that deserve fuller treatments. I urge readers who are interested to pursue the literature cited in the chapters that follow.

Finally, a word on the title, *Natural Experiments.* When I was becoming interested in the topic of EBM, I noted that conservation biologist Reed Noss and others were referring to such initiatives as "experiments" in a cautionary way, to remind observers that it was not clear whether they would, in fact, benefit the environment. Legal scholar Brad Karkkainen (2001/2002), an EBM enthusiast, urged scholars to "take a serious and sustained critical look at each project as a potentially generalizable experiment rather than a purely sui generis local response to a particular problem" (211), noting that the most interesting and robust of these efforts appear to share elements of a common institutional architecture. He advised engaging in "a rigorous process of identifying and evaluating the unique strengths and weaknesses of each project, subjecting each to careful monitoring and comparative benchmarking to isolate factors that appear to contribute to their success, as well as those that are obstacles to success" (211). It is in the spirit of both of these comments that I refer to the cases described in this book as "experiments."

Like every author, I am indebted to an enormous number of people and organizations. I began the research for this book in 2003, during a sabbatical that was generously funded by a Domestic Public Policy grant from the Smith Richardson Foundation. A timely grant from the Humanities and Social Sciences fund at MIT enabled me to finish up the data collection.

At the inception of this project, DeWitt John, Craig Thomas, and the late Steve Meyer gave helpful feedback on my research design. In constructing the cases I interviewed more than 100 people, sometimes taking up several hours of their time. Without their help, I could never have understood how events really unfolded. I also relied heavily on the work of local journalists who have done yeoman's work in documenting these initiatives. Chris Bosso and Steve Trombulak provided pointed advice on early chapter drafts. After I had completed data collection for the original cases, David Fairman encouraged me to consider adding comparison cases with more "hopeful" outcomes. I owe a serious debt to Craig Thomas, who not only commented extensively on the research design but also made suggestions on an early draft *and* provided an extraordinarily focused critique of the final draft. Thanks are also due to the following people, who furnished maps or other hard-to-get material: Art Brandt, Catherine Byrd, Sue Carnevale, Kevin Connally, Mark Hanna, Mary Ruth Holder, and Jan Loftin.

Two anonymous reviewers for MIT Press provided helpful direction for a final round of revisions. Series editors Michael Kraft and Sheldon Kamieniecki were encouraging throughout the process. Clay Morgan and his assistant, Meagan Stacey, shepherded the manuscript through the review process. Copy editor Beth Wilson and senior editor Katherine Almeida attended to the details with great care.

Closer to home, I have been fortunate to have the support of my colleagues at MIT, particularly JoAnn Carmin, Robert Fogelson, Lang Keyes, Larry Susskind, and Larry Vale. In addition, my research assistants at MIT have been extremely helpful. Kate Van Tassel and Sharlene Leurig retrieved data for several of the cases. Jess Burgess, Molly Mowery, and Rachel Henschel searched for images and checked references. Alexis Schulman was a gem: she gathered case material and rounded up missing information with irrepressible good cheer. Xenia Kumph provided all kinds of administrative help, always graciously acceding to my last-minute requests. Students in my seminar on the Politics of Ecosystem Management provided thoughtful reactions to my ideas. More generally, my students have consistently been a steady source of delight and hope.

I want to extend special thanks to those who painstakingly reviewed individual chapters: Maeveen Behan, David Braun, Carolyn Campbell, Jim Canaday, Tony Davis, Mark Hanna, Bill Kier, Mark Kraus, Joseph Koebel Jr., Clif Ladd, Kent Loftin, Sam Luoma, Geoffrey McQuilken, John Ogden, Jerre Stallcup, Tina Swanson, and Lou Toth. The chapter reviewers did not always agree with my interpretation of events, but they were patient in their efforts to set me straight. Despite their best efforts, there are surely inaccuracies that remain, and I am wholly responsible for those.

As always, my family and friends have taken good care of me, made me laugh, and accepted me as I am—which is about as much as anyone could ask for. I am especially grateful to Liz Phillips, who went above and beyond in reading the entire manuscript and smoothing out infelicitous passages. Also, I want to thank two people who came into my life at a difficult time: Jeannine Sudol and Nancy Goldstein. They are living proof of silver linings. Finally, this book is dedicated to the many people who are working to make ecosystem-based management work on the ground. Without them we would really be in trouble.

Natural Experiments

1

Introduction

Since the 1980s the theory and practice of natural resource management have undergone a profound transformation. The environmental movement of the late 1960s and early 1970s spawned an avalanche of federal statutes, as well as state laws and local ordinances, that addressed the environmental problems caused by more than a century of industrialization. By the early 1980s, however, a three-pronged critique of the newly instituted regulatory framework had emerged. Detractors charged that centralized decision making produced uniform rules that did not reflect local conditions, ignored interrelationships among natural system elements, and stifled innovation; top-down, expert-driven regulation prompted local resistance and endless rounds of legislative, administrative, and judicial appeals; and inflexible mandates resulted in minimal compliance and an inability by regulators and the regulated community to learn or adjust to new circumstances. Yet even as these allegations gained political traction, observers were documenting a spate of innovations in environmental problem-solving that, according to their proponents, promised to reinvigorate efforts to mitigate human impacts on the natural world.

Among the most potentially revolutionary of the new approaches was ecosystem management, now more commonly known as ecosystem-based management (EBM).[1] Scholars and practitioners have offered dozens of formal definitions, but most agree that at a minimum EBM entails collaborative, landscape-scale planning and implementation that is flexible and adaptive (Cortner and Moote 1999; Grumbine 1994, 1997; Hartig et al. 1998; B. R. Johnson and Campbell 1999; Karkkainen 2002a, 2002b; Keiter 1998; Meffe et al. 2002; Szaro, Sexton, and Malone 1998). Although EBM shares attributes with many of the other environmental problem-solving approaches that emerged in the 1980s and

1990s—particularly an emphasis on decentralization, holism, collaboration, and flexibility—it is distinct from its various cousins in some important respects, the most important being the scale at which problems are addressed (Cestero 1999) and the nature of government involvement (Koontz et al. 2004).[2] For example, the efforts that Edward Weber (2000, 2003) terms "grassroots ecosystem management" (GREM), which aim to change the culture rather than the rules of a place, are typically initiated by residents of rural, western communities threatened by disputes over natural resource extraction, mostly on public lands. By contrast, EBM—as more commonly defined—tends to be instigated by government officials and seeks to institutionalize new forms of governance to address pollution and resource management problems. EBM initiatives span large landscapes that may encompass marine or other aquatic ecosystems, publicly and privately owned land, and urban as well as rural areas.

The ecosystem-based approach has gained particular prominence in the United States and elsewhere because it promises to coordinate the activities of jurisdictions and agencies with disparate missions, integrate management of public resources with stewardship of the surrounding matrix of private land, and facilitate policy learning and adjustment. It has the potential to resolve the apparently intractable controversies that accompany our ubiquitous sprawling, resource-depleting pattern of development. Because of the concept's broad appeal, during the 1990s a host of nongovernmental organizations, professional societies, federal agencies, and state officials endorsed ecosystem-based approaches to land-use and natural resource policy-making (see, for example, Beattie 1996; Christensen et al. 1996; Dombeck 1996; USEPA 1994; Interagency Ecosystem Management Task Force 1995; NAPA 1995; PCSD 1996; Society of American Foresters 1993; J. W. Thomas 1996; Western Governors' Association 1998). In the 2000s scientists, managers, and advocates began promoting EBM for marine systems as well (McLeod et al. 2005; Pew Oceans Commission 2003; U.S. Commission on Ocean Policy 2004). Yet despite widespread enthusiasm for EBM, scholars have not provided systematic evidence of its efficacy in practice—until recently, few initiatives had existed long enough for evaluators to assess their substantive benefits, and of those few, their complexity and heterogeneity made evaluation particularly challenging.

That said, in recent years scholars have been analyzing aspects of EBM, particularly the effects of stakeholder collaboration on natural resource planning and management. They have ascertained that, consistent with proponents' claims, watershed collaboratives and other participation-intensive problem-solving efforts *do* appear to increase human and social capital, as well as the level of stakeholder agreement (Beierle and Cayford 2002; Gunton, Day, and Williams 2003; Huntington and Sommarstrom 2000; Innes et al. 1994; Leach, Pelkey, and Sabatier 2002; Lubell 2005; Weber 2003). In addition, many participatory schemes have taken concrete steps—such as implementing restoration projects and instituting monitoring and education/outreach programs—toward their environmental goals (Huntington and Sommarstrom 2000; Imperial and Hennessey 2000; Leach, Pelkey, and Sabatier 2002).

Researchers have not discerned a clear relationship between these two central achievements, however; in fact, some analysts suggest that funding levels and the passage of time, rather than trust and social capital, are the keys to successful implementation (Beierle and Cayford 2002; Leach and Sabatier 2005; Raymond 2006).[3] Others argue that the context, rather than the internal characteristics of a collaborative group, largely determines a community's willingness to implement a collaboratively formulated plan (Koontz 2005). More important, scholars have been unable to document a causal relationship between collaboration and improved environmental conditions, despite widespread agreement that the most important measure of success is achievement of on-the-ground environmental benefits beyond what would have occurred anyway (Beierle and Cayford 2002; Born and Genskow 1999; Imperial and Hennessey 2000; Kenney 2000; Leach, Pelkey, and Sabatier 2002; Lubell 2004; O'Leary, Nabatchi, and Bingham 2004). In short, although existing empirical work highlights a small number of variables that appear to be correlated with "success," serious gaps remain in our understanding of whether, how, and under what conditions collaborative governance arrangements yield genuine environmental improvements. Systematic evidence of the efficacy of landscape-scale planning and flexible, adaptive implementation is even more elusive.[4]

In an effort to fortify and build on existing scholarship, this book investigates seven efforts to conserve and restore terrestrial or aquatic landscapes, with the goal of ascertaining whether ecosystem-based management produces in practice the benefits promised in theory. More precisely, it asks: to what extent, how, and under what conditions does EBM yield durable, environmentally protective policies and practices that (1) constitute improvements on the status quo and (2) are likely to conserve and restore ecological health?

I chose ecosystem-based management as a category for exploration for several reasons. Of all the new approaches to environmental problem-solving,

EBM is arguably the most likely to achieve environmentally protective results. Unlike national-level decision making, working at a landscape scale facilitates tailoring remedies to the particular ecological and socioeconomic conditions of a specific region. At the same time, landscape-scale efforts can take into account many of the factors—particularly critical ecological processes or functions, pollution that crosses political boundaries, and features of the larger economic or regulatory context—that typically overwhelm the efforts of local jurisdictions. Moreover, regional initiatives may be more likely than local ones to muster the financial and technical capacity to commission and implement sophisticated scientific assessments, as well as the resources to monitor policy implementation.

On the other hand, there are some potentially significant tradeoffs in moving from a local to a landscape scale. In particular, collaboration among stakeholders seems most likely to produce social and human capital when citizens bound by attachment to a particular place can engage in face-to-face deliberation; by contrast, large-scale projects rely on interest-group representatives, whose capacity to speak on behalf of their "constituents" may be limited (Cestero 1999). The trick for EBM initiatives, then, is to capture the purported advantages of working at a landscape scale while harnessing at least some of the benefits of engaging stakeholders.

The importance—in fact, the urgency—of assessing whether and how EBM is likely to conserve or restore the health of natural systems is clear. No economic or social system can survive in the long run if it destroys the resilience of the ecosystems it depends on (Arrow et al. 1995; Daly 1997; Diamond 2005; Rees 2000). As global climate change advances—bringing with it rising sea levels, changing patterns of precipitation, and more severe storms—landscapes will need to be more resilient, not less so. Nevertheless, we are degrading the landscape at an accelerating rate, and the cumulative effects of human activity are becoming increasingly severe and irreversible (Lubchenco 2002; Millennium Ecosystem Assessment 2003; Noss and Scott 1997; Orians 1995; Pew Oceans Commission 2003; U.S. Commission on Ocean Policy 2004). Although the global pursuit of sustainability is essential, the United States is particularly culpable here, since American lifestyles depend heavily on the appropriation of resources from other countries and future generations (Beatley 1998; Wackernagel and Rees 1996).

In the remainder of this book I provide a detailed analysis that supports the following general conclusion. On the one hand, all seven of the initiatives I examine have generated land-use or natural resource management plans that are more holistic and comprehensive than the piecemeal approaches they replaced. Each also boasts concrete achievements, such as the public acquisition of ecologically valuable land. On the other hand, comparison among the cases reveals that the initiatives whose goals were set in collaboration with stakeholders have produced environmental policies and practices that are less likely to conserve and restore ecological health than those whose goals were set through conventional politics.

The initiatives in which goals were set collaboratively have yielded fewer-than-anticipated environmental benefits for a variety of reasons. Above all, to achieve consensus, planners promised to pursue environmental and economic goals simultaneously. To this end, they reframed problems in ways that allowed them to avoid tackling controversial issues or seriously considering policies that would impose short-run costs on development interests. They also adopted technology- and management-intensive solutions that aim to "expand the pie," in the process imposing substantial risk on the environment. In some cases, efforts to implement plans' provisions exposed disagreements that had been glossed over during the collaborative process, resulting in stalemate and delay. Because of insufficient funding and inadequate margins for error in the plans themselves, flexible policy tools and a rhetorical commitment to adaptive management appear unlikely to compensate for these shortcomings.

By contrast, the initiatives in which goals emerged out of conventional politics have yielded greater-than-expected environmental benefits because political officials—judges, administrators, or elected officials—employed political capital and regulatory authority to promote an overarching, environmentally protective goal. Such pro-environmental leadership, which typically occurred in response to lawsuits or campaigns to raise the salience of an environmental problem, enhanced the influence of precautionary interpretations of science and established strict floors below which plans could not fall. It thereby mitigated the disparity in power between development and environmental interests. It also induced a positive feedback, as environmentally protective policies and practices yielded tangible benefits around which new constituencies formed.

A Road Map

In the chapters that follow, I lay out the empirical basis for and elaborate on the argument summarized above, drawing on evidence from seven cases that all involve efforts to conserve or restore ecosystems but vary in the extent to which they include the three elements—a landscape-scale focus, collaborative planning, and adaptive implementation—that constitute full-fledged EBM. In chapter 2, I describe the impetus for EBM in greater detail and propose two models of EBM—optimistic and pessimistic—derived from the writing of scholars and practitioners. I also explain my criteria for choosing cases and assessing the consistency of the evidence in those cases with each model.

In chapters 3 through 6, I analyze four nationally recognized EBM projects. The first two cases involve efforts to protect and restore *terrestrial* ecosystems in rapidly urbanizing regions of the Southwest: Austin's Balcones Canyonlands Conservation Program (BCCP, chapter 3) and the San Diego Multiple Species Program (MSCP, chapter 4). Both of these cases concern habitat conservation planning initiatives that were sparked by the listing of endangered songbirds, and both were cited by prominent federal officials as exemplars of the ecosystem-based approach. Chapters 5 and 6 describe two highly publicized efforts to conserve or restore *aquatic* ecosystems—the Comprehensive Everglades Restoration Plan (CERP, chapter 5) and the California Bay–Delta Program (CALFED, chapter 6). In all four cases, efforts to appease conflicting interests by meeting all demands simultaneously yielded minimally protective plans and halting implementation. As a result, although each project has produced impressive advances in scientists' understanding of damaged ecosystems and has enhanced localities' ability to raise money for environmental improvements, in their current form none are likely to conserve or restore the landscapes they aim to protect.

In chapters 7 through 9, I analyze three comparison cases. Chapter 7 describes Pima County, Arizona's Sonoran Desert Conservation Plan (SDCP), which, like the MSCP and the BCCP, was triggered by the proposed listing of an endangered species. By contrast with the MSCP and the BCCP, however, Pima County officials, not stakeholders, took the lead in devising the SDCP. Moreover, from the outset they portrayed the plan as primarily a mechanism for conserving Pima County's biological diversity, not simply meeting the legal requirements of the Endangered Species Act; they also imposed strict restraints on development until the plan's details were finalized, despite the vociferous objections of development interests. As a result, the SDCP's habitat preserve hews closely to the configuration prescribed by the county's scientific advisory team.

In the last two cases—the efforts to restore Florida's Kissimmee River (chapter 8) and California's Mono Basin (chapter 9)—proponents pursued ecological restoration through conventional politics, such as salience campaigns, lawsuits, and expert planning. Political officials responded by supporting a single, environmentally protective goal and employing regulatory tools to ensure that goal was met. These two cases affirm that when planners focus primarily on ecological restoration, even if doing so provokes resistance, they can achieve genuine environmental improvements.

Finally, in chapter 10, I examine the similarities and differences among all seven cases to evaluate the benefits and drawbacks of landscape-scale planning, stakeholder collaboration, and flexible, adaptive implementation; situate those findings among the claims of theorists, as well as the conclusions of scholars who have evaluated EBM or its constituent elements; and raise some methodological caveats about which claims I am most and least confident of. I conclude by suggesting some implications of my findings for advocates and policymakers, and proposing policy changes that might improve their ability to promote environmentally beneficial outcomes.

2

Why Ecosystem-Based Management?

Before diving into the cases, this chapter describes how the concept of ecosystem-based management (EBM) emerged in the 1980s in response to perceived deficiencies in the existing environmental policymaking system. In particular, critics decried regulators' focus on individual economic sectors or activities, media, or species, and their consequent inability to address more complex and diffuse "second generation" problems. Many scholars converged on an ecosystem-based approach as a remedy. In general, their definitions of EBM featured three elements: a landscape-scale focus; collaborative planning that engages all stakeholders; and flexible, adaptive implementation of planning goals. Proponents cited numerous benefits of EBM, but after an initial wave of enthusiasm skeptics began questioning whether it would really produce the promised results. Drawing on these competing perspectives, I develop two simple models of EBM: optimistic and pessimistic. I conclude by outlining the methods I used to compare real-world EBM with each of these models.

Drawbacks of the Conventional Regulatory Approach

EBM is, in many respects, a logical outgrowth of frustration with the existing regulatory system, which is an uneasy blend of local and national land-use and natural resource policies. Although the conventional approach has produced notable environmental improvements, many observers argue it is unlikely to solve the nation's remaining environmental problems. Ecologists and conservation biologists complain that the traditional regulatory framework implicitly treats complex, diffuse phenomena as if they are separable into problems that are well bounded, clearly defined, and linear with respect to cause and effect.

Policy analysts point out that although centralized decision making that generates uniform rules accompanied by penalties for noncompliance has been effective at curbing the harmful practices of huge industries, it is unwieldy for addressing problems attributable to the habits of individuals and small businesses. They also draw attention to the political liabilities of the "decide–announce–defend" model, in which decision making is contentious and polarizing, stalemate is common, and the policies that result are poorly implemented and ceaselessly subject to challenge.

The Scientific Foundations of Environmental Regulation

The applied-biology community has been both a source of criticism of conventional regulation and a wellspring of the EBM concept. Even as the federal government was passing the nation's landmark environmental laws in the early 1970s, the scientific ideas on which those laws were based were becoming obsolete. A series of dramatic conceptual and value shifts within applied biology—acceptance of change as natural and of humans as part of natural systems, recognition that ecosystems are open rather than closed, and a focus on biodiversity and ecological integrity rather than commodity production—underpinned a profound critique of conventional natural resource management.

According to the classical, equilibrium-based paradigm in ecology, the natural world is "tightly organized, interdependent, and highly coevolved" (Barbour 1996, 233); furthermore, ecological systems are closed, self-regulating, and subject to a single, stable equilibrium. Ecological change occurs by succession, which leads to stable climax communities; disturbances, which are rare, push succession back to earlier stages; and humans are harmful additions to natural systems (Pickett and Ostfeld 1995). As historian Donald Worster (1994, 202) explains, from the equilibrium-based perspective "the ultimate goal of nature...is nothing less than the most diverse, stable, well-balanced, self-perpetuating society that can be devised to meet the requirements of each habitat." Ecologist Eugene Odum (1972) spelled out the implications of the classical model: nature is fragile, finite, and interdependent, and human intervention poses the main threat to the "homeostasis" toward which ecosystems tend. In the 1960s and 1970s, these ideas—captured by the "balance of nature" metaphor—became entrenched in popular thinking about ecology (Kempton, Boster, and Hartley 1995) and were embedded in federal statutes and regulations (Tarlock 1996).

The classical perspective furnished the rationale for two very different but equally problematic approaches to natural resource management: tight control and benign neglect (Botkin 1990; Holling and Meffe 1996; Pickett and Ostfeld 1995).[1] According to ecologist C. S. Holling (1995, 8), tight control over ecosystem attributes that normally fluctuate leads to "systems [that are] more likely to flip into a persistent, degraded state triggered by disturbances that previously could be absorbed." The consequences of a host of conservation-era policies—including forest, range, and fishery management—reveal the pitfalls of tight control: native species and the ecosystems they depend on have declined precipitously (Holling 1995; Ludwig, Hilborn, and Walters 1993; Noss and Cooperrider 1994). Benign neglect, which is often practiced in nature preserves, has had similarly poor results: exploding wildlife populations, confined within park boundaries, have decimated vegetation and then undergone precipitous declines (Botkin 1990; Chase 1986).

During the 1970s and 1980s, ecologists S. T. A. Pickett and P. S. White pioneered a new ecological perspective based on empirical observations that suggested several important deficiencies in the classical paradigm.[2] According to the "flux-of-nature" view, most ecosystems are *not* self-regulating, but experience important limits from external sources. Stable equilibria occur infrequently, and successions are rarely deterministic; instead, they are affected by the specifics of a particular ecosystem. Disturbances—such as fire, floods, drought, and storms—play a central role in shaping ecosystem dynamics, even though such events may occur infrequently relative to the scale of human lifetimes. And landscapes that have not experienced human influences have been the exception for hundreds, if not thousands, of years (D. A. Perry and Amaranthus 1997; Pickett and Ostfeld 1995). Taking these insights into account led ecologists to adopt a revised view of ecosystems as moving targets that are open to exchange of organisms, material, and energy from outside and whose futures are uncertain and unpredictable (Allen and Hoekstra 1992; Holling 1996; Pickett and White 1985). As Worster (1994, 394) explains: "Nature should be regarded as a shifting landscape of vegetative patches of all textures and colors, a veritable patchwork quilt of living things, changing continually through time and space, responding to an increasing barrage of perturbations."

Scientists caution that although the flux-of-nature view "emphasizes processes, dynamics, and context, rather than endpoint stability" (Meffe and Carroll 1994, 269), not all changes are consistent with maintaining an ecosystem's integrity and resilience—and therefore its long-term health. Fluctuations in the natural world have functional, historical, and evolutionary limits, and human-induced changes can greatly exceed those limits in their rate and extent. For example, species diversity is generally enhanced by disturbances that occur at intermediate levels of frequency and intensity but do not exceed the capacity of the system to recover between disturbances. Diversity is reduced, however, by disturbances to which the species in the system are not adapted or which destroy habitats faster than they can recover (D. A. Perry and Amaranthus 1997).

Parallel to the upheaval within ecology was the emergence in the 1980s of a new discipline, conservation biology, partly in reaction to a growing emphasis in ecology on formal theory, prediction, quantification, and modeling—all with the aim of achieving academic respectability, not real-world problem-solving. By contrast, conservation biology is explicitly interdisciplinary and policy relevant; it combines historical, comparative, and experimental approaches at scales appropriate to policy concerns. Conservation biology also arose out of a sense of urgency about the extinction crisis and the belief that scientists had the responsibility to do more than merely study phenomena (Soulé 1985). Therefore, unlike traditional scientists, its practitioners are more concerned with avoiding Type II errors (rejecting a true or useful hypothesis) than averting Type I errors (accepting a false hypothesis)—that is, they prefer to put the burden of proof on those who would impose environmental risk rather than on the defenders of natural systems (Holling 1995; Schrader-Freschette 1996).

Moreover, unlike its predecessors in applied natural resource management fields, whose goal is to maximize natural resource production within the constraints of natural systems, conservation biology aims to protect biodiversity (Noss and Cooperrider 1994; D. A. Perry and Amaranthus 1997).[3] In the long run, say conservation biologists, efforts to manage for maximum production of food or fiber lead inevitably to brittle, vulnerable ecosystems. By contrast, management that aims to retain or restore biological diversity results in better functioning, more resilient ecosystems (Christensen et al. 1996; D. A. Perry and Amaranthus 1997).

The Limits of Local Land-Use and Water Management

As scientists have become more outspoken in their criticisms of conventional environmental regulation, many planners and regulators have formulated a parallel critique. They point out that the nation's strong tradition of local autonomy in the realms of land-use and water management has thwarted efforts to address sprawl and its attendant environmental and social problems. They observe that local government control over zoning and development rules has led to fragmented, piecemeal land conservation efforts. And they charge that water management has been parochial and shortsighted, driven by a concern with facilitating local urban and agricultural development without regard for its cumulative environmental impacts.

The Futility of Local Land-Use Planning The environmental consequences of the traditional system of land-use decision making are plainly evident: despite pervasive "smart growth" rhetoric, new roads carve up the landscape surrounding cities; suburban development surrounds tiny habitat islands; and wetlands are eradicated to make way for strip malls. John Turner and Jason Rylander (1997, 60) describe a view of America from the air, in which

Cul-de-sac subdivisions accessible only by car—separated from schools, churches, and shopping—spread out from decaying cities like strands of a giant spider web. Office parks and factories isolated by tremendous parking lots dot the countryside. Giant malls and business centers straddle the exit ramps of wide interstates where cars are lined up bumper to bumper. The line between city and country is blurred. Green spaces are fragmented. Only a remnant of natural spaces remains intact.

Such images are not limited to any particular part of the country: between 1970 and 1990, the population of Los Angeles grew 45 percent, whereas the metropolitan land area increased 300 percent (Turner and Rylander 1997). During the same period, in the metropolitan regions of New York, Chicago, Philadelphia, Boston, Detroit, Washington, D.C., Cleveland, and St. Louis, the urbanized area grew by nearly 39 percent, but population grew by only 3.4 percent (Wievel, Persky, and Senzik 1999). As a result of these sprawling development patterns, traffic congestion has worsened, air and water quality have deteriorated, and wildlife habitat has become fragmented and degraded (Beatley 2000; Burchell et al. 2005; Kahn 2000).[4]

In theory, local governments can do a lot to curb sprawl and promote environmentally friendly development: they can craft comprehensive land-use plans that delineate and protect habitat, aquifer discharge zones, and other environmental features; use their zoning powers to encourage dense development of existing urban centers and to protect undeveloped land; and employ tax incentives, impact fees, congestion pricing, and

other financial mechanisms to direct development away from environmentally sensitive areas. In practice, however, few localities engage in formal planning, and even fewer have stringent growth management rules.
So, for example, in his evaluation of local plans in Florida, Samuel Brody
(2003) finds that despite strong interest in ecosystem management at the
state and regional levels, local jurisdictions have not inserted ecosystem
management principles—such as ecological goals, coordination with
other jurisdictions, and policies that promote conservation of ecological
values—into their comprehensive plans.

In any case, local rules may not succeed in curbing sprawl (Logan and
Zhou 1989). For instance, local growth management policies in southern
California have hardly slowed the pace of development because growth
advocates simply have adopted new tactics to resist and redefine restrictions on their operations (Warner and Molotch 2000). More generally,
Anthony Downs (2005) argues that "Political resistance to raising densities in existing neighborhoods and blocking outward extension of future
growth makes it unlikely that a metropolitan area will adopt a broad
smart growth program."

State growth management laws improve local outcomes only marginally, in part because they bear the heavy imprint of concessions made to
developers, local governments, industry, and utilities during their draft-
ing (Popper 1981). In addition, most are discretionary rather than mandatory, and most also contain generous waiver provisions (Keiter 1998).
Even in the handful of states with stringent growth management laws—
such as Oregon, Vermont, and Florida—localities continue to emphasize
development or at least fail to incorporate progressive environmental
provisions into their plans, particularly if they face no sanctions for failing to meet state standards. Richard Norton (2005) finds that even after
more than two decades under the state's widely touted Coastal Area
Management Act of 1974, North Carolina's coastal resources continue
to decline as a result of the cumulative impacts of low-density development. More generally, planning scholars have found that although strict
state growth management policies can improve the quality of local plans,
promote compact development, and reduce sprawl, their effects are modest and may be swamped by other factors, such as population, income,
and household size (Anthony 2004; Burby and May 1997; Carruthers
2002; Wassmer 2006; Yin and Sun 2007).

The main stumbling block to curbing sprawl, even in the context of state
growth management laws, is the overwhelming pressure on local officials

to promote economic growth (Logan and Molotch 1987; P. Peterson 1981). Even the most progressive localities are constrained in the extent to which they can engage in environmentally protective policymaking by their inability to control the flow of capital across their borders. As a result, local officials routinely approve upzoning—increasing the density allowed on a property—which creates value for developers and rewards land speculation. By contrast, inner city construction is often more costly and time-consuming than developing outlying land because of the myriad requirements for compatibility with neighboring uses.

Of course, some local governments do adopt environmentally protective practices, but when each jurisdiction makes decisions independent of its surrounding communities, the results can be perverse. Because municipal borders rarely correspond to ecological boundaries, piecemeal local decisions do not add up to the coherent, landscape-scale protection that biologists contend is necessary for healthy ecosystems. Moreover, since the wealthiest communities are more likely to enact protective bylaws than their poorer neighbors, many critics allege such efforts are motivated by exclusivity, and deplore them because they push unwanted land uses toward those communities with the least capacity to resist them (Bruegmann 2005).

The Detrimental Impacts of Traditional Water Management   Just as local land-use decision making has come under attack, numerous critics deride the traditional approach to water management. Historically, water management in the United States has been driven by and, if anything, has exacerbated the local desire for unfettered growth. Since the nineteenth century local water agencies have fostered agricultural and urban expansion, and water policy—although often administered by state and federal agencies—has reflected that emphasis. As Robert Gottlieb and Margaret Fitzsimmons (1991) explain, in the late 1800s eastern cities such as Boston and New York confronted limits to their local water supplies and petitioned their respective states for rights to distant water sources, as well as subsidies and taxing powers to finance the development and maintenance of those sources. In the South and Midwest, urban and agricultural interests turned to the Army Corps of Engineers to harness local rivers for navigation and flood control. In the West, development interests led by the railroads sought federal support to capture and transport water, prompting the formation of the Department of the Interior's Bureau of Reclamation. According to Gottlieb and Fitzsimmons (1991, 3), however, "federal intervention neither centralized policy decisions nor established the priority of a national agenda. . . . the choices that framed the activities and set the purpose of water development remained locally rooted."

Numerous commentators have documented the environmental impacts of growth-driven water management. Dams, erected to divert water for irrigation or urban use or to generate hydroelectric power, have destroyed stream habitat, changed water temperatures and degraded water quality, blocked fish runs, and prevented the movement of sediment, nutrients, and organisms downstream. In most river systems, dams also interfere with the interaction between the river and its banks—a relationship that provides critical nursery grounds and food sources for many kinds of fish. As a result, "Species of fish once so numerous as to be legendary are now on the brink of extinction" (Grossman 2002, 1). For example, in the early 1800s the Columbia River teemed with Chinook salmon. Although overfishing took a heavy toll on salmon populations, and careless land management practices badly damaged their spawning habitat, the largest single cause of the salmon decline on the Columbia was the hydroelectric dams built in the 1930s and 1940s. Above the Bonneville Dam, which is 145 miles from the Pacific, only 50 miles of the 1,214-mile-long river remain free-flowing; the rest has become "a series of placid, computer-regulated lakes" (Wilkinson 1992, 195). As a consequence, historic Chinook runs have declined 75 to 85 percent, and many stocks are facing extinction.

About 600,000 miles of what were once free-flowing rivers now lie stagnant behind dams, while less than 1 percent of the nation's river miles are protected in their natural state (Grossman 2002). So it's hardly surprising that, according to William Graf (1999, 1309), "The construction and operation of dams has already had greater hydrological and ecologic impacts on American rivers than any changes that might reasonably be expected from global climate change in the near future."[5] Recognition of the ecological and human consequences of dams has spawned intense resistance to the construction of new dams (Leslie 2005; McCully 2001), and in the United States, advocates have seized on relicensing requirements for dams that furnish hydropower to promote ecological restoration (Grossman 2002; Lowry 2003).

Although more attention has focused on dams, critics note that diversions of water for irrigation and urban water supply, as well as river channelization and levee-building to facilitate navigation and prevent seasonal flooding, have also been pernicious (Schneiders 1999). For example, a combination of huge federal water-storage and conveyance projects and myriad smaller dams, irrigation diversions, and groundwater pumps extracts so much water from the Colorado River that in most years it runs dry before reaching the Gulf of California. Moreover, when wastewater is returned to the river, it is often laden with salts, sediments, and chemicals; it also raises water temperatures, rendering streams inhospitable to cold-water fish. Channelization and levee-building on the Mississippi River have drastically altered sedimentation and flow regimes. One result is that the Mississippi River delta wetlands—which constitute nearly 40 percent of the total coastal salt marsh in the lower 48 states—have been disappearing at a rate of 25 square miles per year (UCS 2006). In total, at least 90 percent of the water discharged from U.S. rivers is strongly affected by channel fragmentation caused by dams, reservoirs, diversion, and irrigation (Jackson et al. 2001).

The cumulative impacts of water development are even more serious than the individual effects, as Noss and Cooperrider (1994, 265) explain:

Because water moves throughout landscapes, aquatic systems tend to integrate and reflect all that is being degraded at a regional scale. Excessive erosion of topsoil, salinization of water from irrigation, runoff of pesticides from fields and oil from roads, and influx of inadequately treated wastes and sewage all accumulate downstream where they are compounded and often have synergistic effects.

The impacts of upstream diversion and pollution often culminate in estuaries such as the Chesapeake Bay, which was once among the world's most productive. Today, even after more than 30 years of intensive efforts to rehabilitate the Chesapeake, it remains badly degraded. The 2006 report card by the nonprofit Chesapeake Bay Foundation pegged the Bay's health at 29—only two points above the score it had received for the previous three years—indicating that it continues to be nearly 75 percent less productive than it was prior to European settlement.

The Perceived Limits of Federal Regulation

Although conceived as a way to compensate for deficiencies in local land-use and water management, the federal environmental regulatory framework has come in for its share of criticism as well. While acknowledging that the mandates enacted in the 1970s to protect the nation's natural resources have been responsible for notable environmental improvements, detractors charge they are inefficient and produce unintended consequences (Bardach and Kagan 1982; Crandall 1983; Landy, Roberts, and Thomas 1994; Melnick 1983). Others point out that, substantively, federal environmental statutes are narrowly cast and aim at the largest, most visible offenders. They are therefore poorly suited to problems whose causes are multiple, complex, diffuse, and variable, such as habitat destruction and non-point-source water pollution (Esty and Chertow 1997; Graham 1999; Mazmanian and Kraft 1999). Economists complain that fixed, uniform-rule-oriented mechanisms encourage reacting to, rather than preventing, pollution and resource damage; do not reflect variation in local ecological conditions; and inhibit efforts to go beyond minimum levels of protection (Davies and Mazurek 1997; Hockenstein, Stavins, and Whitehead 1997). Critics also decry the traditional top-down, expert-driven approach to standard-setting, combined with adversarial procedures for resolving disputes. Such processes, they say, provoke contestation and litigation over rules, polarization among contending factions, and government paralysis. Furthermore, the usual "decide–announce–defend" approach—in which agency experts devise rules with only a perfunctory public comment period—stifles creativity and yields solutions that are likely to be resisted locally and challenged by both advocacy organizations and industry (Dryzek 1990; Karkkainen 2002; Mazmanian and Kraft 1999; Moote and McClaran 1997; Susskind and Cruikshank 1987; Weber 1998).

Critics have taken particular aim at the Endangered Species Act (ESA), arguably the nation's most stringent environmental law. The ESA, which prohibits destruction of species or their habitats once species are listed as endangered, boasts some substantial achievements: it has ensured the protection of millions of acres of habitat and has played an important role in rescuing more than a dozen species, including the iconic bald eagle, from the brink of extinction (Barringer 2004). Nevertheless, although they recognize the utility of the ESA as a legal tool, most conservation biologists (and many environmental activists) contend that because it aims to protect individual species rather than biological processes or landscapes, it is not the ideal mechanism for conserving biodiversity (Angermeier and Karr 1994; Meffe and Carroll 1994).[6] In fact, conservation biologist Dennis Murphy (1999, 232) argues that the law's "myopic focus on individual development projects has led to further species attrition and fragmentation of remaining habitat." Making matters worse, federal officials are reluctant to enforce the ESA, particularly on private lands, where many of the nation's endangered species reside. Although the U.S. Fish and Wildlife Service has interpreted section 9 of the act as prohibiting the destruction of a listed species or its habitat, the agency has neither the budget nor the political clout to enforce this provision: it is extremely difficult to get a conviction for damaging habitat, and efforts to do so often result in "expensive, protracted administrative and legal battles" (Dennis Murphy 1999, 232). Moreover, federal officials are well aware of the political sensitivity of trying to enforce the ESA on private land (Beatley 1994; Durbin and Larmer 1997).

Critics have also assailed the Clean Water Act. Like the ESA, the Clean Water Act is responsible for some impressive accomplishments. It has prompted declines in industrial discharges and the widespread construction of sewage treatment plants, resulting in a 50 percent reduction in municipal "loading" (Houck 1999).[7] On the other hand, a study by the U.S. Public Interest Research Group confirmed previous findings by the Environmental Protection Agency (EPA) and Government Accountability Office that many major facilities are in significant noncompliance with their Clean Water Act permits (Leavitt 2006). Furthermore, the Clean Water Act is notoriously ineffective at addressing non-point-source pollution, an ever-larger portion of the total. Section 303—the total maximum daily load (TMDL) provision, which requires every state to establish water quality standards and then devise a plan for achieving them—is the EPA's best weapon for addressing non-point-source pollution. But the TMDL provision is barely functioning: after years of delay, in 2000 the agency issued a draft TMDL rule in response to lawsuits by environmentalists.[8] The rule provoked such a furor that Congress suspended it and asked the National Research Council (NRC) for an evaluation. In 2001, after the NRC released a critical report, the Bush administration postponed the TMDL rule's effective date, and in 2003 rescinded the rule altogether.

The federal government has been similarly reluctant to use section 404 of the Clean Water Act to halt the loss of wetlands. Under section 404 the U.S. Army Corps of Engineers and the EPA jointly administer the program that dispenses permits to fill wetlands. Between 1996 and 1999, the Corps approved 99.7 percent of all permit applications, authorizing 85 percent of those within 14 days. Between the program's inception in 1979 and 1999, the EPA, which can veto a Corps decision, rejected only 11 of 150,000 permit applications (Zinn and Copeland 2001). Because of a slowdown in agricultural conversion since the 1970s, urban and rural development accounted for more than 60 percent of freshwater wetlands loss in the United States between 1998 and 2004.[9]

Finally, the federal government has been loath to challenge state water rights rules, and therefore unwilling to use the Clean Water Act to improve river flows.[10] The goal of the act is "to restore and maintain the chemical, physical and biological integrity of the Nation's waters," but the federal agencies that administer the act have focused almost exclusively on chemical pollution, neglecting rivers' biological health (Postel and Richter 2003). As a result, it is the Endangered Species Act, rather than the Clean Water Act, that has provided the impetus for the most ambitious efforts to protect and restore river flows.

According to detractors, federal efforts to regulate environmentally damaging development of land and water are not only ineffective, they have also generated a rancorous anti-environmental reaction. In the early 1990s, that backlash came to a head as property rights organizations teamed up with Wise Use groups to challenge federal enforcement of the ESA and Clean Water Acts (Layzer 2006; Switzer 1997). Although few of the anti-environmentalists' legislative and judicial efforts succeeded, the movement did manage to intimidate federal agency officials and create a regulatory climate hostile to new efforts at environmental protection. Moreover, the politics surrounding land-use and natural resource decision making became deeply polarized as adversaries challenged one another continuously, in turn creating policy instability, political stalemate, and local resistance to implementation.

One Remedy: Ecosystem-Based Management

Even as the conventional environmental regulatory framework was coming under fierce attack, new strategies were emerging that many believed could address the panoply of concerns raised by critics of the status quo. One remedy was an approach that has become known as ecosystem-based management—a set of practices that does not supplant, but rather is added to the existing regulatory framework. Whereas the conventional system of land-use and natural resource policymaking involves implementing a disjointed and often conflicting mix of local and national regulations, EBM entails purposefully bringing stakeholders together on a subnational but supralocal scale to devise a single, holistic plan that is tailored to the conditions within a specific region. Furthermore, unlike the conventional approach—which critics describe as hierarchical and rigid—federal, state, and local partners implement EBM cooperatively, using "flatter," more flexible and adaptive governance arrangements. As a result, according to its proponents, EBM promises not only to end the political stalemate that afflicts efforts to address environmental problems, but also to yield more effective and durable solutions than the existing combination of national-level regulatory schemes and local-level land-use and water management.

Not everyone has been sanguine about EBM, however. Skeptics worry that it will yield lowest-common-denominator solutions whose implementation is likely to be stymied by the same obstacles that have impeded efforts at environmental protection to date: a systemic bias toward short-run economic considerations and a related unwillingness to acknowledge the fundamental lifestyle changes required to attain ecological health.

Defining Ecosystem-Based Management

Since it emerged in the 1980s, the concept of EBM has evolved as scholars and practitioners have become increasingly concerned with implementation. Early proponents of ecosystem management—many of whom were scientists—adopted an ecocentric view that emphasized the overarching goal of ecological health or integrity. Based on his authoritative surveys of the literature on ecosystem management in 1994 and 1997, environmental educator R. Edward Grumbine (1994, 31) concluded that "Ecosystem management integrates scientific knowledge of ecological relationships within a complex sociopolitical and values framework toward the general goal of protecting native ecological integrity over the long term." More specifically, according to Grumbine, ecosystem management aims to (1) maintain viable populations of all native species in situ; (2) represent, within protected areas, all native ecosystem types across their natural range of variation; (3) maintain evolutionary and ecological processes; (4) manage over periods of time long enough to maintain the evolutionary potential of species and ecosystems; and (5) accommodate human use and occupancy within these constraints.

Similarly, forest ecologist Jerry Franklin (1997, 27) asserted that ecosystem management involves managing ecosystems to assure their environmental sustainability, defined as "the maintenance of the potential of our terrestrial and aquatic ecosystems to produce the same quantity and quality of goods and services in perpetuity." Policy analyst Chris Wood (1994) put it more bluntly: "To embrace the ecosystem management concept is to accept that ecological factors, such as maintaining biological diversity, ecological integrity, and resource productivity dictate strict limits on social and economic uses of the land" (7).

Shortly after the concept of ecosystem management emerged, however, proponents concerned about local resistance began incorporating an explicit element of stakeholder collaboration into their definitions (Duane 1997), although most continued to assume that a shared interest in ecological health would emerge from that process (Cortner and Moote

1999). For instance, according to the federal Interagency Ecosystem Management Task Force (1995, 3), although ecosystem management is applied within a geographic area defined primarily by ecological boundaries, its goals should emerge from a "collaboratively developed vision of desired future conditions that integrates ecological, economic, and social factors." Similarly, planners Szaro, Sexton, and Malone (1998, 3) advised: "Overall, the mandate should be to protect environmental quality while also producing resources people need. Therefore, ecosystem management cannot simply be a matter of choosing one over the other." And Jamie Clark (1999), director of the U.S. Fish and Wildlife Service, cautioned: "It's crucial to work with the people who live in a place when devising solutions; that's the only way to foster a new relationship with the land."

According to a 1998 review by legal scholar Robert Keiter, by the mid-1990s most definitions of ecosystem management shared six principles: (1) goals are socially defined through a shared-vision process that incorporates ecological, economic, and social considerations; (2) management requires coordination among federal, state, tribal, and local entities, as well as collaboration with other interested parties; (3) planning and management are based on integrated and comprehensive scientific information that addresses multiple resources; (4) management aims to maintain and restore biodiversity and sustainable ecosystems; (5) management occurs at large spatial and temporal scales to accommodate the dynamic and sometimes unpredictable nature of natural processes; and (6) management is adaptive—that is, it entails establishing baseline conditions, monitoring, reevaluating, and adjusting to new information. Although subsequent definitions have generally been consistent with Keiter's, many scholars and practitioners now prefer the term "ecosystem-based management," noting that the phrase "ecosystem management" conjures up an image of managing ecosystems rather than the human beings who live in those systems (McLeod et al. 2005).

To make my analysis tractable, I focus on three core attributes of EBM that, taken together, should facilitate the conservation and restoration of ecosystem health. First and foremost, EBM involves *addressing problems at a landscape, or regional, scale*. As forest ecologists David Perry and Michael Amaranthus (1997, 49) explain, "The critical role of landscapes and regions in buffering the spread of disturbances, providing pathways of movement for organisms, altering climate, and mediating key processes such as the hydrologic cycle means that the fate of any one piece of ground is intimately linked to its larger spatial context." Nature does not provide a rigid system for classifying or demarcating ecosystems or landscapes, however. Furthermore, as noted earlier, ecosystems are not closed with respect to exchanges of organisms, matter, and energy. Drawing boundaries for EBM is therefore a pragmatic exercise that depends on both an area's natural features and the issues that prompt the problem-solving effort (Yaffee et al. 1996). All that said, a crucial feature of EBM is its emphasis on planning and managing for multiple ecological elements and processes, which typically involves coordinating activity across political boundaries.

Second, EBM entails *collaborative planning*, in which public officials, private stakeholders, and scientists assemble voluntarily to seek consensus on a solution that promises joint gains (Susskind and Cruikshank 1987).[11] In most collaborative planning processes, participants deliberate with the aim of reaching consensus, generally defined as a willingness by all to accept the decision of the group.[12] The theoretical appeal of consensus is that because each party has a veto, power is equalized and openness is encouraged (Foster 2002; Pellow 1999); the unanimity requirement lowers the risk of discussing problems and considering corrective actions (Kenney 2000).

Third, EBM relies heavily on *flexible, adaptive implementation* of planning goals. A flexible implementation strategy is one that employs information, incentives, performance standards, and voluntarism, rather than prescriptive rules and deterrence (Fiorino 2004). Adaptive management, in its ideal form, begins with the establishment of baseline conditions and the identification of gaps in knowledge about a system; next, scientists devise management interventions as experiments that test clearly formulated hypotheses about the behavior of the system and monitor the results of those interventions; finally, managers modify their practices in response to information gleaned from monitoring (Gunderson, Holling, and Light 1995; Holling 1978; K. N. Lee 1993; Meffe et al. 2002).

The Hypothetical Benefits of Ecosystem-Based Management: An Optimistic Model

Proponents have amassed a host of persuasive arguments that comprise an optimistic model of how these three attributes of EBM, working in concert, should yield effective and durable solutions (see table 2.1). According to this model, EBM produces environmental improvements because a landscape-scale focus facilitates a comprehensive approach and coordination among agencies and jurisdictions. Collaboration with stakeholders inspires innovative plans that are both durable and grounded in the best available

Table 2.1
The Optimistic Model of EBM

EBM Attribute	Hypotheses	Observable Implications
Landscape-Scale Focus • Planning for multiple ecological elements and processes at a supra-local but sub-national scale.	Integrative scientific assessment→awareness of relationships among ecosystem elements and processes→develop ment of a comprehensive plan.	Planners use more holistic, less parochial language over time; plan addresses main causes of ecosystem decline.
	Coordination among agencies and jurisdictions→Consistent goals and actions.	Representatives of agencies and jurisdictions meet regularly, consult with one another, engage in projects, issue joint reports, and/or establish common environmental objectives.
Stakeholder Collaboration • Allowing stakeholders to establish rules of engagement, define issues, design data collection and analysis, and help develop solutions.	Face-to-face deliberation →trust among stakeholders→transformation of interests→ innovative plan.	Stakeholders perceive other participants as trustworthy; stakeholders perceive common interests; plan adopts novel or rarely used measures.
	Use of local knowledge combined with agreement on plan's scientific basis→ feasible plan grounded in the best available information about the ecosystem.	Scientists solicit opinions of local naturalists; stakeholders agree on scientific basis for the plan; the plan is consistent with the recommendations of scientists.
	Stakeholders' involvement in crafting plan→perceptions of fairness and legitimacy→plan that is likely to be implemented without challenge.	Stakeholders perceive the plan as fair and legitimate; implementation proceeds without legislative, administrative, or judicial challenge.

Table 2.1
(continued)

EBM Attribute	Hypothesis	Observable Implications
Flexible, Adaptive Implementation • Relying on flexible policy tools and adaptive management.	Emphasis on flexible mechanisms→efforts by managers and stakeholders to go beyond legally required minimum levels of environmental protection.	Managers and stakeholders adopt more environmentally protective practices than what is legally required.
	Adaptive management→ adjustment in the face of new information→implem entation is consistent with the best available understanding of the ecosystem	Scientists synthesize monitoring data and communicate results to managers; managers modify practices based on information garnered from monitoring; practices reflect the current scientific understanding of the ecosystem.

information about a particular ecosystem. And a reliance on flexible, adaptive implementation facilitates responsiveness to new information and encourages efforts to go beyond the legally required minimum.

A Landscape-Scale Focus

According to proponents of an optimistic model of EBM, addressing problems at a landscape scale offers two main environmental benefits. First, by prompting an integrative scientific assessment that captures the interrelationships among an ecosystem's main elements and functions, landscape-scale planning raises policymakers' and stakeholders' awareness and knowledge of critical ecological processes. As a result, they are more likely to design solutions that are holistic and comprehensive—and therefore more effective at conserving biological diversity—than are uniform national-level policies (Christensen et al. 1996; Meffe and Carroll 1994; Dennis Murphy 1999; Noss 1983; J. M. Scott et al. 1999).

Second, addressing problems at a landscape scale forces managers in different agencies and jurisdictions to make their actions consistent with one another (Keiter 2003; C. Thomas 2003). Such coordination, in turn, alleviates the problems that arise when federal and state agencies managing resources within a single ecosystem pursue inconsistent policies (Freeman and Farber 2005). It also averts the "death by a thousand cuts" that occurs when localities make decisions that disregard spillovers across jurisdictional boundaries and that facilitate urban sprawl (Beatley and Manning 1997; Beatley 1998; Drier, Mollenkopf, and Swanstrom 2001; Orfield 1997).

Stakeholder Collaboration

According to the optimistic model, collaborative planning offers several environmental benefits beyond those that accrue as a result of landscape-scale problem-solving. In the conventional regulatory approach, professionals in centralized bureaucracies develop rules with only perfunctory citizen involvement—a process that fosters distrust and cynicism. By contrast, collaborative processes promise to transform participants' views of their adversaries and of their own interests. Over time, as they engage repeatedly in face-to-face deliberation, participants discover shared values and come to trust and respect one another (Axelrod 1984; Barber 1984; Dryzek 1990; Innes and Booher 1999; Kemmis 1990; Larmer 1996; Susskind and Cruikshank 1987). In turn, stakeholders come to understand that their well-being is intimately tied to that of the community (Barber 1984; Healey 2006; Weber 2000). Central to this view is the notion that

communicative planning has the potential to transform material conditions and established power relations through the continuous efforts to "critique" and "demystify"; through increasing understanding among participants and hence highlighting oppressions and "dominatory" forces; and through creating well-grounded arguments for alternative analyses and perceptions—through actively constructing new understandings (Healey 1993, 243–244).[13]

In theory, by building trust and transforming interests, and hence power relations, collaborative interactions enable participants to discover unanticipated alternatives and develop innovative solutions (Brunner 2002; Dryzek 1990; Innes 1996, 1999; Sabel, Fung, and Karkkainen 2000; Wondollek and Yaffee 2000). As Phil Brick and Ed Weber (2001, 18) explain: "Instead of a system premised on hierarchy, collaboratives devolve significant authority to citizens, with an emphasis on voluntary participation and compliance, unleashing untapped potential for innovation latent in any regulated environment."

In addition, according to the optimistic model collaboration yields solutions that are more effective at solving environmental problems because the process incorporates more and better information, and does so more thoroughly, than top-down approaches. Unlike decision making by narrowly trained experts, collaboration draws not only on traditional science but also on local knowledge, which is based on extended, close observation of how an ecosystem behaves (Berkes 1999, 2004; Brunner et al. 2005; Fischer 2000). In doing so, it helps filter out the biases and broaden the perspectives of experts, while simultaneously enhancing the technical expertise of citizens (Brick and Weber 2001; Hillman, Aplin, and Brierley 2003; Ozawa 1991; Sabel, Fung, and Karkkainen 2000; Susskind and Cruikshank 1987; Wondollek and Yaffee 2000). Moreover, participants in a deliberative forum are likely to consider the full variety of available information because reasoning—rather than tactics—predominate (Andrews 2002; Barber 1984; Ehrmann and Stinson 1999; Freeman 1997; Ozawa 1991). By contrast, adversarial techniques "create barriers to a full airing and reconciliation of disputed scientific and technical points and contested political claims and, in fact, encourage a distortion of the issues and debates" (Ozawa 1991, 28).

A third theoretical benefit of collaboration is that plans are more likely to be implemented fully and without challenge than plans devised using a more conventional approach. Whereas federal rules often fit awkwardly within a particular local context, plans devised by stakeholders are likely to be feasible in practice because they incorporate firsthand knowledge of local socioeconomic and cultural conditions (Knopman, Susman, and Landy 1999; Wondollek and Yaffee 2000). Furthermore, those who must implement a collaboratively devised plan are likely to feel a sense of ownership and perceive it as legitimate, and hence be committed to its implementation (Gunton, Day, and Williams 2003; Innes 1999; Wondollek and Yaffee 2000). As Gordon and Coppock (1997, 44) point out: "The inclusiveness of the process broadens the base of support, making it harder for diehard opponents to overturn agreements as soon as they see a political advantage." By contrast, locals tend to perceive mandates issued by federal officials as unfair and illegitimate, and hence to resent and resist them (Bardach and Kagan 1982; Susskind and Cruikshank 1987; Sax 2000).

Flexible, Adaptive Implementation
According to the optimistic model, flexible, adaptive governance arrangements promise at least two major environmental benefits. In theory, flexibility fosters a sense of stewardship among regulated entities, increasing the likelihood they will take protective measures that exceed minimum legal requirements (Fiorino 2004; Gunningham 1995). By contrast, say critics of the status quo, command-and-control regulation appears unreasonably burdensome and arbitrary, so provokes resistance or efforts to circumvent

the rules. Those who do comply are likely to engage in the minimum legally necessary remediation (Fiorino 2004; Freeman 1997; Ruhl 1995).

Adaptive management improves problem solving by promoting continuous learning, which is essential given the inherently unpredictable nature of dynamic ecosystems, the unanticipated effects of management interventions, and our limited ability to understand natural processes (Holling 1995, 1996; Holling and Meffe 1996; K. N. Lee 1993). The expectation is that through the continuous infusion of new information, managers will, over time, be able to reduce the risk and uncertainty associated with managing natural systems (Stankey et al. 2003). By contrast, the uniform rules promulgated in the conventional mode impede efforts to respond to new information about the environment or even to discern the effect of a particular management intervention.

The Potential Pitfalls of Ecosystem-Based Management: A Pessimistic Model

Notwithstanding the avalanche of positive press that has accompanied EBM initiatives, skeptics fear that many such efforts are at best a waste of time and money and at worst an alternative means by which development interests can dominate policymaking (see table 2.2). According to the pessimistic model, EBM will not yield environmentally protective solutions because the mechanisms touted by proponents rarely, if ever, work as promised. Instead, EBM undermines environmental protection efforts by drawing limited resources away from or disabling the tools—such as administrative appeals, lawsuits, and public relations campaigns—that historically have been environmentalists' most effective weapons.

A Landscape-Scale Focus
According to the pessimistic model, trying to solve problems at a landscape scale is unlikely to result in environmentally effective plans; instead, the benefits of greater comprehensiveness may be swamped by the effect of forces emanating from beyond the region, such as transboundary air pollution, population growth, and economic globalization (Bradshaw 2003; Paehlke 2001). Further undermining plans' protectiveness, say critics, economic interests exert more influence at state and local levels than at the national level, especially in times of recession (Bradshaw 2003; Layzer 2002). For example, scholars have found that national forests are better protected than state forests because historically states have responded

Table 2.2
The Pessimistic Model of EBM

EBM Attribute	Hypotheses	Observable Implications
Landscape-Scale Focus • Planning for multiple ecological elements and processes at a supra-local but sub-national scale.	Development interests dominate→avoidance of measures that impose costs on development interests.	Planners reject proposals that impose costs on development interests.
	Institutional and other barriers→failure to cooperate by agencies & jurisdictions.	Representatives of agencies and jurisdictions act independently; do not regularly meet, consult, produce joint reports, or undertake joint projects.
Stakeholder Collaboration • Allowing stakeholders to establish rules of engagement, define issues, design data collection and analysis, and help develop solutions.	Effort to reach consensus→ marginalization of extremes→lowest-common-denominator solutions.	Extreme views or alternatives are suppressed or not considered; planners disregard ambitious but potentially contentious solutions.
	Elevation of local socio-economic considerations→dilution of precautionary science→ minimally protective plan.	Planners reject precautionary measures out of concern for short-term economic considerations.
	Disagreement on plan specifics, dissatisfaction among those who did not participate, or changes in the political context→legislative, administrative, and legal challenges that impede implementation.	Stakeholders disagree on plan specifics during implementation; changes in the political context prompt defections; efforts to implement the plan are challenged in legislatures, agencies, and/or the courts.
Flexible, Adaptive Implementation • Relying on flexible policy tools and adaptive management.	Flexible mechanisms→ evasion of plan's protective measures by laggards.	Reluctant managers and stakeholders do not implement plan's provisions.
	Adaptive management→ resistance by managers or capture by development interests.	Managers resist adjusting practices in the face of information suggesting more protective measures are needed; development interests resist efforts to make management practices more environmentally protective.

to their own revenue needs in managing their lands (Scheberle 2004). Skeptics also point out that institutional barriers—such as disparate missions, different organizational interests or cultures, incompatible statutory mandates, and a lack of incentives and rewards for cooperating—may stymie efforts at interagency and interjurisdictional coordination (Keiter 1998; Moote and McLaran 1997; C. Thomas 2003).

Stakeholder Collaboration

Skeptics are even more dubious about the environmental benefits of collaboration. George Cameron Coggins (2001) contends that most natural resource disputes arise because some losses for current users are inevitable if environmental problems are addressed with genuine, long-term solutions. But collaboration undermines efforts to depart dramatically from the status quo because, in an effort to attain consensus, planners exclude or marginalize those with "extreme" views, skirt contentious issues, focus on the attributes of the ecosystem that are easiest to control, and avoid considering solutions that impose costs on participating stakeholders (Beierle and Cayford 2002; Coglianese 2001; Eckersley 2002; M. Peterson, Peterson, and Peterson 2005). As a result, consensus-based collaboration is likely to yield lowest-common-denominator, rather than innovative, solutions (Coglianese 2001). Or, as Thomas Stanley, Jr. (1995, 261), charges:

As currently espoused, [ecosystem-based management]...promises the impossible—that we can have our cake and eat it too. Worse, however, it addresses only the symptoms of the problem and not the problem itself. The problem is not how to maintain current levels of resource output while also maintaining ecosystem integrity; the problem is how to control population growth and constrain resource consumption.

Some skeptics believe that collaboration actually exacerbates the power imbalance between environmental and development interests, and therefore holds the potential for *worse* outcomes than the conventional approach (Fainstein 2000, 2005; Tewdwr-Jones and Allmendinger 1998). They point out that local environmentalists rarely have the skills, experience, or money to participate effectively in collaborative processes; at the same time, collaboration consumes time that might otherwise be spent on activism and watchdogging, rewards moderation rather than principled opposition, and delegitimizes or contains conflict—which has been an effective means of mobilizing support (Amy 1990; Coggins 2001; Lange 2001; McCloskey 1996; Pralle 2006; Savitz 1999; Stahl 2001; Steinzor 2000). Furthermore, although giving everyone a veto theoretically empowers disadvantaged interests, powerful interests can force agreement by threatening failure of the process; in this way "more powerful and knowledgeable participants are able to co-opt dissident viewpoints that may be critical to seeking more creative and just solutions" (Foster 2002, 494). Finally, as legal scholar Stephen Nickelsburg (1998) observes, by increasing the difficulty of making large changes, unanimity rules advantage those who benefit from the status quo, typically business and property owners.

The pessimistic model also challenges the assertion that collaboratively devised plans are more likely to be implemented than those developed under the conventional approach. Because collaborative processes tend to deal with the most tractable—rather than the most serious—problems, the resulting plans are likely to be vague or vacuous, and conflict is simply deferred until implementation (Beierle and Cayford 2002; Born and Genskow 1999; Coglianese 2001; Finnigan, Gunton, and Williams 2003; Mansbridge 1980). In other words, implementation may proceed as long as it consists of "picking the low-hanging fruit," but may not persist as the thorny issues that prompted the problem-solving effort in the first place arise again. Moreover, if the political context changes, participants who feel they can get a better deal in another venue are unlikely to remain loyal to the negotiated solution, and instead may pursue their objectives in the legislature or the courts. Those who were excluded for being "extreme" are particularly likely to challenge collaboratively devised plans (Lange 2001). Further undermining the prospects for successful implementation is the fact that even if collaborative solutions are perceived as fair and legitimate by those at the table, the broader public may not share those views (Coglianese 2001).

Flexible, Adaptive Implementation

The pessimistic model also conveys doubt that flexible, adaptive governance arrangements will yield environmentally protective results. Critics point out that, historically, reliance on flexible implementation has resulted in capitulation to development interests, and they worry that the combination of vague plans and provisions for adaptive management is likely to result in similar concessions during implementation to attentive and well-financed development interests (Coggins 2001; Lowi 1999; Steinzor 2000). Research on corporate social responsibility suggests that optimists' expectations that flexibility will promote stewardship and efforts to go beyond the legal minimum may be misplaced. Because voluntary solutions can be evaded without consequence, and given the many short-term economic incentives the planning process does not address, many—and perhaps most—participants may continue to engage in business as usual rather than embracing new ways of operating (Coglianese and Nash 2002; Harrison 1999; Morgenstern and Pizer 2006; Rivera, de Leon, and Koerber 2006; D. Vogel 2005). The pessimistic model also suggests that, despite the theory that adaptive management allows for adjustment to new information, in practice learning may not occur because of turnover in government personnel, resistance by managers and stakeholders to changing established practices, and an unwillingness by some to undertake activities that threaten core values (B. L. Johnson 1999; C. Walters 1997; Stankey et al. 2003).

An Empirical Investigation

In short, for many observers ecosystem-based management promises not only to end the political stalemate over the environment but also to yield genuinely protective policies; for others, EBM at best perpetuates the status quo and at worst threatens to subvert or undermine existing environmental safeguards. This book aims to shed some empirical light on this debate by addressing the question: *to what extent, how, and under what conditions does EBM yield durable, environmentally protective policies and practices that (1) constitute improvements on the status quo and (2) are likely to conserve and restore ecological health?* To the extent that EBM can bring about environmental improvements, what attributes of the process or context are primarily responsible, and how do they operate in practice? Which can be replicated, and which are place-specific? On the other hand, if EBM fails to yield environmental improvements, which aspects of the process or context impede progress?

To address these questions, I investigate a series of EBM "experiments" in depth, using a focused-comparison method to systematically compare the predictions of the optimistic and pessimistic models with the evidence in each case (George and Bennett 2005).[14] Although each case is idiosyncratic, since each is "invented" in the place where it arises, my purpose is to identify some underlying procedural or contextual regularities that enhance or impede environmental progress, and thereby determine the extent to which insights from individual cases may be transferable across regions and projects. Exploring a small number of cases qualitatively enables me to focus on the process by which inputs produce outcomes in different contexts. Looking at cases over a long period of time allows me to document dynamic relationships and feedbacks. By contrast, snapshots can be misleading, because outcomes change over time; some apparently moribund initiatives get resuscitated, whereas others that appear successful at one point subsequently fall apart.

Selecting Cases

To maximize the likelihood of detecting the benefits of EBM, I began by identifying cases that scholars, journalists, and public officials consistently have touted as exemplary collaborative efforts to design landscape-scale plans whose implementation relies heavily on flexible, adaptive management. In two cases—Austin's Balcones Canyonlands Conservation Plan (BCCP) and San Diego's Multiple Species Conservation Plan (MSCP)—rapidly expanding urban areas in the southwestern United States have tried to preserve habitat for multiple species despite severe pressure to develop the region's remaining private land. In two others—the Comprehensive Everglades Restoration Project (CERP) and the California Bay–Delta (CALFED) program—policymakers aim to restore aquatic systems in the face of a burgeoning demand for water by urban, rural, and environmental interests. (I also collected data on a fifth case, the Chesapeake Bay Program, which involves both land and water, but decided not to include it in the final analysis because its scale—it comprises a region that spans six states—made comparison with the other cases unwieldy.)

I anticipated that examining such prominent experiments would increase the likelihood of discerning the positive relationships among procedural elements and environmentally protective outcomes posited in the optimistic model: each has had sufficient resources and capacity to generate technical and logistical support for the planning effort, has been in existence long enough to have achieved some of its objectives, and has faced high-level scrutiny and pressure to produce positive results. On the other hand, all are in regions facing serious growth pressures, and so constitute a hard test of the optimistic model of EBM.[15]

Although broadly similar in terms of several key attributes, the cases are sufficiently different with respect to the nature of the problem and the context of the initiatives that comparisons among them are likely to shed light on the factors that cause variations in environmental protectiveness. For example, there are good reasons to distinguish between efforts to address problems in terrestrial and aquatic ecosystems. As conservation biologists Reed Noss and Allen Cooperrider (1994, 282) note, conserving biodiversity in aquatic systems poses some unique challenges. First, aquatic systems are linear and branched, so that the flow of water forms a continuum from head-waters to sea (or sink, in the case of landlocked systems). Thus, upstream events such as pulses of pollution can have effects far downstream. Second, few reserves have been designated for aquatic resources...Furthermore, except for a few small coastal watersheds, no river systems exist that have not been severely modified by humans and that might serve as controls or benchmark aquatic systems. Finally, since aquatic systems are inherently connected, it is difficult to establish down-stream reserves or refugia that are reasonably protected or buffered from both upstream and downstream influences, much less atmospheric influences.

Similarly, although all four cases are in regions that face major population growth pressures, they are in areas with different political cultures: San Diego, California; Austin, Texas; northern California; and southern Florida.

It eventually became apparent, however, that despite some variation among inputs there was very little variation in the environmental protectiveness of outputs. In each of the exemplary cases, EBM yielded some environmental benefits, such as land acquisition and habitat restoration, but appeared unlikely to conserve or restore ecosystem health. So, to improve the validity of my inferences about which aspects of the process were particularly crucial in shaping outcomes, I chose three comparison cases that were similar in terms of location and problems being addressed but appeared to be producing more substantial environmental improvements: Pima County, Arizona's Sonoran Desert Conservation Plan (SDCP), the Kissimmee River Restoration in central Florida, and California's Mono Basin Restoration (see table 2.3).[16]

Analyzing the Cases

In analyzing the cases, one objective is to ascertain whether and how EBM has yielded more effective environmental protection than the traditional regulatory approach. This entails comparing the policies and

Table 2.3
Case Selection

	Minimal Environmental Benefits	Substantial Environmental Benefits
Terrestrial Ecosystem	Austin BCCP San Diego MSCP	Pima County SDCP
Aquatic Ecosystem	CERP CALFED	Kissimmee River Restoration Mono Basin Restoration

practices (outputs) and environmental improvements (outcomes) produced by EBM with those that likely would have resulted from a continuation of the status quo, as well as with outputs and outcomes in the comparison cases. A second objective is to elucidate whether and how EBM has yielded policies and practices that are likely to result in genuine environmental improvements. This involves comparing the outputs and outcomes of the planning process in each case with a set of generic criteria derived from the literature and with what scientists say would be sufficient to conserve or restore the target ecosystem. A third objective is to ascertain the extent to which the individual elements of the EBM process have contributed to observed outputs and outcomes. To do this, I compare the evidence of what has happened in the four cases of EBM with the empirical implications of both the optimistic and the pessimistic models described above. I garner evidence from extensive, semistructured interviews with key informants in each case, as well as from documents produced by each program, secondary-source assessments, and newspaper reports. I triangulate among multiple data sources whenever possible to avoid the pitfalls of relying too heavily on participants' perceptions, which are likely to reflect the strong psychological effects of long-term involvement (the "halo effect").[17]

Measuring Environmental Outputs and Outcomes

In recognition of both the complexity and the uncertainty of efforts to link trends in environmental quality to relatively recent changes in planning and management, I focus on assessing the likelihood that an EBM initiative will yield environmentally beneficial outcomes over time (see table 2.4). I begin by investigating the extent to which the plan and its implementation have resulted in policies and practices that are likely to enhance the target ecosystem's biological integrity. I ask: does the plan contain measurable objectives, such as performance standards, that are consistent with the best available scientific understanding of what it would take to conserve or restore critical ecological elements, functions, and processes?

Second, I gauge the extent to which the policies and practices specified in the plan are likely to achieve those objectives. Does the plan contain provisions for protecting representative ecosystem types or areas of concentrated biodiversity? Does it prescribe conserving and, where necessary, restoring processes that have been critical to the system's evolution—such as nutrient cycling, disturbance–recovery regimes, and

Table 2.4
Criteria for Evaluating Outputs and Outcomes

Outputs

Does the plan specify measurable objectives that are consistent with those prescribed by scientists as necessary to conserve or restore critical ecological elements, functions and processes?

Are the policies and practices in the plan necessary to achieve these objectives?

• Are representative ecosystems protected?
• Are evolutionary processes conserved/restored?
• Are terrestrial reserves designed according to state-of-the-art principles?
• In aquatic systems, are pre-diversion hydrological cycles restored?
• Have planners attended to land use in the surrounding matrix?
• Are there provisions to enhance implementation?

Is the plan precautionary—that is, does it avoid imposing the risk of failure on the natural system?

• Does it avoid actions that may impose irreversible damage and prescribe actions that promise environmental benefits?
• Does it employ buffers or otherwise provide latitude for increasing protection?
• Does it try to reduce the intensity of management over time?

Is the plan being implemented in an environmentally protective fashion?

• Are the plan's environmentally beneficial provisions being implemented?
• Have agencies and jurisdictions instituted environmentally beneficial practices more generally?
• Are managers adjusting their practices in response to new information?
• Are managers and stakeholders committed to greater environmental stewardship?

Outcomes

Have there been measurable environmental improvements in species, habitats, and ecological processes targeted by the plan?

---

predator–prey dynamics—and then managing them so that they continue to operate within their historic range of variation? In the case of terrestrial systems, do planners follow the state-of-the-art principles of reserve design by assembling large reserves with appropriate area-to-perimeter ratios; maintaining corridors of sufficient quality and size to ensure wildlife movement; and minimizing barriers to movement and intrusions into preserves from roads, power lines, and houses? In the case of aquatic systems, do planners try to restore the prediversion (or preobstruction) hydrologic cycle—including the quality, quantity, timing, distribution, and flow of water; conserve upland habitat; protect and restore adjacent wetlands and riparian vegetation; eliminate exotic invasive species; and reintroduce or augment populations of native species? In both terrestrial and aquatic systems, do planners attend to land-use practices in the surrounding matrix, where much biodiversity is located? Beyond its substantive elements, does the plan contain provisions that enhance the prospects for implementation by, for example, designating a reliable funding stream, requiring monitoring of what scientists believe are the key indicators of ecosystem health, and establishing penalties for non-compliance?

I consider plans more protective if they adopt approaches that scientists regard as conservative or precautionary—that is, they refrain from prescribing actions that may impose irreversible damage and do prescribe actions that are environmentally protective, even if the benefits are uncertain. A precautionary approach to EBM involves employing buffers to accommodate uncertainty—for example, by setting aside more land or water for the natural system than the bare minimum—and minimizing and mitigating edge effects, which occur at the intersection of natural habitat and human-modified habitat and are generally associated with harmful impacts on wildlife and plants (Meffe et al. 2002; Scott et al. 1999).[18] Plans that aim to reduce the intensity of management over time are more precautionary than those that rely on engineering and intensive management to achieve their goals (Angermeier and Karr 1994; Noss et al. 1999).[19] Philosophically, perpetuating intensive management reinforces the illusion that humans can control ecological systems and reduces the likelihood they will recognize natural limits (Francis 1993; Hillman, Aplin, and Brierley 2003; Wallace et al. 1996). Practically speaking, approaches that rely on intensive management leave little room for error and, because they are costly to maintain—particularly as energy costs skyrocket—are susceptible to changes in financial priorities over time. In short, as Reed Noss and his coauthors (1999, 118) argue, "By letting natural processes prevail to the maximum practical extent, optimal conditions for the maintenance of biodiversity are provided at minimal costs in hands-on management. To minimize errors of commission, direct interventions are to be avoided wherever possible."

Third, I assess the extent to which each initiative has actually funded and carried out the environmentally beneficial projects specified in its plan and whether, in fact, implementing agencies and jurisdictions have adopted more protective practices. For example, have they instituted policies that restrain environmentally harmful behavior or land-use practices? Have they undertaken habitat restoration or species recovery projects, acquisition of development or water rights for conservation, investment in the repair or maintenance of infrastructure to eliminate waste, or removal of invasive species? Are they adjusting their practices in response to new information gleaned from monitoring? Importantly, I try to ascertain whether staff are committed to a more protective approach and have sufficient knowledge and resources to implement it.

This analytic approach, which focuses on outputs, is indisputably a second-best solution to the quandary of evaluation because even fully implemented projects may not achieve their ecological goals (Koontz and Thomas 2006). For example, one analysis found a low success rate among the fully implemented stream restoration projects that had been evaluated to date (Kondolf and Micheli 1995). The authors cite a study of in-stream aquatic habitat enhancement structures installed in Alberta, Canada between 1982 and 1990 in which one-third were completely ineffective. They describe a second study of 161 habitat enhancement structures on 15 streams in western Oregon and Washington in which 18 percent had failed outright, and 60 percent were damaged or ineffective.

Ideally, then, I would conclude by assessing environmental outcomes; in particular, I would calculate the degree to which an EBM initiative actually enhanced an ecosystem's health by restoring its biological integrity, defined as its "ability to generate and maintain adaptive abiotic elements through natural evolutionary processes" (Angermeier and Karr 1994).[20] To this end, I have assembled the best available evidence of improvement in species, habitats, and processes targeted by each plan—such as upward trends in the populations of keystone species, restoration of native vegetation communities, and reestablishment of natural processes.

Unfortunately, although the particulars vary from case to case, definitive evidence is generally scant: it is difficult to measure environmental outcomes because monitoring in even the most prominent EBM experiments is sporadic, and more systematic approaches have only recently been put in place. It is even harder to attribute observed effects to a particular intervention, either because baseline data are lacking or because of an abundance of other, potentially confounding variables. For example, scientists may be unable to explain with any certainty fluctuations or trends in the population of a migratory bird species within a period of less than a decade. Similarly, the effect on an estuary of reductions in the amount of nutrients seeping into groundwater may not show up for years, and variations in rainfall rather than pollution reduction activities may be responsible for much of the short-term variation in the estuary's water quality. To reduce the likelihood that my judgment of a plan is an artifact of overly strict criteria, I err on the side of giving the plan credit for any documented environmental improvements. On the other hand, I argue that—given these initiatives' primary purpose of conserving and restoring ecological integrity—if ecosystems are not recovering in discernible ways, or if the declines that prompted the initiative continue, there is reason to believe the plan is insufficiently protective.

Detecting Causal Relationships

In addition to formulating a judgment about each initiative's environmentally protective outputs and, to the extent possible, outcomes, I try to ascertain how each of the three main elements of EBM has contributed to the observed results (see tables 2.1 and 2.2). To determine the impact of working at a landscape scale, I look for evidence that planners began using more holistic, less parochial language and devised a plan that addresses the main causes of ecosystem decline. I also assess whether agencies and jurisdictions are consulting with one another more frequently, are issuing joint reports, or have established common environmental objectives.

To evaluate the impacts of collaborative planning, I determine the extent to which informants regarded other participants as true to their word, came to think differently about their interests, and chose implementation mechanisms that were novel or had not been widely adopted. I also investigate whether scientists solicited the opinions of local naturalists and others knowledgeable about local ecology; whether stakeholders agreed on the scientific basis for the plan; and the degree to which the plan is consistent with the recommendations of scientists. I also seek evidence that stakeholders perceive the plan as fair and legitimate and decline to appeal it.

To detect the effects of flexible, adaptive governance, I collect evidence about whether flexible mechanisms have prompted stewardship and efforts to go beyond legal minimum levels of environmental protection. I also look for the adoption of a systematic monitoring program, the availability of a forum for synthesizing and analyzing monitoring data, the existence of clear lines of communication between scientists and managers, and indications that managers are modifying their practices in response to information from monitoring.

To increase confidence in my results, I try to account for threats to the validity of the causal inferences I have drawn based on the evidence in each case. I consider whether the effects I observe might have happened anyway, even in the absence of EBM, as a consequence of forces that were already in place. I draw on several lines of evidence to make this judgment: the project's environmental impact statement, which explicitly compares the chosen alternative to a "no project" scenario; counterfactuals, which involve considering at crucial junctures what might have gone differently or did happen but did not affect outcomes; and, where possible, simultaneous treatment of similar problems by different means. At the same time, I recognize that unaccounted-for changes in the larger ecological, political, or economic context may be largely responsible for the dynamics I observe. Alternatively, because each EBM experiment was initiated in response to a crisis, improvement may simply reflect the tendency of any extreme outcome to be followed by one that is less extreme, simply because fewer extreme random factors influence outcomes on successive occasions (Geddes 2003).

3

Setting Aside Habitat for Songbirds, Salamanders, and Spiders in Austin, Texas

The 1990s saw the completion of the nation's first landscape-scale multiple-species habitat conservation plan, the Balcones Canyonlands Conservation Plan (BCCP). Preparation of the BCCP began in 1988, when a series of endangered species listings prompted the city of Austin and Travis County to assemble a stakeholder group charged with advising policymakers on the elements of a regional conservation plan. Eight years later, in May 1996, the U.S. Fish and Wildlife Service (FWS) approved the BCCP, which aimed to create a 30,428-acre preserve comprising a small number of closely spaced habitat blocks—an achievement Interior Secretary Bruce Babbitt called "visionary" (Haurwitz 1996) and "the best example in the nation of a new and broader approach to endangered-species protection" (Haurwitz 1993a).

The BCCP has yielded several tangible benefits for Austin's rare songbirds, salamanders, and spiders. It has attracted funding from the federal government as well as from local taxpayers, and officials have used that money to acquire large tracts of biologically valuable land. By contrast, the conventional regulatory approach—project-by-project permitting— would likely have yielded a patchwork of open spaces with little cumulative ecological value. In addition, preparation and implementation of the BCCP have raised awareness of and concern about habitat conservation among city and county staff, and given them a rationale to manage set-aside lands for biological diversity.

The BCCP's preserve system is unlikely to conserve the Texas Hill Country's biological diversity in the long run, however. It comprises fewer acres than the bare minimum prescribed by scientists, and set-aside lands are both fragmented and insufficiently buffered from urban encroachment. Moreover, despite a rhetorical commitment to adaptive management, there is no central entity that can synthesize monitoring data across preserve sites or generate recommendations for revising management practices. In any case, there are few options for adjusting the preserve in the face of information suggesting it is inadequate: as Travis County has struggled to acquire its portion of the preserve's acreage, many of the most biologically valuable sites have been developed; Austin does not have sufficient resources to manage the land it has, even as the city's rapid growth has created new threats to its open space.

The BCCP is likely to fall short of its biological goals because the legal and political context in which the plan was formulated heavily favored development interests. Despite the popularity of environmental protection in the Austin area, environmental interests did not have adequate resources or land-use expertise to negotiate as equals with developers, nor were they sufficiently cohesive to compel the region's political leaders to take a stand on behalf of conserving biological diversity. Even if local officials had been inclined to take a pro-environmental position, they were—and continue to be—hampered by a dearth of regulatory levers with which to extract concessions from developers or raise money for acquiring and managing preserve land: the city of Austin had the authority to zone and impose impact fees on developers, but much of the vulnerable habitat lay in the unincorporated county, which did not. The only real pressure to make the plan more protective came from the threat that the FWS would not approve a plan that did not promise to meet the requirements of the Endangered Species Act. But federal officials were reluctant to deny permits during the planning process or to impose stringent biological conditions on skittish local officials. Constrained by this pro-development context, the collaborative process that yielded the BCCP focused on minimizing the financial burden on developers and assuaging property rights concerns. The plan's reliance on flexible, adaptive implementation has only exacerbated the challenges of assembling and managing a biologically viable preserve. Property owners defect from the BCCP when doing so serves their economic interests, monitoring and management are only minimally coordinated, prospects for adaptation are limited, and the preserves face threats from urban encroachment and recreational use.

The Origins of the BCCP

The origins of the BCCP date back to the 1970s, when Travis County entered a period of explosive economic expansion. The city of Austin devised a plan to manage growth, but developers easily found ways to undermine its intent, and the metropolitan area continued to sprawl westward into the environmentally fragile Texas Hill Country. In the late 1980s, however, an endangered species "crisis" prompted local officials to establish a collaborative process to formulate a landscape-scale multiple-species habitat conservation plan.

The Impetus for Habitat Conservation Planning

As is often the case, Austin's extraordinary natural environment, which features unusual geological and biological diversity, is part of what makes it so appealing to developers. The city sits on the eastern edge of the Edwards Plateau and straddles the upthrust of the Balcones Fault, which separates the blackland prairie to the east from the rolling hills of the Edwards Plateau to the west. In 1991, the Nature Conservancy named the Texas Hill Country one of 12 "Last Great Places" on Earth for its unique and endangered natural features: the area's craggy hills, steep-walled canyons, and artesian springs sustain a variety of plant communities, each of which supports its own wildlife. Beneath the city lies the Edwards Aquifer, a vast limestone formation of honeycombed rock that stretches northeastward from Kinney and Uvalde counties to Travis County. This phenomenally productive aquifer stores the cool, clear water that nourishes the area's springs; it is also riddled with caves and sinkholes that are home to a host of rare cave (karst) invertebrates (Collier 1992d; Haurwitz 1993d).

The Hill Country's unusual geology and abundant clean water attracted human settlers as far back as 12,000 years ago. The area was sparsely populated, however, until the 1960s, when the region's growth accelerated rapidly: the population of the Austin metropolitan area grew nearly 35 percent between 1960 and 1970, and nearly 50 percent in the subsequent decade, reaching 536,688 by 1980 (Butler and Myers 1984). In the early 1980s, lured by the region's well-educated residents and picturesque natural setting, a handful of high-tech companies relocated to Austin, prompting Chase Econometrics to forecast that over the course of the decade the city would experience the nation's highest employment growth rate (Butler and Myers 1984).

Austin responded to the surge in construction that accompanied its burgeoning population with a concerted effort to restrain sprawl: in 1979, the city completed a comprehensive growth management plan, *Austin Tomorrow*, which recommended sharply restricting growth in the west, directing it instead toward a "preferred growth corridor" running north–south through the heart of the city. Two factors combined to thwart this strategy, however: first, Austin voters repeatedly rejected proposals to issue bonds to fund capital improvements, thereby eliminating a powerful incentive for inducing growth in desired locations; second, the Lower Colorado River Authority (LCRA) agreed to provide water services to developers who had been denied by the city, thereby enabling them to build in outlying areas.[1] Lacking the regulatory leverage to enforce the comprehensive plan, city officials were compelled to manage the region's rapid growth through case-by-case negotiations with proponents of major developments, and by the mid-1980s the city was sprawling westward, carving up the western Travis County landscape.

In the late 1980s, however, Austin's expansion encountered a series of challenges when the development boom collided with the needs of the region's wildlife. In October 1987 the FWS listed the black-capped vireo, a diminutive songbird, as endangered. Over the next year the FWS brought to an abrupt halt several road projects, as well as a massive housing project on the 4,500-acre Steiner Ranch, because they threatened to destroy vireo habitat. Then, in September 1988 the agency listed as endangered five species of karst invertebrates (soon to be followed by a sixth) in response to pleas from scientists and a highly publicized cave occupation by the environmental group Earth First!.

Setting Up a Collaborative, Landscape-Scale Planning Process

Faced with this volatile situation, Assistant City Manager Austan Librach began exploring the possibility of crafting a regional multiple-species habitat conservation plan (HCP), a novel mechanism for protecting endangered species on private land that environmentalists Bill Bunch and Barbara Dugelsby had brought to the city's attention.[2] As Librach soon discovered, however, the city had to forge its own path: officials in the Arlington FWS office knew little about HCPs; moreover, the FWS had approved only a handful of HCPs since 1982, and none of those plans had tried to conserve habitat for multiple species in a rapidly growing urban area. After ascertaining that members of the development and environmental communities were willing to participate in this enterprise despite this uncertainty, in late 1988 Librach proposed to newly inaugurated Mayor Lee Cooke that the city set up a steering committee to formulate recommendations for what was originally called the Austin Regional Habitat Conservation Plan (Librach 2003).

About 100 people showed up in response to the city's announcement, and two-thirds of them wanted to be members of the new committee. In order to streamline the planning process, Librach formed a smaller group, asking each community (development, environmental, and government) to choose its own emissaries. The resulting 15-member Executive Committee included four developer representatives, four environmental advocates, and three local and three state officials.[3] David Braun, of the Nature Conservancy, agreed to chair the committee's twice-monthly meetings, all of which were open to the public, and an FWS official served as an *ex officio* member.

One of the Executive Committee's first tasks was to set up a biological advisory team (BAT) that could develop a credible scientific basis for the plan. The committee took pains to create a team that could produce results that would be unassailable, particularly within the development community: to this end, they asked zoologist Doug Slack of Texas A&M, one of the state's most conservative academic institutions, to chair the BAT. The remainder of the team consisted of ten scientists affiliated with a variety of academic and government institutions and two prominent local experts, who served as advisers.[4] The committee also selected consultants to collect land-use and economic data; synthesize the biological, legal, and economic data; and prepare the HCP itself. Again, the committee made sure the consulting firms they chose—Kent S. Butler & Associates and Espey, Huston & Associates—were acceptable to the development community (Braun 2003).

Collaborative, Landscape-Scale Planning

Over the next seven years a dizzying series of collaborative efforts were made to design a landscape-scale preserve system within the severe constraints posed by a context inhospitable to land conservation. Consistent with the optimistic model, the BAT produced an integrative assessment of the Hill Country's biological diversity and laid out a set of minimum measures necessary for conserving it. More consistent with the pessimistic than the optimistic model, however, jurisdictions struggled to cooperate in assembling a preserve. Also consistent with the pessimistic model, collaboration with stakeholders neither brought about trust, transformation, or innovation, nor yielded agreement on scientists' recommendations. Instead, planners whittled down the acreage of the recommended preserve and compromised on its configuration in order to assuage developers, property rights activists, and recalcitrant county officials.

Biologists Formulate the Scientific Basis for the Plan

With little idea of what the outcome would be, the Executive Committee asked the BAT to come up with a set of recommendations for a preserve system that would protect the amount and configuration of habitat necessary to ensure the viability of the region's endangered species. To this end, the science team began by constructing a list of nearly 160 species, based on recommendations from the FWS, the Texas Parks & Wildlife Department, the Texas Natural Heritage Program, the Texas Organization for Endangered Species, and local experts. The team then narrowed its focus to nine rare species—two migratory songbirds, five karst invertebrates, and two plants—based on three considerations: the presence of a significant population in the study area, the existence of serious threats to the species' survival, and the ability of a regional habitat conservation plan to materially affect its prospects (BAT 1990).

In January 1990, after a year of study, analysis, and synthesis, the BAT released an assessment that focused attention on ecosystem functions and urged planners to adopt a precautionary approach. The team introduced the report by defining the problem in a way that contrasted sharply with the prevailing "endangered species block development" view. "Habitat destruction," wrote the BAT, "is the underlying reason that the species encompassed by the [habitat conservation plan] are in danger of extinction." Furthermore, the authors pointed out, human activity is the primary cause of habitat fragmentation, and "everyone living in the study area bears some degree of responsibility for [the] current plight" of the region's rare and endangered species (BAT 1990, 2–3). The discussion that followed emphasized the importance of focusing conservation efforts on whole ecosystems rather than individual species. "Fundamentally," said the BAT (1990, 31–32):

the species to be protected are not isolated entities, but are components of an ecosystem which is a complicated network of interacting organisms. Maintaining an intact ecosystem will reduce management costs and will also reduce the possibility of currently unknown factors thwarting the goals of the [plan]. Thus, in addition to the following specific recommendations on preserve design, the BAT strongly recommends that the system of preserves be designed under the overall goal of maintaining an intact ecosystem, as opposed to several highly managed populations.

Finally, the team pointed out that, to be effective, the preserve network would have to conserve large blocks of contiguous habitat. The report warned:

The more fragmented a system of preserves is, that is, the smaller the individual preserves are and the greater the distance between them, the greater will be the total area required. Significant fragmentation of the preserves could easily increase the total area required by a factor of ten above the recommendations given below. (BAT 1990, 32)

On the other hand, BAT members—although technically independent of the Executive Committee—were well aware of the political and economic considerations that would shape the reception of the report and therefore based their acreage recommendations on estimates of the minimum necessary preserve for the particular species of concern (Pease 2006). For the black-capped vireo, for example, the BAT noted that surveys conducted in 1989 had found only 59 breeding pairs across the study area, which included Travis County and portions of Williamson, Hays, and Burnet counties, and that the population had been declining by 25 to 55 percent annually. Estimating that 500 to 1,000 breeding pairs would be needed to maintain a viable population of vireos, the BAT prescribed a habitat preserve comprising no fewer than 123,500 contiguous acres and noted that if the preserve consisted of smaller parcels, as many as 864,500 acres could be required. In recognition of the serious threat posed by edge effects, the BAT also specified the attributes of a properly configured preserve, saying that "less than 5% of the area of any preserve should be within 100 meters [330 feet] of the preserve boundary or any large human disturbance" (BAT 1990, 33).

For the golden-cheeked warbler, the BAT recommended establishing two separate preserves of at least 29,640 acres each, in order to maintain a minimum population of 500–1,000 breeding pairs, and emphasized that each preserve should be contiguous and unfragmented because warblers are extremely sensitive to incursions on their habitat. The team noted that a single 123,500-acre preserve would serve to protect both the vireo and the warbler, but it also observed that the only remaining habitat fragments that met its reserve configuration criteria were in and around the South Post Oak Ridge area. The BAT concluded its discussion of the warbler on a precautionary note, saying that because the Travis County population was central to the preservation of the entire species, the plan should aim to protect more than the minimum amount of habitat (BAT 1990, 37).

With respect to the karst invertebrates, the BAT noted that simply protecting individual caves would be inadequate because such a strategy would neglect the water, energy, and nutrient needs of karst ecosystems.

Therefore, the team advised employing a host of land-use restrictions, such as protecting the water recharge zone for the karst, avoiding activities that involve using pesticides or changing the water flow regime, protecting the mantle of natural vegetation that provides energy and nutrients to the karst ecosystem, and creating a large, undisturbed preserve to prevent intrusion by nonnative fire ants. Finally, the BAT identified specific locations where endangered plant species should be conserved.

Overall, the BAT's recommendations reflected scientists' judgment about what would constitute a comprehensive approach to protecting Austin's biological diversity in the face of considerable uncertainty. According to Chuck Sexton, a biologist for the city of Austin and a BAT member, "What we wanted to do was to protect large enough, viable preserve systems in a few blocks to protect the processes. That means the large vegetation patches the birds needed for all of their life functions, as well as things like the groundwater and the watersheds for these cave ecosystems" (Collier 1990a). Defending the team's integrative approach, FWS biologist Joe Johnston explained: "Each species is an integral part of our environment, and we don't know all the consequences of losing even one of them. When do we come to the point where we've reached the straw that broke the camel's back?" (Collier 1990b).

Stakeholders Craft a Landscape-Scale Plan

After the BAT released its report, the Executive Committee turned over the job of designing a preserve system to a small group of consultants led by Kent Butler, a planner at the University of Texas, and Clif Ladd, a biologist with Espey, Huston & Associates. Planning proceeded haltingly, with policymakers reconstituting the collaborative entity several times and consultants proffering increasingly modest preserve designs. Collaboration did not produce trust or transformation among its participants largely because the context in which negotiations occurred was heavily weighted in favor of development interests. Although represented in equal numbers, developers and environmentalists were not equally powerful. Developers were represented by veteran attorneys well versed in the convoluted process of permitting and land-use law, and they were abetted by both a hostile state legislature that repeatedly denied Travis County the authority to enact land-use controls and local officials worried about their tax base. By contrast, the environmental representatives were mostly volunteers; only one, Bill Bunch, was a lawyer and professional activist. Making matters worse, the Austin area's environmental activists did not unite behind a single agenda; as a result, despite Austin's relatively pro-environment city culture, political leaders had little incentive to press for a more protective approach.[5]

The Executive Committee Agrees on a Plan   Developers were taken aback by the BAT's recommendation for a 123,500-acre preserve but were resolute about moving forward with an HCP, which they regarded as an opportunity to alleviate the obstacles posed by Endangered Species Act enforcement. For example, attorney David Armbrust said: "We're facing a development moratorium if we don't work through this. It would just absolutely stop development in northwest and southwest Travis County for decades" (Collier 1990a). Throughout the spring of 1990, development interests and their allies continued to emphasize the federal government's ability to block construction, as well as the expense, delay, and uncertainty associated with getting individual permits (K. Martin 1990; Stanush 1990a, 1990b).

The urgency of formulating a regional plan increased dramatically in March 1990, when the FWS recommended listing the golden-cheeked warbler as endangered. The agency pointed out that the warbler, the only bird that breeds exclusively in Texas, had lost about 40 percent of its habitat in western Travis County during the 1980s; its remaining habitat—closed canopy oak/juniper woodland—was rapidly being fragmented, resulting in cowbird parasitism (USFWS 1990). Because warblers inhabit mature forests, the location of preserve land was relatively inflexible; most of it was close to the city and highly prized for its real estate value (Pease 2006). Also, as the BAT had noted, warblers are highly sensitive to intrusion, so it is essential to buffer their habitat from development.

Like developers, the mainstream media portrayed the regional plan as essential to the Austin area's economic well-being. For instance, in late March consultants released maps showing the potential habitat of the plan's target species within the planning area: 60,000 acres for the black-capped vireo, 67,000 acres for the golden-cheeked warbler, 4,000 acres for the plants, and 150,000 acres for the cave bugs. Reporting on the maps, journalist Bill Collier (1990c) observed that protecting habitat blocks would facilitate development:

Without the regional plan...every developer or individual property owner wishing to disturb the habitat of a protected species must go through the expensive and time-consuming process of seeking a federal permit. Once a federal permit is approved by the wildlife service, all public and private projects in the planning area that are consistent with its restrictions will be allowed to proceed.

In early May 1990, when the FWS announced it was emergency-listing the warbler as endangered, a headline in the *Austin American-Statesman* blared: "Protected Bird Halts Development on 67,000 Acres." According to development lawyer and Executive Committee member John Joseph, the warbler listing promised to "further destroy what limited demand there is in the Austin market for anything." Moreover, he added, "Anyone who owns land out there is impacted—farmers and ranchers, even people who own homes or individual lots" (Collier 1990c).

In the midst of this media rhetoric, consultants Butler & Ladd scrambled to come up with a preserve design they believed would be both biologically defensible and politically palatable, given existing land-ownership patterns and the constraints on local officials' regulatory authority to steer development away from sensitive habitats. In mid–June they unveiled maps that identified two 60,000-acre tracts west of Austin where conservation efforts would focus, noting that they would not prescribe acquiring all 120,000 acres but would aim to acquire sufficient acreage within those areas to preserve the target species (Banta 1990). They declined to identify acquisition sites for karst invertebrates, retreating from the BAT's suggestions for land-use restrictions and instead recommending that jurisdictions in the planning area require a cave survey as part of their development approval processes. At the end of June, the consultants presented a system of three habitat blocks totaling 77,600 acres, of which 51,500 acres would have to be acquired (the remainder was already publicly owned). They estimated the cost of purchasing land at $86 million and annual operating costs at $2 million (Collier 1990d).

Despite its modesty relative to the BAT's prescriptions, Butler and Ladd's proposal failed to quell the opposition to the HCP concept that was brewing among political officials in many of the jurisdictions within the original planning area, some of whom feared that "undeveloped land in their jurisdiction [would] be roped off as a preserve and that they would be forced to impose development fees and turn over tax money for a cause many of them [did] not support" (Vierebome 1990). In early December 1990, in response to criticism of their initial proposal, the consultants presented an even more conservative draft of what was now known as the Balcones Canyonlands HCP (BCHCP). This time, instead of three large habitat blocks, the consultants designated six "macrosites," whose locations were based on tracts that were publicly owned or whose development potential was already strictly limited by Austin's comprehensive watershed protection ordinance. They pared down the preserve acreage from 77,600 acres to 64,202 acres, of which 45,412 acres would have to be acquired; 6,640 acres were already publicly owned, and 12,150 acres would be protected through mitigation agreements with landowners (Collier 1990f). To reduce the local obligation, they also recommended that the federal government purchase the bulk of the land—some 30,000 acres—for a national wildlife refuge. To limit the amount of property that would have to be acquired, consultants proposed surrounding core areas with buffers where land use would be restricted.

Despite these concessions, critics immediately assailed the consultants' proposal, and in late February 1991 the Executive Committee approved a revised draft of the BCHCP. It did so, however, only after a series of meetings yielded amendments that mollified its detractors. The new plan accommodated some of biologists' concerns by incorporating suggestions to better protect the warbler, but most of the changes aimed to placate developers. The committee agreed to reduce the amount of land outside the preserves that would serve as buffers; spread acquisition costs by financing part of the preserve with a real estate transfer fee and making development impact fees due later in the process; and protect habitat exclusively through land acquisition, rather than through a combination of acquisition and land-use controls—even though doing so raised the plan's cost to $113 million. The committee also took pains to emphasize that the document was "Not Just a Habitat Conservation Plan." To this end, it dropped the word "habitat" from the plan's title (it was now the Balcones Canyonlands Conservation Plan, or BCCP) and focused the text on the economic and water quality benefits—a purposeful shift in emphasis aimed at rallying public support and defusing opposition.

The Plan Languishes   The draft plan still needed the approval of the local entities that would have to implement it, as well as permission from the state legislature to implement some of its provisions; despite the modifications it faced resistance on both fronts. For instance, the consultants had proposed that Travis and Williamson counties take the lead on financing the plan, but Williamson County Judge John Doerfler—who, along with Georgetown Mayor William Connor, had joined the Executive Committee in early 1991—had cast the lone dissenting vote on the plan, expressing serious reservations about its size and cost.[6] Similarly, Travis County Judge Bill

Aleshire had told the Executive Committee that his ability to support the plan was complicated by an unfinished road project for which county taxpayers owed $22 million on bonds issued by the Southwest Travis County Road District No.1.[7]

As they strove to mitigate resistance, the plan's proponents continued to make pragmatic arguments that it would provide open space and water quality benefits, free up for development hundreds of thousands of acres encumbered with endangered species restrictions, and save money in the long run. For example, a March 4 editorial in the *Austin American-Statesman* explained that the BCCP was not just for "birds and bugs," but was also an opportunity to preserve and enhance water quality at a reasonable cost while freeing land for "appropriate development." A March 9 editorial emphasized that the BCCP should be regarded as "a means to put predictability into land development while protecting endangered species, water quality and our quality of life." Glenn West of the Greater Austin Chamber of Commerce reiterated the idea that "Without a conservation plan, it doesn't appear the Endangered Species Act is going to permit any reasonable level of development in western Travis County." Michael Spear, the regional FWS director in Albuquerque, added: "I hope people understand very clearly the alternative," noting that developers and landowners wishing to modify property in western Travis and Williamson counties would have to get their own permits from the FWS—an expensive and time-consuming process. "It will be burdensome to those people," Spear said. "The thought that every one of them would have to get an individual permit is mind-boggling" (Collier 1991a). Consultant Kent Butler assured skeptical neighborhood leaders in Austin that the plan would save money in the long run because land freed for development would produce tax revenue (J. Wilson 1991).

In a last-ditch effort to increase the plan's political appeal, in April 1991 the consultants released a revised draft of the BCCP that explicitly touted its economic benefits. The text emphasized the problems with the status quo, including declining property values because of a real estate market that was nervous about endangered species listings; inconvenience to ordinary people because of delays in construction permits and lending; delay and uncertainty for developers; increases in the cost of public works projects; and, finally, uncertain prospects for endangered species themselves because of fragmented conservation (City of Austin 1991). Notwithstanding proponents' efforts to accommodate the plan's critics, as legislators considered proposals to give county officials the regulatory clout they needed to raise money and prod developers to participate in the plan, opposition surfaced from a variety of quarters. The Texas Association of Realtors, the Texas Farm Bureau, the village of Lakeway, and property owners from Travis and Williamson counties all lobbied hard against the measures. The legislative battle further intimidated reluctant local officials and disrupted the already fragile dynamics of the BCCP's Executive Committee: Maury Hood, who had been representing the Texas Capitol Area Home Builders Association, resigned from the committee after representatives of the Texas Association of Home Builders testified against the enabling bill. Ultimately, the state legislature declined to give Travis County the authority to charge development fees and create zoning districts and building codes.[8]

Resuscitating the Plan, Round 1   Although the legislative setback severely constrained options for funding the BCCP, in June 1991 newly elected Austin Mayor Bruce Todd (formerly the Travis County representative on the Executive Committee) set out to revive the plan. Taking over from the Nature Conservancy's David Braun as chair of the Executive Committee, Todd vowed to have agreement on the plan within three months. Throughout the summer of 1991, as the Executive Committee wrestled with criticism of the plan's cost and doubts about its biological sufficiency, backers continued to try to enhance its public appeal by stressing its economic importance. As one editorial explained:

No, birds won't stop singing and bugs won't quit chirping if there is no plan. Quite the contrary, protected species will remain protected by federal law. Landowners, however, won't be able to in any way disturb those habitats without permission from the U.S. Fish and Wildlife Service. Big landowners can and have petitioned the agency for permission to develop their property, but it is a costly and time-consuming process. Smaller landowners will find it virtually impossible to thread their way through the bureaucracy on their own. Besides, a comprehensive overall preservation plan promises an efficient and orderly way to accommodate limited development while protecting the endangered species. (Anon. 1991)

Similarly, a July 17 column titled "Growth Depends on Habitat Plan's Success" emphasized that the BCCP was "a development plan, a project that would allow construction in areas now paralyzed by federal environmental laws" (M. Kay 1991a). The author went on to argue that the name of the plan, which contained the word "conservation," falsely implied that its organizers represented only environmental interests. Extending the economic theme, Mayor Todd attributed the drop in land values in western Travis County to enforcement of the Endangered Species Act, which he claimed had depressed land values, causing the tax revenues in both Austin and Travis County to decline, and suggested the BCCP would provide a remedy (Martinez 1991). (In fact, most of the drop in property values was a consequence of both the sluggish economy, which depressed demand for new development, and the low prices offered by the Resolution Trust Corporation, or RTC, which owned large swaths of land in the area.[9]) Plan proponents also pointed out that the preserve system would draw tourism and businesses looking to relocate to places where employees would have access to the outdoors (M. Kay 1991b).

The arguments failed to sway Bill Aleshire, the outspoken Travis County judge who was now a member of the Executive Committee, having taken the spot vacated when Bruce Todd became mayor of Austin. Aleshire continued to express his ambivalence about the plan. "I am not fighting the plan," he said, "but I am afraid that there are folks out there who are caught up with the vision and are not asking due diligence questions." In a letter to Mayor Todd, Aleshire asked: "Is there proof that the project is necessary or, at least, desirable? Is it practical and affordable? Do our people support it?" (M. Kay 1991b). Aleshire also continued to raise concerns about the proposed acquisition of two parcels in particular—the Uplands and Sweetwater tracts, both owned by the RTC—which county officials had hoped would be developed, thereby adding to the tax base of the county's troubled road district. Aleshire worried that if the RTC sold the land to a nontaxable entity, the county would run the risk of defaulting on the road district's bonds. Finally, Aleshire insisted he would withhold his support for the plan until he saw an economic study documenting that its financial benefits would outweigh its costs.

In late November 1991, Pat Oles, an LCRA board member and chair of a task force appointed by Mayor Todd to resolve the plan's remaining biological and legal issues, endorsed another series of modifications aimed at quelling opposition: (1) further reducing the size of the plan's preserves to 29,100 acres, in order to cut local costs; (2) increasing the size of the proposed federal wildlife refuge from 30,000 to 41,000 acres to address biologists' concerns, while dropping it from the BCCP to avoid dealing with opponents of the refuge; (3) having the FWS rather than the county collect development fees—a move that would placate Aleshire and eliminate the need for legislative clearance; (4) formally dropping the disgruntled Williamson County from the plan; (5) emphasizing that jurisdictions would acquire land only from willing sellers and would not use condemnation; and (6) reducing the land area shown on maps as "potential preserves" to defuse hostility from landowners (Collier 1991b).[10]

Despite these additional concessions, property-rights activists formed a coalition called the Texas Alliance for Property Rights to fight the plan on behalf of an estimated 40 groups whose members felt their interests had not been represented in the process of formulating the BCCP (Collier 1991c). In hopes of propitiating landowners who felt they had been excluded, in January 1992 Mayor Todd belatedly appointed two new representatives to the Executive Committee: Robert Brandes, who owned 150 acres on Lake Travis and had been an outspoken critic of the plan, and Steve Gurasich, part owner of a proposed 1,100-acre development (Collier 1992a). Property-rights advocates were not appeased, however, pointing out that they had not been consulted on the appointments, which still did not include any rural ranchers, and noting that they were included only when the committee's work was nearing completion.

Although the public controversy continued, in early February 1992—after several more adjustments—consultants submitted a final draft of the plan for public review and Executive Committee approval. The new version proposed conserving 29,160 acres at a projected 20-year cost of between $138 million and $143 million. It proposed requiring mitigation for development outside the preserve but within the planning area of $3,000 per acre of habitat destroyed or $600 per acre in a proposed development (Collier 1991b). (No fees would be assessed if owners could show their land contained no occupied habitat.) Working with Oles' proposals, consultants projected development fees would yield $62.3 million over 20 years. Other revenues would include a $40.6 million surcharge on water rates, $27 million from property taxes, $7.5 million from the state, $5 million from an assessment on public projects within the planning area, and $1 million from visitors to the preserve. Rather than creating a new regional authority to manage and monitor the network of set-aside lands—a suggestion that had provoked resistance among some local officials—the consultants suggested that a committee comprising representatives from Austin, Travis County, the LCRA, and the Texas Parks & Wildlife Department coordinate preserve management (Collier 1992b).

Ongoing divisions among stakeholders continued to thwart the Executive Committee's efforts to agree on a plan, however. Environmentalists complained that the draft was biologically inadequate because it did not address the habitat requirements of the Barton Springs salamander, for which two University of Texas biologists had filed an endangered species listing petition. The salamander, they noted, depends on the quantity and quality of water pouring from Barton Springs, which is fed by a recharge zone that stretches from Austin southwest into Hays County. The major chronic threat to the salamander is runoff of silt, oil, and chemicals from construction, roads, and yards; if listed, the salamander could trigger a federal mandate for tougher water quality measures throughout the 354-square-mile recharge zone (Collier 1992b; Haurwitz 1993e).[11] Judge Aleshire announced that he, too, opposed the BCCP in its current form. Aleshire reiterated his chief complaints: first, by removing preserve land from the tax rolls, the plan would impede the county's ability to pay off its debt for the Southwest Parkway and therefore hurt its bond rating; second, the plan did not provide details of how a proposed bond issue for the preserve would be repaid (Collier 1992c).

In order to keep skeptics on board, Mayor Todd decided to skirt the remaining areas of discord among committee members by crafting a resolution that recommended the plan as a general foundation, not a blueprint, for a final product to be submitted to the FWS. In late February the Executive Committee approved Todd's resolution 16–1 (environmentalist Bill Bunch cast the dissenting vote), at which point the Parks & Wildlife Department assumed responsibility for shepherding the plan through the process of gaining approval from local governments and the FWS. In May and June 1992 the department convened staff from the city, county, and LCRA to generate suggestions for revising the plan before submitting it for final approval by each of those entities' governing boards. The plan soon ran aground again, however, foundering on the very questions the Executive Committee had dodged: funding and biological adequacy.

Despite continuing opposition, the plan's supporters persisted, and over the summer two events breathed new life into the BCCP. First, in early June consultants released an economic study aimed at silencing skeptics that concluded the plan's benefits would far outweigh its costs. Then, in August—following a campaign orchestrated by many of the city's business and civic leaders—Austin voters propelled the plan forward by approving (65 percent to 35 percent) a $22 million bond issue for land purchases. The sales pitch for the bond had focused on the BCCP's recreation and water quality benefits, rather than species conservation, as well as on the bargain available to the city as a result of a deal brokered by the Nature Conservancy to acquire nearly 9,633 acres of land from the RTC. "The main message is the water quality benefits as well as the fact that this property is never going to be any cheaper," explained John Scanlan, an attorney and co-chair of Texans for the Economy and Nature (TEN), the coalition that promoted passage of Proposition 10 (Collier 1992e). Bolstering proponents' arguments, shortly before the bond election the FWS had pronounced the BCCP biologically sound and likely sufficient to lift building prohibitions in western Travis County. Shortly after the election, the LCRA, the Austin City Council, and the Travis County Commissioners Court signed off on the preliminary plan. (The Texas Parks & Wildlife Commission, which oversees the Parks & Wildlife Department, withheld its endorsement. It cited lack of funding—the plan had assumed a contribution from the department of $7.5 million—and asked for more specifics.)

In late February 1993, after the Austin City Council voted unanimously (7–0) to submit the biological design of the 29,160-acre preserve design to the FWS for approval, planners set out to resolve the problems with the financing and administrative portions of the permit application. By this time, consultants were estimating the total land acquisition costs at $165 million over 30 years. They suggested some of that could come from development fees: a countywide surcharge on building permits and a mitigation fee of $1,000 per acre would cover acquisition costs during the program's first three years, or until the preserve was completed; after that, the building permit surcharge would cover preserve operations and maintenance. The plan hit yet another roadblock, however, when—after another bruising battle—the state legislature again declined to pass a bill that would authorize the city and county to levy habitat mitigation fees.

Another Setback: Travis County Declines to Fund the Plan   In early 1993, with the advent of a sympathetic administration in the White House, BCCP proponents hoped that the federal government would step up to provide funding and regulatory backbone that the state and some recalcitrant county officials refused to supply. They were encouraged when, in March, newly installed Interior Secretary Bruce Babbitt visited Austin and praised the initiative, calling it "the best example in the nation of a new and broader approach to endangered-species protection" (Haurwitz 1993a). Then, in August, Secretary Babbitt announced the Interior Department would buy 5,000 acres of habitat, at a cost of about $5 million (in addition to the 41,000 acres it had already committed to acquiring for the Balcones Canyonlands National Wildlife Refuge) as long as local entities came up with funding for the BCCP (Haurwitz

1993f). (The FWS, which had raised questions about the BCCP's adequacy for the warbler, anticipated that the additional acreage would improve the bird's prospects.)

Even as federal officials strove to compensate for local resistance, however, the FWS continued to grant individual permits that were chipping away at the potential preserves and undermining the possibility of a biologically viable system. By the summer of 1993 the real estate market was heating up once again, and developers were filing individual permit requests at an increasing rate, saying they could not wait for a habitat plan that might never materialize. Between 1988 and mid-1993 the agency had approved three highway projects, one shopping mall, a research center, and five subdivisions in Travis County. By late June 1993 proposals for ten new subdivisions in western Travis County were awaiting FWS approval, and the agency expected dozens of additional proposals in the coming year and a half, as the economy—and hence demand for new housing—rebounded (Haurwitz 1993c).

As individual permits chiseled away at the proposed set-asides, the environmental coalition—which had always been tenuous—fractured over the biological merits of the BCCP's preserve system (Haurwitz 1993b). Although the mainstream groups, such as the Austin Sierra Club and the Travis Audubon Society, continued to back the plan, they acknowledged that the acreage protected was a bare minimum in terms of the biological requirements of the covered species. As George Avery of the Sierra Club explained, everyone in the biological and environmental communities had concerns about the preserve's biological sufficiency, but his group thought a plan was better than no plan. Tom McCuller of Travis Audubon had a similar perspective: "If we don't have a regional plan," he said, "we don't have any hope at all of saving the birds" (Haurwitz 1993b). But Bill Bunch insisted the plan simply did not save enough habitat. Similarly, although they recognized that project-by-project approvals would lead to further fragmentation of the remaining land, Earth First! activists regarded the BCCP as a sop to developers and called it "the incredible shrinking habitat plan."[12]

Despite the environmental community's ambivalence, in September a group of business executives, moderate environmentalists, and public officials led by the Nature Conservancy established a political action committee, the Texas Legacy Committee, to lobby county voters to approve a $48.9 million bond issue to acquire the 12,000 acres of habitat needed to complete the preserve. An interim coordinating committee comprising six local, state, and regional officials backed the plan, whose cost consultants now estimated at $180 million (Haurwitz 1993g). Ultimately, virtually every major local group endorsed the bond issue; only three organizations—Earth First!, the Texas Capitol Area Home Builders Association, and a group calling itself the Travis County Taxpayers—expressed outright opposition. As the election neared, Babbitt weighed in as well, holding a press conference in front of the Travis County courthouse at which he called the BCCP "the flagship" of habitat conservation plans, a milestone in ecosystem-scale planning under the Endangered Species Act (Haurwitz 1993h).

Despite its high-profile support, in November 1993, county voters rejected the BCCP bond by a narrow margin, dealing the plan what observers thought was "a stunning and possibly fatal blow" (Haurwitz 1993i). Low turnout (17.3 percent of registered voters went to the polls) largely determined the 52 percent to 48 percent outcome, since fiscally conservative voters tend to prevail in low-turnout elections.[13] Also likely contributing to the defeat was the environmental community's tepid support and its corresponding unwillingness to mobilize voters. Making matters worse, immediately after the bond issue was defeated, negotiations between environmentalists and builder Freeport–McMoran that would have solved the Southwest Road District problem and freed up the Uplands and Sweetwater tracts for acquisition collapsed. On November 6 an editorial in the *Austin American-Statesman* pronounced the BCCP dead.

Resuscitating the Plan, Round 2   The next two years were a roller-coaster ride for the BCCP, as its supporters proposed innovative funding mechanisms they hoped would reinvigorate the plan. After a failed effort by Mayor Todd to generate interest in an approach called "Conserve as You Grow," in June 1994—at the request of the Texas Capitol Area Home Builders Association—Interior Secretary Babbitt stepped in to try and jump-start the HCP process. Six months later, after a series of private meetings with local officials and some stakeholders, Babbitt announced a proposal that called for a 30,428-acre preserve system—slightly more land than the previous version required, to compensate for the macrosites' increasing fragmentation. He stipulated that land purchases be financed using mitigation fees paid by developers ($5,500 per acrein known habitat; $2,750 per acre in potential/unoccupied habitat), in exchange for which they would get a mitigation certificate and certainty that they could develop. This approach reduced the mitigation ratio required of developers from 3:1 to 1:1. Each certificate would mitigate one acre of habitat (as determined by aerial photographs) and would sell at a price determined by the market, with the initial price set at the prevailing price for raw land in Travis County (Librach 1995).

To flesh out the details of the latest option, dubbed the Shared Vision Plan, Babbitt convened a new stakeholder group—this one chaired by County Commissioner Valarie Bristol. The group's primary aim was to devise a funding scheme, not to ensure the plan's protectiveness; in fact, organizers specifically excluded environmentalist Bill Bunch, who had consistently demanded more stringent species protection measures during his tenure on the BCCP Executive Committee. By May 1995 the newly constituted working group had agreed on an approach that simplified participation and dramatically eased the financial and logistical burden on developers, reducing their share of the funding to 26 percent (Librach 1995). The latest version of the BCCP relied on a combination of the mitigation certificates proposed by Babbitt and tax increment financing, a mechanism suggested by assistant city manager Joe Lessard.[14] It also shortened the time line for land acquisition to between 10 and 20 years, depending on the pace of development. To placate disaffected property owners, the working group agreed to charge small landowners, ranchers, and farmers a greatly reduced certificate fee and emphasized that participation in the BCCP was voluntary; developers remained free to apply for individual permits through the FWS rather than signing on to the regional plan.

The Balcones Canyonlands Conservation Plan

In the spring of 1995, the city of Austin and the Travis County Commissioners Court signed off on the BCCP, and in early May 1996 the FWS approved a joint permit for the city and county. By the time the BCCP was finalized, the permit area encompassed 561,000 acres, including 87 percent of Travis County (but excluding Williamson County entirely). The permit grants coverage for the two migratory songbirds, six karst invertebrates, and 25 additional species of concern. The planned preserve areas are located within seven macrosites—five large ones (Bull Creek, Cypress Creek, North Lake Austin, South Lake Austin, and Upper Barton Creek) and two small ones (West Austin and Pedernales) (see figure 3.1). The macrosites are slated to comprise 30,428 acres, of which 7,347 acres are already publicly owned when the planning process began. Planners estimated the BCCP's 30-year cost at $160 million.

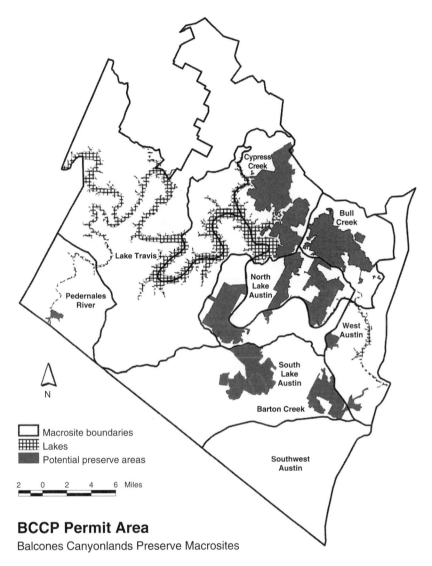

Cypress
Creek

Bull
Creek

Lake Travis

North
Lake
Austin

Pedernales
River

West
Austin

N

South
Lake
Austin

Barton Creek

▢ Macrosite boundaries
▦ Lakes
■ Potential preserve areas

Southwest
Austin

2   0   2   4   6   Miles

BCCP Permit Area
Balcones Canyonlands Preserve Macrosites

Figure 3.1
Austin's Balcones Canyonlands Proposed Preserve Areas

The plan relies on a variety of flexible tools to raise money to acquire preserve land. The primary instrument allows those who want to build outside the preserve to purchase participation certificates based on the total acreage of different habitat zones within the tract they plan to develop. Initially, the cost of a participation certificate was $5,500 per acre of occupied songbird habitat, $2,750 per acre of habitat whose status is unknown, and $55 per acre of karst habitat. Landowners seeking to build a single-family home or other low-density development on 15 acres or less outside the preserve boundary can purchase a special provisions certificate for $1,500; agricultural and ranching operations also get special permit status. Landowners within the planned preserve areas retain the option of selling their property to the city or county or getting a permit from the FWS to develop a portion of it.

The approach established by the BCCP is superior to the trajectory the city of Austin and Travis County were on. Consistent with the optimistic model of EBM it is relatively comprehensive because it conserves habitat in large blocks. By contrast, if the FWS had continued to grant development permits, habitat would have been even further fragmented. Parcels of land containing endangered species would have been saved, but they would have been isolated from one another, and unoccupied but potential habitat would not have been conserved at all. According to the plan's environmental impact statement (EIS), such fragmentation would have imposed "potentially severe adverse long-term impacts on the viability of the species and the supporting ecosystems in the area" (City of Austin and Travis County 1996, 2).

On the other hand, consistent with the pessimistic model, the plan imposes substantial risk on the region's natural systems. It aims to protect only a small portion of covered species' habitat: the EIS estimates that 30,000 to 60,000 acres of land will be developed over the 30-year life of the permit, including up to half of occupied vireo habitat, 71 percent of potential warbler habitat, and nearly 85 percent of potential karst invertebrate habitat (City of Austin and Travis County 1996). The plan thereby allows the "take" of 55 percent of the black-capped vireo population and anticipates that four of the nine known populations of bracted twistflower will be lost, adding that the plan does not adequately protect this plant species.[15]

In addition to conserving only minimal acreage, the planned preserve areas do not adhere to the configuration recommended by the BAT. For example, the plan allows 20 percent of designated preserves to be within 330 feet of the preserve boundary or other type of edge, whereas the BAT advised allowing less than 5 percent of the preserve to be in that category (Hood 1998). Furthermore, the plan fails to designate buffer areas. According to biologist Chuck Sexton (2003), when consulting biologists agreed to the 30,000-acre figure for the preserve, they stated the set-asides would need to be buffered from urban encroachment by lower-density, restricted development. But development interests adamantly opposed buffers, or, for that matter, any land-use controls. The end result was a hard boundary with development allowed to come right up to the edge. Biologists concluded that the only way to make the approach adopted by the BCCP work was to manage the preserve intensively—precisely the approach the BAT had advised against.

Even as it imposes risk on Travis County's natural systems, the BCCP assures landowners that if they meet the terms of the permit, they will not be responsible for any additional land-use restrictions or financial contribution, even if new information reveals a covered species is inadequately protected. Under the plan's "no surprises" provision, if the FWS determines that additional mitigation measures are necessary to conserve a species, the agency—not the landowners or permit holders—has the primary obligation for undertaking and paying for mitigation measures. The only exception is for "extraordinary circumstances," in which case the agency can ask the permit holder to do the absolute minimum.

Implementing the BCCP

Despite its tortured origins and some ongoing challenges from disaffected landowners, the BCCP concept has proven durable. This is less a function of buy-in from stakeholders, as the optimistic model of EBM suggests, than it is commitment by dedicated city and county staff to acquiring preserve lands and managing them primarily to conserve biological diversity. Also contrary to the prediction of the optimistic model, the plan's flexible implementation has not spurred efforts to exceed the legal minimum level of conservation. Because participation in the plan is voluntary, land within the planned preserve areas has extremely high development value, and the permit holders have minimal leverage to spur participation, landowners have continued to obtain individual permits when they believe they can cut a more favorable deal with the FWS. As a consequence, the preserve has been fragmented well beyond what scientists had deemed acceptable. Moreover, despite concerted efforts to coordinate management and monitoring, there is little capacity for adaptation because there is no central entity that can synthesize and disseminate information gleaned from monitoring; in addition, the combination of the "no surprises" clause and rising land prices virtually precludes acquiring additional land despite information showing the existing preserve is biologically inadequate. In any case, managers are struggling—with limited staff and funding—to maintain the biological value of the existing preserves in the face of urban encroachment and escalating demands for public access.

Assembling the Preserve
In mid-July 1996, Travis County issued the first mitigation certificate under the BCCP for a single-family house on a 3.5-acre tract on the southern shore of the Colorado River. The event hardly signified smooth sailing for the plan, however: within a year of its approval, the Texas legislature was considering a bill that would slash the BCCP's funding, revise its biological underpinnings, and impose new requirements for review and approval of HCPs in Texas (Haurwitz 1997a). Legislative sponsors crafted the bill—dubbed the Death Star by BCCP supporters—in response to vociferous complaints by landowners who were furious that their land had been designated for acquisition but had not yet been purchased. Although the legislation failed, complaints by those who claimed they were being treated unfairly created additional pressure on both the jurisdictions and the FWS to accommodate their concerns.

Even more serious, development in some of the highest-quality warbler habitat was undermining the biological integrity of the preserve: by the fall of 1997, 21 single-family lots, two subdivisions, and one water-line project had enrolled in the BCCP; meanwhile, however, developers were seeking an ever-larger number of individual permits from the FWS for land that was supposed to have been acquired for the preserve. Some large landowners outside the preserve were choosing to work directly with the FWS as well. As lawyer Alan Glen explains, they did this in order to customize their permits, an option that was not available through the BCCP (Haurwitz 1997b). (Others suggest that development lawyers advised their clients to work with the FWS rather than the city or county because doing so was more lucrative for the lawyers.) In addition, some landowners were simply grading land without permission from either the jurisdictions or the FWS.[16] As a result, actual and potential habitat was disappearing, and the city and county were not raising the money they needed to assemble the preserves. In hopes of increasing participation in the BCCP, the permit holders agreed to drop the price of participation certificates for songbird habitat from $5,500 per acre to $3,000 per acre.

Despite these handicaps, thanks largely to aggressive fund-raising by Travis County, as of the fall of 2007 the Balcones Canyonlands Preserve (BCP) included 44 cave sites and was only 2,522 acres short of its target. On the other hand, with revenues from participation certificates lower than expected and land values escalating, the prospects for completing the reserve system were receding. (The Balcones Canyonlands National Wildlife Refuge was suffering the same problem as the BCP: although the federal government had acquired some 20,000 acres, the FWS had expanded the refuge boundary several times, as parcels within the initial target area were developed before the government could acquire them.)

Even more important than the number of acres set aside is their configuration, and the BCP is far more fragmented than scientists on the BAT had hoped it would be. Rather than the contiguous blocks that scientists recommended, the preserve system is a patchwork of disconnected properties; the FWS has allowed development on more than 1,300 acres of high-quality habitat that planners had targeted for preservation. Finally, the preserve system is not buffered from urban encroachment. Although scientists had insisted that warblers would need a 330-foot buffer around core preserve areas, in most places development comes right up to the edge and trails crisscross the remaining open land. There are no land-use regulations governing activities in the matrix surrounding the BCP; in fact, landowners often demand brush clearance at the wildland-urban interface for fire protection, which further erodes the preserve's biological value (Koehler 2003).

Management and Monitoring

The BCP's minimal size and substantial fragmentation have made intensive management and monitoring essential to its long-term biological viability. Responsibility for these activities is divided among the local participants: the permit holders, Austin and Travis County, and their cooperating partners—the LCRA, the Nature Conservancy, and the Travis Audubon Society. The partners meet quarterly to discuss trends and produce joint annual reports that document site-by-site monitoring activities as well as extensive management, including fencing, prescribed burns, clearing of weeds and invasive species, and trail management. There is no formal mechanism, however, for synthesizing monitoring information into a preserve-wide assessment or for feeding the results to managers. Moreover, although the partners communicate frequently, the relationship between the city and county remains uneasy: the city has completed its acquisition but does not have the resources to monitor and manage its holdings; meanwhile, the county has a steady stream of money to devote to monitoring and managing its preserves but is struggling to meet its acreage target. Periodically, disagreements flare over whether and how to count land toward preserve acreage.[17]

In addition, preserve managers face intense challenges because of the preserves' proximity to the urban area. Despite managers' best efforts, infringement on the preserves poses a serious and growing hazard. Blue jays, cowbirds, fire ants, feral hogs, and an exploding white-tailed deer population threaten the songbirds' nests. The cats and dogs of nearby residents, as well as nonnative garden plants, crowd out more sensitive species. Illegal dumping, property encroachment, and unauthorized public use degrade the land as well (J. Johnson 2000; Mottola 2005; Stiffler 2005). Urban encroachment continues to undermine the best efforts of preserve managers because the city regularly grants variances to its relatively strict environmental standards and because many developments are building out under approval granted prior to the imposition of stricter development standards enacted in the 1980s (Duerksen and Snyder 2005).

The most visible management challenges arise with respect to golden-cheeked warbler habitat. According to their permit, the partners are supposed to manage the BCP to "control human activities to eliminate or mitigate any adverse impacts of human activities to the Warbler." To this end, managers have limited access to the preserve, and they close some areas completely during the six months that warblers breed there (Mottola 2005). Curbing recreational activities has been controversial, however; for example, in the spring of 1999, park officials' proposal to ban mountain biking and horseback riding on BCP lands prompted a public outcry. By 2005 managers were allowing mountain biking and running in about 25 percent of the BCP and hiking in 32 percent (though in some sections groups of more than three were prohibited), mainly on land that was already public parkland prior to adoption of the BCCP (Mottola 2005). To reduce the chance that public access would harm conserved species, preserve managers devised strategies to increase visitors' sensitivity; for example, the city offers classes in nature conservation that grant permits to visitors to hike, bird-watch, and photograph year round in some preserves (Alford 2000; Connally 2006). The hope is that such programs will encourage people to volunteer for data collection and erosion-control duties to supplement the limited resources available to managers.

Despite efforts to educate the public about the preserves' fragility, demand for recreational access continues unabated. Controversy has arisen in part because proponents touted the BCCP to the public as a way of gaining open space, not protecting biological diversity. In an effort to ensure its adoption, Interior Secretary Babbitt compared enacting the BCCP to the creation of Central Park, propagating the belief that the preserve system would be accessible for recreation. Similarly, the Sierra Club's George Avery (1993) stated that the preserve lands would be available for outdoor activities (subject to certain restrictions), including nature hikes, walking, jogging, camping, and swimming. As important, managers feel caught between their mandate to put species' needs first and their desire to increase public support for the preserve system. Willie Conrad, the BCP manager for Austin's water utility, which has jurisdiction over the city's preserves, reminds critics that the primary purpose of land management is to ensure that public access does not harm the creatures the preserves were established to protect. Steve Windhager, a member of the BCP's Scientific Advisory Committee, defends such conservatism by pointing out that many studies have shown that nesting ability for endangered birds declines in areas with trails running through them. He adds that although the impact of humans is poorly understood, managers must "err on the side of caution, keeping bird sanctuary sections closed until it can be shown that human access is safe." But Ted Siff, the BCP's Citizen Advisory Committee chair, counters that "For the BCP's long-term acceptance and appreciation, citizens should be encouraged to understand its purpose by getting them on the land" (Mottola 2005).

Conclusions

The process of preparing and implementing the BCCP has yielded a variety of policies and practices that promise significant environmental benefits. As predicted by the optimistic model of EBM, a landscape-scale perspective prompted the city and county to acquire land in a more biologically sound configuration than would have been possible as a result of project-by-project enforcement of the Endangered Species Act. The process of formulating the BCCP also spurred the creation of the Balcones Canyonlands National Wildlife Refuge, for which the FWS has acquired nearly 21,500 acres through fee-title acquisition and conservation easements. Moreover, the plan has enabled the permit holders to raise more than $100 million, nearly half of that from the federal government, which continues to fund

Travis County's land acquisition (Connally 2006, 2007). Jurisdictions are more aware of the region's biological diversity and are coordinating their monitoring and engaging in activities they might not otherwise have undertaken, such as cowbird management. Moreover, although preserve managers feel compelled to offer some public access, particularly on land that was parkland prior to the designation of the BCP, they are making a concerted effort to manage most of the land for its biological values.

Although the BCCP marks an improvement over the status quo, it is unclear how well the covered species are faring as a result of the plan. A status review of the black-capped vireo released in May 2006 revealed more sightings of the bird but concluded this was a result of more comprehensive surveying rather than an indication of a growing population. The study's authors also pointed out that the threats to the vireo are the same or increasing in the Edwards Plateau region, where Travis County sits. As of late 2007, the FWS was still working on a status review for the golden-cheeked warbler. Although the existing evidence on species' conditions is ambiguous, biologist Chuck Sexton (2003) contends that, had there been no BCCP, "I'm absolutely convinced the cave critters would be essentially extinct and those systems degraded to nothing. The warbler would be substantially on its way to local extirpation in the heart of the species range."

Though it may have slowed the decline of its target species, the BCCP is unlikely to conserve Austin's biological diversity in the long run. Consistent with the pessimistic model of EBM, the plan's acreage target is the bare minimum that biologists would sign off on, and its configuration does not meet the minimum standards established by the BAT. A lack of buffering and land-use restrictions in the surrounding matrix exacerbates the preserves' vulnerability to urban encroachment. The prospects for addressing the preserves' weaknesses through adaptive management are limited by the inability of the permit holders to discern problems across the BCP. The declining availability and rising cost of undeveloped land in the macrosites, combined with the "no surprises" clause, constrain permit holders' flexibility to adjust in the face of information indicating the BCP is biologically inadequate. Moreover, a controversy in 2006–2007, in which the city of Austin proposed putting a water treatment plant on BCP land, provides a clear reminder that even set-aside land is not guaranteed protection (Coppola 2007a).

The biological weaknesses of the BCCP are primarily a result of the pro-development context in which stakeholder collaboration and flexible implementation have occurred. Austin has a powerful and unified business community (Glen 2003), which is tenaciously supported by a state legislature that is reluctant to grant localities the authority to regulate land use. Although the development community was briefly in disarray at the end of the speculative boom of the 1980s, when the BCCP process began, it quickly regrouped during the 1990s, as over 800 high-tech companies moved into the region and Austin's population grew from 450,000 to over 650,000 (Duerksen and Snyder 2005). The development community worked hard to ensure that the plan employed a voluntary approach that relied on incentives rather than land-use regulations. To this end, it drove a wedge between the FWS and local officials by charging them with collusion if they discussed particular properties (Librach 2003; Vosler 2003).

At the same time, Austin's environmental community was ill-prepared to play an effective role in formulating the BCCP: it was not cohesive, and therefore did not speak with a single voice. As a result, despite the centrality of environmental concerns in local culture, political leaders were unwilling to challenge the dominant view of the problem as endangered species hampering development, or of the BCCP as a way to revitalize the regional economy and avert the dire consequences of the Endangered Species Act. Similarly, although some of Austin's journalists worked hard to convey the biological richness of the Hill Country and, in particular, the uniqueness of the species covered by the BCCP, most adopted the more typical framing: the Endangered Species Act was blocking development, and an HCP would end the gridlock. Given its highly constrained context, the stakeholder process yielded neither trust nor transformation among its participants; instead it reproduced the existing balance of power between developers and environmentalists.

David Braun, who chaired the Executive Committee during the plan's formative years, has attributed the plan's weaknesses in part to a lack of public awareness, saying that it took most of the ten years spent getting the plan in place simply to raise public awareness of the endangered species and their needs (Braun 2003). Polls taken in the early 1990s detected strong support for a regional conservation plan, however. For example, an April 1991 survey by RPC Market Research found that more than 80 percent of the 400 Travis County voters polled would support a program of land acquisition for environmental protection, and more than half said they would support higher taxes to fund a bond issue to finance the program (M. Kay 1991b). Similarly, a citywide poll taken in April 1991 asked voters whether they would support a $50 million bond issue to protect endangered species west of the city, as well as Barton Springs water quality, and 58 percent said yes (Collier 1991b). These results suggest that the potential was there for strong public support for a genuinely protective preserve, had environmentalists been able to mobilize it.

Apologists for the BCCP argue that an unfavorable economic situation limited planners' options. But the economy had a mixed impact over the life of the planning process. On the one hand, the original plan was formulated during a lull in the area's explosive growth, which reduced the pressure to develop the remaining habitat. And, thanks to the economic downturn that prompted the savings and loan collapse of the 1980s, the RTC controlled large parcels of land that it could sell at below-market prices for conservation purposes. On the other hand, the earlier boom period had fueled unrealistic expectations about rising land values and left legacies of debt, such as that incurred by the Southwest Travis County Road District. Moreover, declining land values not only reduced property tax revenues but also created caution around spending money to acquire habitat. By the early 1990s the economy had begun to recover, creating the potential to raise money for acquisition; by then, however, land values—and hence the incentive to develop—were rising much faster than tax collection or participation in the plan. In short, the economy had countervailing effects throughout the process.

The federal government had a similarly mixed role in the BCCP. On one hand, the FWS opened an office in Austin to support the plan's development; the RTC made large tracts of inexpensive land available, enabling the city to reach its target acreage more easily; and Interior Secretary Babbitt provided highly visible political support for the plan. Congress designated a new national wildlife refuge in the region and repeatedly provided funding for acquisition of both refuge and BCP land. On the other hand, the FWS has issued permits for development within the planned preserve areas throughout the process, justifying its actions by pointing out that the agency lacks the authority to deny individual permits if an individual project does not jeopardize a species (Vosler 2003).

In short, there is no simple explanation for the BCCP's shortcomings; public support, economic conditions, and the choices made by the federal government all affected the shape of the final plan and its subsequent implementation. But policymakers' reliance on stakeholder-based planning and flexible, voluntary implementation greatly increased the likelihood of a minimally protective plan. Without local political leadership backed by adequate regulatory leverage to offset the power of the development community, there was little chance that the BCCP would achieve its biological goals.

4

Saving San Diego's Coastal Sage Scrub

Shortly after planning for the BCCP began, a similar process got under way about 1,300 miles to the west. In 1991, a handful of jurisdictions in southern California, led by the city of San Diego, established a working group whose charge was to create a plan that would conserve biodiversity in the midst of one of the fastest-growing urban areas in the country. Six years later, the U.S. Fish and Wildlife Service (FWS) and the California Department of Fish and Game (DFG) signed off on the Multiple Species Conservation Program (MSCP), which allowed local officials to grant permits for development that would destroy the habitat of federally listed endangered or threatened species; in exchange, the signatories pledged to acquire, manage, and monitor habitat for those and other species. Interior Secretary Bruce Babbitt hailed the result as "the jewel of habitat conservation plans," a model for the nation that would "preserve the most environmentally sensitive pieces of the San Diego landscape" (Cone 1997).

In terms of environmental protectiveness, the MSCP is an improvement over what had been the status quo in San Diego. Over a 30-year period the plan aims to assemble a 172,000-acre preserve network of biological core areas and wildlife corridors. Although existing regulations would have protected a comparable number of acres, permitting individual projects on a parcel-by-parcel basis without regard for regional impacts almost certainly would have resulted in a more fragmented and less biologically valuable patchwork of open space. The MSCP has also enabled the city and county of San Diego, which together are responsible for more than 90 percent of the preserve, to raise millions of dollars from state and federal coffers, much of which they have used to acquire thousands of acres of habitat.

Despite the plan's achievements, however, there is little reason to believe it will actually conserve San Diego's unique biodiversity. Because of concessions to development interests, the plan allows jurisdictions to meet their goals by including some patches of land that have little biological value while permitting construction on important habitat; in addition, it contains no provision for buffering conserved areas from surrounding urban lands. As a result, even if fully assembled, the resulting preserve will be less protective than what scientists who advised on the plan recommended as being minimally adequate. To compensate for shortcomings in the preserve's design, the wildlife agencies emphasized that adaptive management would be essential to ensuring its biological viability. But jurisdictions' inability to agree on a regional funding mechanism has impeded efforts to manage set-aside tracts and gather information on their biological effectiveness. In any case, the dearth of additional land, combined with legal assurances that limit landowners' obligations, sharply constrain managers' ability to adjust preserve boundaries given information suggesting that species covered by the plan are in trouble. In short, the MSCP imposes the risk of failure squarely on the natural system.

The main reason the MSCP is only minimally protective is that negotiations occurred within a context that enabled the region's development community to define the problem to be addressed and dominate negotiations among stakeholders during critical stages of the planning process. San Diego's growth machine—including developers, builders, and large landowners, often with the backing of local officials—characterized the problem as the obstacles to development caused by endangered species listings, and the main purpose of the plan as streamlining the regulatory process and preventing future listings. Throughout the planning process they focused on meeting legal requirements and minimizing the cost of assembling the preserve, staunchly resisting efforts to adopt a precautionary (or environmentally risk-averse) approach. An overmatched environmental community failed to build a broad coalition or mobilize the public, and so was unable to compel local political leaders to support an environmentally protective plan. Even environmentalists' ostensible allies in the wildlife agencies became progressively less willing to take a hard line and focused more on completing a plan as the negotiations drew to a close. The result was a plan that sought to ensure development while conserving, to the extent possible, the region's biological diversity. A reliance on flexible, adaptive implementation has done little to ameliorate the plan's weaknesses because funding shortfalls have severely curtailed management and monitoring efforts, and local officials have resisted imposing stringent limits on development.

The Origins of San Diego's MSCP

The MSCP planning process occurred at the tail end of a decades-long development whirlwind in San Diego that began in the 1970s. Environmentalists had campaigned persistently for growth control during that era, but they had only limited success against the city's extremely well-oiled growth machine: although proponents of growth management gained recognition and clout in the 1980s, the region's sprawl continued virtually unabated. In the early 1990s, however, the threat of an endangered species listing gave environmentalists an apparent trump card and compelled developers to participate in an entirely different kind of planning process—a landscape-scale effort to conserve the region's remaining biodiversity—that promised to achieve much more environmentally beneficial results.

San Diego's Sprawl

When the first white settlers arrived in the San Diego region, they encountered a landscape full of plant and animal species that had evolved in the region's sunny, temperate Mediterranean climate. San Diego's regional ecosystem was dominated by coastal sage scrub—a mix of low-growing, drought-tolerant shrubs such as sage, coastal sagebrush, California buckwheat, and lemonadeberry, as well as succulents like prickly pear and cholla cactus that grow nowhere else in the United States (USFWS 1993a). Coastal sage scrub communities were embedded in a shifting mosaic of other habitat types, including grasslands, chaparral, oak woodlands, and riparian systems. Those plant communities provided mating, nesting, and foraging habitat for thousands of species of insects, birds, frogs, lizards, and mammals, many of which were endemic to southern California.

Well into the twentieth century the landscape remained relatively undisturbed—San Diego was a sleepy, slow-growing city whose economy depended primarily on a handful of military installations. In the 1970s, however, drawn by the region's weather and topography, people began flocking to the city. In 1974, as this population boom was getting under way, environmental planners Donald Appleyard and Kevin Lynch issued a 51-page report titled *Temporary Paradise?* that warned San Diegans their city was losing its best qualities. The authors urged residents and city officials to recognize that "This bold site, its openness, its sun and mild climate, the sea, the landscape contrasting within brief space are (along with its people) the wealth of San Diego. They are what have attracted settlers to the place and still attract them. They must not be destroyed" (4). But Appleyard and Lynch's advice had little impact. Thanks to San Diego's parochial and relatively conservative political culture, strong Republican majority, and long history of city and county governments controlled by developers and their allies, the politics of growth prevailed. Between 1980 and 1990 San Diego County's population grew 34 percent, reaching 2.5 million, and the process of accommodating the newcomers followed the sprawling pattern common throughout the United States: developers built a spiderweb of roads leading to subdivisions and malls far from downtown.

In the mid-1980s some environmentally concerned residents— infuriated by a series of developments that undermined the city's existing growth management plan—mobilized to force the city to control growth and experienced a measure of early success.[1] In 1985, after a hard-fought battle in which pro-growth forces outspent environmentalists by a factor of ten, San Diegans approved a ballot initiative (Proposition A) that required a popular vote to change the status of property in the designated "future urbanizing area." The following year, residents elected Mayor Maureen O'Connor, a proponent of slow growth, and in 1987 she proposed an interim development ordinance (IDO) that aimed to put the brakes on the city's accelerating expansion. Faced with the threat of an even more draconian, citizen-sponsored growth-limiting ballot initiative, the City Council agreed to pass the mayor's measure. But the City Council's subsequent unwillingness to enforce the IDO provoked yet another round of ballot initiatives in 1988. This time—in response to a barrage of negative publicity by the building industry, combined with the complexity of the initiatives themselves—voters rejected all four options. Recognizing the depth of public concern about growth, however, both the City Council and the County Board of Supervisors enacted resource-protection ordinances in the summer of 1989.

Although it gave local officials some tools to regulate growth, this incremental series of regulations did little to curb sprawl in San Diego: both the city and the county routinely granted exceptions to and exemptions from local ordinances and plans in approving new projects; in practice, mitigation requirements were flexible, upzoning was common, and review under the California Environmental Quality Act (CEQA) was perfunctory (USFWS 1993a). Even when state or federal approval of a project was required because of wetlands or endangered species on the site, regulators focused solely on the individual parcel, not the cumulative impact of development. As a result, by the early 1990s developers had bulldozed between 70 and 90 percent of the region's original vegetation, including more than 85 percent of its coastal sage scrub, to make way for highways, houses, businesses, and shopping malls (USFWS 1993b). The habitat that remained was highly fragmented and often badly degraded, and many of the creatures that relied on coastal sage scrub for shelter and foraging were in precipitous decline.[2]

The Impetus for a Multiple-Species Habitat Conservation Plan

Fortuitously, an opportunity to conserve what remained of San Diego's biodiversity arose when entrepreneurial officials in the FWS's Carlsbad office urged the city's Clean Water Program to mitigate the biological impacts of a proposed sewer upgrade and expansion by designing a multiple-species habitat preserve for the 582,243-acre (900-square-mile) area covered by the sewer system.[3] Following the FWS's recommendation, in March 1991 the Clean Water Program brought together a small group of stakeholders to advise policymakers on the features of a regional preserve and possible mechanisms for assembling it, and in July that group began its work in earnest. Providing impetus for developers to come to the table was the proposed endangered species listing of a small songbird, the coastal California gnatcatcher (*Polioptila californica californica*), which had become vulnerable to parasitism and predation as its low-elevation coastal sage scrub habitat became sparser and more fragmented. In late 1990, with the help of the Natural Resources Defense Council, ornithologist Jonathan Atwood—whose work had prompted the subspecies designation for the California gnatcatcher—petitioned both the FWS and the state to list the bird as endangered.[4]

For San Diego's environmentalists, the problem was simple: unfettered development was gobbling up the region's open spaces and, as a result, destroying its natural heritage along with residents' quality of life. As biologist and county park ranger Robert Patton explained, preserving the gnatcatcher and its habitat was a moral imperative:

The only reason it's in danger of extinction is because of man's activity and man's economic growth. How do you balance the life of a whole population of an entire species against us building homes for a species that's overpopulated the planet and possibly destroying it? To me, that's a real moral dilemma. (Rodgers 1991)

For developers and many local officials, however, the problem was the arduous, duplicative, and costly regulatory process that ensued when they proposed building in the habitat of state- or federally listed species. In the outcry that followed the listing proposal, the building industry challenged the FWS's assertion that developers had razed 90 percent of the region's coastal sage scrub, putting the figure instead at 65 percent; ridiculed scientists' suggestion that gnatcatchers strongly prefer low-elevation habitat; predicted that listing the bird would spark lawsuits; and commissioned a study that forecast that protecting it would prompt an economic catastrophe, costing as many as 212,000 jobs and more than $20 billion in business activity (Silvern 1991a). Painting environmentalists as misanthropic extremists, builders dismissed arguments for listing the gnatcatcher as "a no-growth agenda bordered in green" (Coddon 1991).

Concerned about an impending collision between developers and environmentalists in southern California, in the spring of 1991 the state stepped in. At the urging of the Irvine Company, one of southern California's largest developers, Governor Pete Wilson proposed a new state initiative, the Natural Communities Conservation Program (NCCP), under which landowners would voluntarily conserve some of their land as habitat on the condition that endangered species designations be put on hold. From its inception, the NCCP aimed to reconcile continued development with environmental protection; as Douglas Wheeler (1996, 8), secretary of the state's Resources Agency, explained: "The goal is to anticipate and prevent the controversies and regulatory gridlock caused by the single species approach, and substitute for it the long-term stability of complete ecological systems."

The NCCP Act, which was signed by the legislature in October 1991, authorized a pilot program in a 6,000-square-mile region of southern California that would focus on preserving the coastal sage scrub ecosystem. In the meantime, although the state declined to list the gnatcatcher (giving the new NCCP Act as its rationale), the FWS kept the pressure on San Diego by announcing in September 1991 that a federal listing was warranted and promising to make a final determination within the year. Then, in early December, the state and federal governments signed a memorandum of understanding indicating that southern California's pilot NCCP plans could also serve as habitat conservation plans (HCPs) under the Endangered Species Act.[5] Shortly thereafter, San Diego enrolled its inchoate MSCP planning process in the new state/federal endeavor. In December 1993, in an effort to bolster participation by landowners and jurisdictions, the FWS enhanced the incentive to enroll in the MSCP by issuing a "special rule" that listed the coastal California gnatcatcher as "threatened," licensed the incidental take of gnatcatchers in return for the preparation of a combined NCCP/HCP, and permitted participating jurisdictions to allow the destruction of 5 percent of their remaining coastal sage scrub while the plan was being prepared.

Collaborative, Landscape-Scale Planning

In some respects the MSCP's collaborative, landscape-scale planning process was consistent with the optimistic model of ecosystem-based management (EBM). Despite the relatively short time allotted and many remaining uncertainties, consultants developed a sophisticated scientific assessment of San Diego's remaining biological diversity. Planners subsequently shifted their focus from individual development projects to biological resources. Furthermore, although the various stakeholders in the working group had different values and competing interests, over the course of the multiyear planning process relationships among them improved. On the other hand, more consistent with the pessimistic model, the collaborative process was not transformative: rather than agreeing on a common vision, participants struck pragmatic bargains on some issues, failed to reach consensus at all on others, and adopted vague language that concealed underlying disagreements. Nor did collaboration precipitate agreement on the science: developers consistently challenged the technical basis for planning, commissioning competing studies and eventually proffering their own preserve design. Thanks to a development-friendly context and environmentalists' inability to generate countervailing political power, developers ultimately were able to exert superior power in negotiations over the preserve design.

The MSCP Working Group
The original advisory committee established by the Clean Water Program in March 1991 consisted of public agency officials and large developers. Within two months, however, the city—hoping to enhance the group's legitimacy—contacted several environmentalists and invited them to select representatives to participate. The 26-member MSCP Working Group put together by the city's Clean Water Program included representatives from the large landowner/development and environmental communities; officials from the city and county of San Diego, as well as

from special districts and nine other jurisdictions covered by the sewer system; and state and federal wildlife agency officials. In 1992 the group added several new members, including a representative of the Endangered Habitats League (EHL), a moderate environmental organization formed specifically to participate in NCCP planning, and a representative of the San Diego County Farm Bureau.[6] To chair the group, the Clean Water Program selected landscape architect Karen Scarborough, president of the Citizens Coordinate for Century 3, a nonprofit organization dedicated to promoting urban planning. Scarborough received the unanimous approbation of the group, and by all accounts she was a determined and capable leader who was dedicated to reaching agreement on a plan. As co-chair, the group approved Jim Whalen—an affable and politically adept representative of a newly formed consortium of developers that called itself, somewhat disingenuously, the Alliance for Habitat Conservation. The city contracted with Ogden Environmental & Energy Services, a highly regarded local firm, to take the lead in conducting the biological assessments and land-use mapping that would form the technical basis of the preserve design. Ogden was also responsible for incorporating the consensus positions of the Working Group into the language of a draft plan that would serve as a basis for policymakers' deliberations. Ogden subcontracted with consultant Rick Alexander to mediate Working Group discussions and prepare a series of issue papers on policy questions around which consensus would be built.

The Working Group met at least once a month, and sometimes more frequently. At first, relations among participants were strained, as each stakeholder struggled to establish his or her position. Working Group meetings were generally civil, however, and over time tensions eased; according to most participants, a fragile trust eventually developed among many former adversaries (Grunewald 1998; Katz 2000; J. Whalen 1999). The group achieved this result, however, partly by marginalizing those who were not perceived as "reasonable." Jim Whalen recalls: "Outlandish positions tended to be dismissed and even ridiculed in the deliberation process of the working group as a whole" (1999, 259). In addition, both developers and environmentalists lobbied policymakers throughout the planning process, and in doing so they periodically charged their adversaries with disingenuousness and subversive tactics, suggesting that participants remained wary of one another. For example, a September 1995 letter sent by Michael Beck (1995) of the Endangered Habitats League to the city's Natural Resources, Culture & Arts Committee, at the tail end of the planning process, described development interests' claim that the Working Group had adopted a 1:1 mitigation standard as "absolutely inaccurate," and characterized the development coalition's support for an early ballot measure on MSCP funding as "a new and disturbing tactical position on funding."

Moreover, despite participants' willingness to cooperate, there was little evidence that their perceptions of their interests were transformed; in fact, letters and memos addressed to policymakers reveal that stakeholders' positions on important issues changed little over the course of the five-year planning process. From the outset, the developers' main objective was to gain regulatory certainty and streamlining while minimizing their financial contribution to the preserve. They recognized that maintaining some open space made good business sense, but they remained adamant that economic and biological considerations should be on equal footing. Environmentalists, by contrast, wanted to maximize—or at least increase substantially—the likelihood that the region's dwindling plant and animal species would survive in the long run, and they remained convinced that biology ought to be the primary consideration in preserve design. These divisions were evident as late as 1996, in a discussion among representatives of key stakeholder groups televised by UCSD-TV, in which builders' and developers' representatives sparred with an environmentalist and a wildlife agency official over the cost of assembling the preserve, the need for regulatory certainty, and equity among current and future homeowners—the very issues stakeholders had disagreed about from the beginning (UCSD-TV 1996).

Because participants differed fundamentally on key issues, the Working Group was unable to resolve a host of important questions; as a result, many of the hard decisions were made outside the group, in subcommittees or in closed-door meetings between local officials and representatives of the wildlife agencies. Both developers and environmentalists remarked repeatedly on the Working Group's tendency to paper over or set aside thorny issues. For example, the Alliance for Habitat Conservation wrote (J. Whalen 1992b): "The Alliance has been most concerned with the possible inability of the Working Group to address more difficult issues, and the undue focus on 'compromising' in order to produce an Issue Paper." The Alliance went on to observe: "By minimizing discussion and resolution of the contentious issues, each Issue Paper becomes little more than a statement of generic concerns and positions without resolving any of the real issues." As a consequence, it said, "The language of the Issue Papers allows for substantial interpretation, and the creation of policies that are diametrically opposed, each of which could well fit within the intent of a given Issue Paper, depending upon a party's viewpoint." Similarly, the Endangered Habitats League and the Sierra Club repeatedly admonished consultants for mischaracterizing issue paper language as consensus statements of the Working Group.

As important as its inability to resolve conflict on major issues was the Working Group's propensity to reflect rather than ameliorate existing differences in power among environmentalists and developers. At the outset, environmentalists had some modicum of credibility. The events of the 1980s—particularly voters' approval of the 1985 growth-limiting ballot initiative—had made manifest their political appeal (Adams 2005); the gnatcatcher listing strengthened their hand. But their resources were inadequate in many important respects: above all, most were unschooled in the intricacies of local zoning rules and resource-protection ordinances (Rolfe 2003). Developers, by contrast, were intimately familiar with local land-use regulations and practices and, unlike environmentalists, had the resources to hire consultants to develop their own scientific, economic, and legal analyses. The emphasis on being "reasonable" in Working Group negotiations further advantaged development interests: because virtually unfettered development was the status quo in San Diego, and reasonableness is—by definition—avoidance of positions that depart from convention, only those who conceded the inevitability of development were treated as legitimate contributors. Over time, developers' superior resources effected a shift in participants' commitment to the process. Developers initially came to the table reluctantly and environmentalists more enthusiastically, but as time elapsed and they saw their concerns being addressed, developers became more committed, and some environmentalists became disaffected (Greer 2003).

Laying a Scientific Foundation

Although developers would dominate over time, from environmentalists' perspective the MSCP process began auspiciously with the preparation of a rigorous scientific assessment of the region's habitat and a set of state-of-the-art preserve design criteria. At the outset, wildlife agency officials pressed Ogden to focus on conserving regional biodiversity. According to the Working Group's Scope of Work, which was taken directly from the sewer system project environmental impact statement (EIS) drafted by the FWS:

The MSCP will be designed to identify, evaluate, and delineate a network of lands that, if acquired and properly managed, would conserve habitat and provide for wildlife movement on a large scale. The network of managed lands is intended to enhance the long-term biodiversity of the greater San Diego area by conserving habitat and, thereby, preserving sensitive species of wildlife. (USFWS 1991, 1)

FWS officials reiterated this protective language as Ogden began preparing the plan's scientific basis. For example, in response to a proposed set of biological criteria for the preserve that—consistent with the charge from the Working Group—focused on listed species, FWS's Nancy Gilbert urged Ogden's lead biologist, Jerre Stallcup, to emphasize that an "essential function of the reserve network is not only to conserve species proposed for listing as endangered or threatened but to conserve biological diversity" (Gilbert 1992, 1). Such a network, said Gilbert, would "preserve most extant species in self-maintaining landscapes which sustain ecological processes and maintain evolutionary opportunities. It is essential," she added, "that the preserve network ensure that natural processes are maintained and human influences minimized" (1).

The development community fundamentally rejected the notion of assembling a biologically based preserve, however, and insisted that doing so would leave insufficient land to accommodate anticipated demand for new development. The Alliance for Habitat Conservation portrayed the development industry as a victim of excessive regulation whose fortunes were tightly linked to those of San Diego. "The industry is now at a breaking point," it said, "where the addition of any new costs will result in further significant reductions in housing starts, jobs, and government revenues, together with a continuing decline in housing affordability" (J. Whalen 1993c, 4). A better approach, said the Alliance, would be to shift the Working Group discussion from biology to financing and allow a preserve system to emerge once the funding constraints had been established.

Nevertheless, the "Biological Objectives and Criteria for Identifying Preserve Planning Areas," prepared by Ogden Environmental & Energy Services in consultation with local and national experts, reflected the FWS's interest in establishing a strong biological foundation for the MSCP while retaining an emphasis on listed species in deference to some members of the Working Group. Specifically, the document stated that the biological goal of the MSCP was "to maximize and enhance biological diversity in the region and to conserve viable populations of endangered, threatened, and key candidate species and their habitats within the MSCP Study Area, thereby preventing local extirpation and ultimate extinction" (Ogden 1992, 1–2). It laid out a set of criteria to be used in identifying land for set-asides, including the extent and richness of high-quality habitat and corridors for species covered by the plan, and the density and richness of covered species within those areas. It also established a set of considerations based on state-of-the-art conservation biology principles, such as large size, vegetative diversity, shape that mini-mizes edge-to-area ratio, the ability to be connected by adequate corridors to adjacent patches, adequate distribution throughout the MSCP area, connectivity with patches outside the MSCP area, buffering by limited-intensity development, and the ability to be managed for the desired use. These criteria were nearly identical to those simultaneously being devel-oped by the NCCP's prestigious Scientific Review Panel (SRP 1993).

Outside experts who reviewed Ogden's work were generally compli-mentary, but expressed concern that the technical analyses on which the firm based its biological criteria were, if anything, insufficiently precau-tionary. For example, biologist Michael Gilpin (1992) characterized the population viability study for the gnatcatcher as the best that had been done to date. At the same time, he commented, "The conclusions of the report, which I believe to be optimistic based on our current understand-ing, are alarming for the fate of the California gnatcatcher" (3). Similarly, ecologist Peter Kareiva (1993) of the University of Washington wrote that Ogden's population viability analysis for the gnatcatcher was one of the better ones he had seen, but noted it nevertheless "could well severely underestimate the threats to the gnatcatcher..." (2). The FWS affirmed that Ogden's population viability analysis was "likely to give an overly optimistic view of gnatcatcher viability" (Office Supervisor 1992, 1). Notwithstanding the tenor of the reviews by outside experts, develop-ment interests disparaged Ogden's analyses as unduly precautionary and hired their own experts to prepare competing studies.[7]

In hopes of creating a planning framework that all the stakeholders could live with, in the early fall of 1992 the Working Group appointed a Biological Task Force comprising biologists from several local consulting firms, the county, and the wildlife agencies to prepare a set of guidelines that would ensure consistency among all the HCPs prepared for western San Diego County. In its draft Biological Standards and Guidelines (Bio Guidelines), released in October 1992, the Task Force acknowledged that although it could provide biological tools, the final plan for the preserve would also take into account land-use, economic, and political considerations. This caveat prompted ornithologist Jonathan Atwood (1992) to worry:

On a more philosophical level...the document emphasizes that..."economic and political feasibility of acquisition" may receive equal weight to the biological criteria. At this point, the whole concept starts to fall apart, and what we end up with is local jurisdictions and landowners only agreeing to conserve those areas that they find economically acceptable. Instead of the process identifying and protecting key parcels of habitat and allowing development to proceed outside the bounds of this reserve, it looks to me like we ask the developers to show us the areas they're willing to give up, and then do the best that we can to put together a reserve system from the "leavings."

At the same time, despite the Task Force's assurances, developers and jurisdictions objected to the Bio Guidelines as too protective and eventually succeeded in diluting their influence on the planning process. At a meeting with wildlife agency representatives in late January 1993, landowners and jurisdictions expressed frustration with the potential for overreaching on the preserve system under consideration, since it was not legally required to be comprehensive. According to the meeting summary (Anon. 1993), "There was strong feeling from the jurisdictions that land use and equity considerations were not getting nearly enough emphasis" (2). When the FWS and biological consultants asked why landowners and jurisdictions were so averse to the Bio Guidelines, the landowners responded that they worried about "zealous junior planners." Each of these "local Aldo Leopolds" would see it as their "mission in life" to assure that the standards set in the guidelines were met, "exacerbating an already untenable entitlement situation" (3). According to Jim Whalen's description of the same meeting, the Working Group's Bio Guidelines Subcommittee agreed that "The bio-guidelines will not be used to plan preserves, and the language in the document should be changed to remove the absolutist wording." He added: "Tradeoffs must be permitted to address jurisdictions' general plans, the value of land, and the other non-biological issues, like equity between jurisdictions, which cannot be ignored" (Whalen 1993a, 1).

Land-use analyses made manifest the likely political challenges of developing a preserve based on a holistic and precautionary interpretation of the available science, as advocated by the FWS. To identify land eligible for inclusion, Ogden devised a habitat evaluation model and, working with technicians from the San Diego Association of Governments (SANDAG) and San Diego State University, conducted a "gap analysis." These exercises entailed digitally mapping the distribution of habitats and species targeted for conservation, identifying the highest-quality habitat, and comparing their distribution to the configuration of land already in public ownership or otherwise protected. They revealed that about 41 percent of the 582,243-acre MSCP study area was urbanized (developed or disturbed), 5 percent was in agriculture, and 54 percent (315,940 acres) was covered by a variety of habitats, most of which were sensitive or rare (City of San Diego 1998). Within the region's remaining habitat, Ogden identified 16 core biological resource areas and associated linkages totaling 202,757 acres.[8] Land-use maps showed that only 17 percent of the habitat in the study area was already preserved as biological open space, however, and nearly two-thirds (194,563 acres) was privately owned. Complicating matters, according to adopted general and community plans, about 40 percent was slated for low-density residential development (City of San Diego 1998).

As it had with the biological analyses, the Alliance for Habitat Conservation criticized Ogden's habitat evaluation model and composite map, which it said reflected "subjective assumptions that are seriously flawed" (J. Whalen 1993b, 1). Instead, the burden of proof ought to be on advocates of protection, the Alliance argued, asking "If, in fact, there are 12 [core areas], what studies exist to suggest that preserving 11, or even fewer, won't provide the long-term viability of the species?" (2). Developers argued that rather than taking proactive measures, the MSCP should *not exceed* existing legal minimum requirements set by the state or federal government, particularly for wetlands or nonlisted species (Birke 1993; McKinley 1993). If it did, warned Leonard Frank (1993) of Pardee Construction, it could prompt a "tidal wave of 'taking' lawsuits."

Mapping the Preserve

Despite their vocal objections, developers were unable to discredit the technical basis for the MSCP preserve. They were, however, able to stymie agreement in the Working Group and get discussions around key decisions moved to more hospitable forums: the individual jurisdictions. Although environmentalists generally retained the backing of the biological consultants and the wildlife agencies, development interests held more sway with local officials, who were the primary decision makers. As a result, developers managed to advance an overall preserve design that ignored important biological considerations. They also prevented mitigation requirements from exceeding historic levels and fended off demands for buffers.

Overall Preserve Design   Initially, the Working Group asked the consultants to develop a map depicting the region's core and linkage areas based on the Habitat Evaluation Model and other input. The resulting map included most, but not all, publicly owned land, as well as some military land, depending on its biological value. The Working Group then asked Ogden to design two preserves within that map: a multiple-habitats (MH) alternative and a developer-backed coastal sage scrub (CSS) alternative that would set aside the bare minimum acreage capable of supporting the gnatcatcher.[9] Disregarding its limited mandate, Ogden proffered an additional, biologically preferred alternative, which—although constrained by an acreage limit imposed by the Working Group—preserved more land than either the CSS or MH option and, more important, protected most of the core and linkage areas. The wildlife agencies concluded that the biologically preferred alternative was, in fact, the minimum configuration likely to conserve San Diego's remaining biological diversity, given the extensive destruction and fragmentation of habitat that had already occurred (Eng and Kobetich 1994b).

Environmentalists, emboldened by the existence of a biologically preferred option, pressed for a "compromise," the result of which was a modified multiple-habitats alternative that set aside more acreage and a greater variety of habitats than the CSS alternative but was less comprehensive in protecting core and linkage areas than the biologically preferred approach. Appalled by the amount of private land included in the modified multiple-habitats proposal, in late 1993 the Alliance for Habitat Conservation proffered what it called "a solution to the MSCP dilemma": the public lands alternative (PLA). To come up with this option, developers' consultants began not with the biological core and linkage areas but with all parcels that were already publicly owned, regardless of whether they were large or small, contiguous or isolated. They also included all the region's military bases—a move that prompted the Navy to withdraw from the Working Group. From there, they added the minimum private lands necessary to get legal coverage for the ten species that were at some stage in the listing process. Finally, they incorporated linkages among core areas based on those included in the CSS alternative, disregarding the fact that the wildlife agencies had rejected that option partly because of its inadequate linkages. According to its authors, the primary goal of the PLA was not to conserve biological diversity but to "minimize cost, business disruption, and the need for costly and controversial major revisions to existing long-range plans" (AHC 1994, 1). The wildlife agencies made it clear, however, that the public lands alternative did not achieve the MSCP's objectives (Eng and Kobetich 1994a).

In December 1993 the Working Group's draft report presented four alternatives for the MSCP—the coastal sage scrub, multiple-habitats, biologically preferred, and public lands alternatives—but the group could not agree on a single approach. Hoping to break the deadlock, in April 1994 San Diego Mayor Susan Golding proposed her own "compromise" option, the Multi-Habitat Preserve Area (MHPA). The MHPA would aim to conserve a target of 155,000–165,000 acres configured to capture 70–80 percent of the biological core areas and 50–60 percent of the corridors identified by Ogden. Like the public lands alternative, the MHPA began with publicly owned properties and built up from there; consistent with developers' preferences, it kept the amount of affected private property—and in particular the amount of land that would have to be acquired—to a minimum. Mayor Golding offered an economic rationale for her approach, saying, "I started with the [developers'] public lands alternative, not because there is no valuable habitat on it, but because obviously it is owned by the public and part of the problem with this whole process is how do you pay for it" (LaRue 1994). The Department of Fish and Game criticized the MHPA, pointing out that many publicly owned lands were too small or isolated or degraded to serve as core or linkage lands, and the assumption that planned open space lands would be available or appropriate for inclusion in the preserve was unwarranted. Nevertheless, with the mayor's imprimatur, the MHPA quickly gained political momentum: the San Diego City Council adopted it unanimously, and it became the focus of wildlife agency evaluations and subsequent Working Group negotiations.

Preserve Boundaries   Development interests also held sway with respect to the contentious and even more critical issue of whether and how to draw a boundary around the preserve—a move that would limit local officials' discretion in permitting development. Unable to reach consensus on this issue, the Working Group decided to allow each of the 11 jurisdictions covered by the plan to delineate its own boundary. Environmentalists complained that such a bottom-up approach would defeat the purpose of comprehensive planning and increase the risk of failing to meet environmental goals, and they pressed hard for the MSCP to include performance standards that would limit jurisdictions' flexibility (Adams 1993; Silver 1993a, 1993b). But the Working Group could not agree on endorsing that approach, and policymakers chose instead to establish the preserve's border through independent subarea planning, much of which consisted of informal negotiations among developers, local officials, and wildlife agency representatives.[10]

In subarea plan discussions with the city of San Diego, most large landowners succeeded in placing all or most of their property outside the MHPA, with the understanding that they would direct any required mitigation toward the preserve. The city, in consultation with a San Francisco law firm, decided that landowners whose property was wholly within the MHPA boundary would be allowed to develop up to 25 percent of their property—approximately what they would have been allowed under the city's existing resource protection ordinance (Greer 2003), and those whose lands straddled the preserve would be allowed to develop up to 40 percent.[11] The city calculated that, taken together, these measures would enable it to conserve 90 percent of its portion of the MHPA.

The county faced a more serious challenge because in the midst of its subarea planning process, several groups emerged—including the San Diego County Business Coalition, the San Diego Association of Realtors, and the newly formed Citizens for Private Property Rights—and began issuing vague threats of "takings" lawsuits if the plan were adopted. The Farm Bureau also actively opposed the plan. Hoping to quell this rebellion, the county supervisors made approval of their subarea plan conditional on the wildlife agencies' agreement with a series of "deal points" that included minimizing interference by the wildlife agencies after plan approval, protecting private property rights, avoiding regulatory duplication, including landowners in the process of deciding which land to conserve, relying on public land to the maximum extent practicable, and providing for the development of infrastructure adjacent to and across preserved land. Anxious to placate county officials, the wildlife agencies acceded to virtually all of their conditions.

Ultimately, the county divided its plan into three segments: in the Lake Hodges and South County segments, officials drew preserve boundaries in consultation with large landowners, just as the city had. When they were unable to reach agreement on a particular parcel, the county designated the property as a plan amendment area—meaning landowners would have to negotiate any future development plan with the county and wildlife agencies. In the Metro–Lakeside–Jamul segment, the largest of the three, the county adopted a soft-line preserve whose boundaries would emerge over time as county staff imposed mitigation requirements on development projects and acquired environmentally sensitive lands. The county designated a preapproved mitigation area (PAMA), essentially equivalent to an MHPA, in this segment.

Mitigation Ratios   In addition to dominating decisions about preserve design and boundaries, developers eventually managed to ensure that mitigation requirements under the MSCP did not exceed historic levels. One of the developers' preeminent goals was to avoid having to foot the bill for assembling, managing, or monitoring the preserve. They argued that mitigation ratios should not depart from those required historically by the city and the county, and they firmly opposed giving local officials the discretion to adopt a precautionary approach to mitigation (J. Whalen 1992a). Simply ascertaining historic mitigation ratios was complicated, however, by the fact that they had been determined in case-by-case negotiations among developers and local, state, and federal officials. In the face of uncertainty about past practices, developers proposed a uniform 1:1 mitigation ratio for all habitat types, regardless of location. The Endangered Habitats League responded that a blanket 1:1 ratio would be well below current practice and would not reflect the sensitivity of dwindling habitat. It recommended ratios from 1:1 to 3:1, with 2:1 being appropriate in most cases (Silver 1995). Although developers acknowledged that local staff members were currently using a 2:1 or higher mitigation ratio as a rule of thumb under the FWS's interim 4(d) rule, they contended such a practice was onerous for developers and should not be accepted as part of the MSCP (Groth et al. 1995).

After extensive debate, the Working Group decided to leave the issue of mitigation ratios, like that of a preserve boundary, to individual jurisdictions. Local officials in turn adopted a flexible approach that steered mitigation toward the preserve while minimizing the departure from historic practices. They began by dividing habitat into tiers, depending on its sensitivity and regional value, and then created incentives for property owners to mitigate within the preserve. For example, the city established that mitigation for Tier 1 habitats developed outside the MHPA is 1:1 if mitigation is carried out within the MHPA and 2:1 only if mitigation is done outside the MHPA. The county adopted a similar approach in its Metro–Lakeside–Jamul segment: mitigation for Tier 1 habitat is 3:1 only if development within the PAMA is mitigated elsewhere, 2:1 if development in the preserve is mitigated within the PAMA, and 1:1 if development outside the PAMA is mitigated inside. As Dan Silver (2003) explains, the net result is that developers are not required to do more than they would have done under existing rules, but the mitigation is more effective. County MSCP program chief Robert Asher (2003) confirms, "We were able to put the system together in a way that mimicked very closely the prior process." Importantly, the rules leave decisions in developers' hands by discouraging but not prohibiting development within the designated MHPA.

Buffers   Finally, developers opposed designating buffers or requiring brush management outside the MHPA, and they wanted mitigation credit for brush management on property adjacent to preserved land; on this they prevailed as well.[12] Consistent with prevailing conservation biology principles, the draft Bio Guidelines noted that "The nature of the surrounding habitat matrix is critical to the viability of the preserve area" (Biological Task Force 1992, 7). Biologists pointed out that without buffers, edge effects would degrade the preserve, and that giving mitigation for brush management—which typically involves clearing as much as half the vegetation—was nonsensical. The Endangered Habitats League observed, "Biologically, the counting of fuel modification zones—that is, purposely degraded vegetation—as full mitigation for impacts to intact habitat is patently absurd. Removing 50% of the vegetative cover from a site *constitutes an impact which itself requires mitigation*, rather than the reverse!" (Silver 1995, 2). The Alliance for Habitat Conservation countered that "Limiting uses adjacent to the...preserve line, or requiring brush management outside the MSCP open space is not acceptable and is simply a different way of requiring buffers" (J. Whalen 1995, 4). The alliance demanded that any reference to "edge effects" be eliminated from the MSCP, on the grounds that the existence of such effects had not been proven.

Despite the well-documented biological importance of buffers, particularly in urban areas, developers prevailed on both counts.[13] Thus, only when preserves are established in areas without adjacent development does new development have to maintain brush management buffers outside the preserve boundary. (Local planners encourage, but do not require, less intensive development—such as parks, golf courses, or streets with housing on only one side—alongside newly created preserves.) Where preserves are established in areas of existing development, however, there is essentially no buffer—or the buffer is inside the preserve boundary. In this situation brush management is allowed inside the preserve, but other active land uses are prohibited (Stallcup 2006).

The MSCP: A Minimally Protective Plan

In 1996, the wildlife agencies approved the MSCP plan, and in 1997 both the San Diego City Council and the County Board of Supervisors voted to adopt it.[14] The plan sets a target of 171,917 acres, to be carved out of a 194,318-acre MHPA (see figure 4.1).[15] Of the preserve's total area, 81,750 acres were already in public ownership: 36,510 acres were federal and state land, and 45,240 acres were locally owned parks and preserves. Planners anticipated that about 63,170 acres of private land would be conserved through development regulations and mitigation requirements; that federal, state, and local governments would acquire another 27,000 acres; and that 18,960 acres would be developed. Between them, the city and county of San Diego are responsible for assembling 153,280 acres, nearly 90 percent of the preserve. The plan estimates the 30-year costs of the MSCP at between $339 million and $411 million in 1996 dollars, including between $262 million and $360 million for land acquisition, and management and monitoring costs totaling $120 million (City of San Diego 1998). Because neither the Working Group nor policymakers had been able to agree on a funding source, the wildlife agencies accepted the plan without a financing mechanism in place but gave the jurisdictions three years to submit a regional funding source to voters for approval.

Because it designates target areas that were contiguous or connected by corridors, the MSCP is considerably more comprehensive than the status quo—consistent with the optimistic model of EBM. Prior to the MSCP, as well as during the planning process, San Diego's landscape was rapidly disappearing: an examination of 15 projects undertaken between 1985 and 1990 in the city of San Diego revealed a 97 percent loss of coastal sage scrub; between January 1991 and March 1993, while the FWS was deliberating over whether to list the gnatcatcher, developers razed another 2,400 acres of undisturbed, low-elevation coastal sage scrub, and local officials permitted the destruction of thousands more acres of other habitat types to make way for golf courses, condos, and gated communities (USFWS 1993a).[16] Project-by-project protection of endangered species certainly would have slowed the rate of loss of gnatcatcher habitat. According to

Figure 4.1
(Southwest) San Diego MSCP Preserve Planning Area

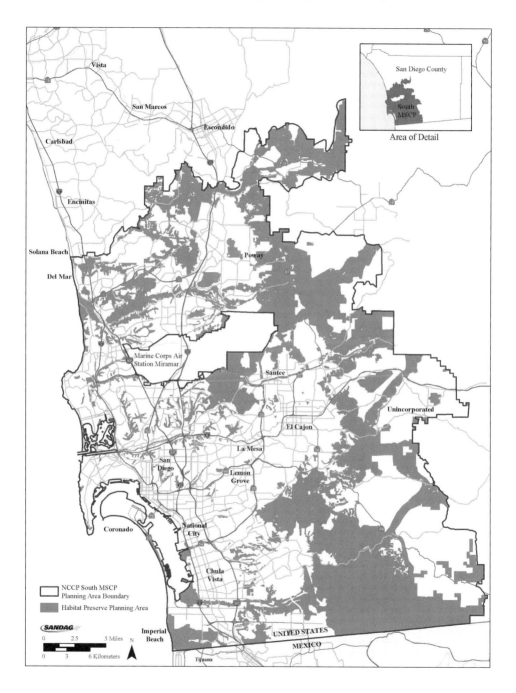

the environmental impact study for the MSCP, the "no project" alternative would have conserved about the same number of acres as the MSCP preserve, but it would have done so in piecemeal fashion, leaving an even more fragmented landscape, and would not have conserved endangered plants on private land. By contrast, the MSCP purports to conserve "both the diversity and function of [the southwestern San Diego County] ecosystem through the preservation and adaptive management of large blocks of interconnected habitat and smaller areas that support rare vegetation communities..." (City of San Diego 1998, 1–5).

On the other hand, as forecast by the pessimistic model, the plan imposes substantial risk on San Diego's plant and animal species and reflects, above all, developers' insistence on equivalence between biological and economic considerations. Consistent with developers' emphasis, the first of the plan's seven objectives is to "establish and maintain a workable balance between preservation of natural resources and regional growth and economic prosperity." Despite the NCCP Scientific Review Panel's recommendation to pursue "no net loss of habitat value," which was based on the recognition that almost all of the region's natural landscape had already disappeared, the MSCP conserves less than two-thirds of the coastal sage scrub and just over half the total habitat in the planning area. Most important, despite a stated desire to adhere to state-of-the-art conservation biology principles, the plan itself notes that much of the MHPA consists of small habitat patches adjacent to existing or proposed development areas. (It asserts that intensive management will minimize the potential biological effects of development along these "interfaces.")

Adding to the risk, the plan covers (and therefore allows "take" of) a large number of species while giving them only tenuous protection. The 1995 draft plan provided legal coverage for 57 species, but as the planning process drew to a close, developers privately threatened to withhold their support unless more species were covered. To assure their continued participation, Fish and Game's Ron Rempel added another 31 grassland- and wetland-dependent species, raising the total to 88. The wildlife agencies subsequently deleted coverage for several of the newly added species, but the rationale for the 28 that remained was vague and, ultimately, controversial (Spencer 2003).[17] As researcher Dan Pollak (2001b, 39) observes, "The plans and related documents created by the local governments and the Department of Fish and Game provide little explanation of how they went about analyzing the needs of each covered species to determine if they were adequately conserved." Pollak notes that in some instances, the decision to declare a species adequately conserved, despite high or unknown risks posed by the plan, was premised on the promise of future management measures that were not specifically guaranteed by the MSCP.

Even as it imposed substantial risk on species, the MSCP provided firm assurances for landowners. During the planning process, development interests had focused on ensuring that the commitments in the plan would be binding; their mantra was "A deal is a deal." They demanded a guarantee that they would be allowed to proceed with a project as long as it met preestablished mitigation standards. Developers also insisted on a clear end to the involvement of the wildlife agencies once the MSCP was approved. Above all, they wanted an unqualified guarantee that landowners would never incur additional conservation obligations once they had complied with the MSCP, even if new species were listed. In 1994 developers got their wish when the Interior Department formalized the "no surprises" clause, which releases landowners from any new obligation to commit additional land, comply with new restrictions, or provide any more money once they have complied with an existing HCP, even if new information indicates the measures previously taken were inadequate.[18]

Implementing the MSCP

Since the adoption of their subarea plans, the city and county have been acquiring parcels of open land and steering mitigation toward the preserve. In addition the state and federal governments have contributed substantial amounts of money and land to implementation. As a result, preserve assembly is ahead of schedule: as of mid-2007, about two-thirds of the land slated for inclusion in the MHPA had been conserved, primarily through dedication of publicly owned land, but also through acquisition, development regulations, and mitigation requirements (City of San Diego 2007; County of San Diego 2007). On the other hand, several major development projects approved at the tail end of and subsequent to the planning process have confirmed environmentalists' concerns that, consistent with the pessimistic model, subarea plans' vague language and heavy reliance on local officials' voluntary willingness to limit development provide insufficient protection for narrowly distributed species. In turn, those projects triggered the litigation that collaborative planning was supposed to avert. Moreover, the optimistic model's theoretical expectations about the benefits of adaptive management have yet to be borne out: because the jurisdictions continue to struggle to fund management and monitoring, the preserve design leaves little room for adjustment, and there is minimal coordination among jurisdictions. At present it appears highly unlikely that adaptive management will ensure the plan's biological success.

Assembling the Preserve

The signal achievement of the MSCP is that it has enabled the city and county to obtain millions of dollars in state and federal grants, which they have used to acquire important parcels of high-value habitat, parts of which otherwise would have been developed. For example, using $7.7 million in grants from the state Wildlife Conservation Board and the San Diego County Water Authority, the city purchased Montana Mirador, a 540-acre parcel on which a developer could have built up to 575 homes. In addition, the state and federal governments independently have preserved thousands of acres of land. In response to pressure from environmentalists, the federal government bought a parcel on Mount San Miguel that was slated for a 27,000-home development. The developer scaled back his plan, and the area became the hub of the planned 43,000-acre San Diego National Wildlife Refuge, established in 1997. South of Mount San Miguel, the federal Bureau of Land Management purchased 3,900 acres of Otay Mountain. And environmentalists persuaded the state to buy another site that had been targeted for a subdivision and a golf course in east San Diego and turn it into the 2,640-acre Crestridge Ecological Preserve.

Beyond acquiring land outright, jurisdictions are supposed to be minimizing development in and steering mitigation toward the designated preserve. The MSCP returned land-use authority to the jurisdictions, which gained sole responsibility for interpreting the plan's biological guidelines as they make decisions on development proposals. The hope was that discretion plus a heightened awareness would encourage local staff to go beyond the minimum in protecting environmentally sensitive land: Marc Ebbin (1997, 705) predicted that in San Diego County "Conservation objectives will now deeply influence local land-use planning and zoning decisions." But although both city and county staff are more vigilant than before about administering local resource protection ordinances, most are not trained biologists. Furthermore, critics worry that staff turnover, combined with the structural factors that have always favored development at the local level, can easily undermine a jurisdiction's commitment to interpreting regulations in an environmentally protective way. Several episodes during the plan's formulation and implementation confirmed these fears and prompted legal action by disgruntled environmentalists.

One major project under consideration in the midst of the MSCP planning process shook environmentalists' faith in the city's willingness to take strong action to limit development of parcels controlled by large developers. In 1993 the Pardee Company submitted for City Council approval a revised version of a development plan for Carmel Mountain that it had first put forward in 1987. Biologists had identified Carmel Mountain, which was one of southern California's last privately held coastal mesas, as a biological core area; because the plan threatened serious environmental impacts, the City Council asked Pardee to work with the wildlife agencies to come up with a blueprint that would be consistent with the MSCP. Although environmentalists recognized that the laws and regulations in place prior to the MSCP probably would not have resulted in much conservation—the Endangered Species Act does not protect plants on private land; mitigation under CEQA is discretionary; and upzoning of agricultural land is routine—they hoped the nascent MSCP would give local officials both the awareness and the willingness to use their leverage to insist that the tract be protected (Rolfe 2003). Instead, after protracted negotiations, in 1998 the city agreed to a plan that set aside nearly 150 acres on the mountaintop but allowed developers to cover about one-third of the mountain with 440 homes (T. Davis 2003d; Rolfe 2000). UCSD biologist Isabelle Kay predicted that the isolated preserve would suffer from massive edge effects (C. Chase 1997).

A second episode that suggested jurisdictions' commitment to conservation was tenuous came in December 1998, shortly after the MSCP plan was approved, when the city allowed developers to bulldoze 65 vernal pools to make way for a mall. The city had updated its zoning and land development code simultaneously with the development of its subarea plan, in the process replacing its Resource Protection Ordinance with an Environmentally Sensitive Lands (ESL) Ordinance that emphasizes species and habitats, consistent with the focus of the MSCP. At the time, environmentalists and several local biologists had raised concerns that the ESL would provide inadequate protection for wetlands because rather than setting encroachment limits, it simply requires that "Impacts to wetlands, including vernal pools in naturally occurring complexes, shall be avoided to the maximum extent practicable." As they had feared, given such discretion, the city decided to grant a deviation to Cousins Market Centers, Inc., after the developer argued that it would be economically infeasible to avoid the vernal pools. The decision prompted 14 environmental groups to sue the city.

Many environmentalists were also disappointed with the outcome of planning for the 22,500-acre Otay Ranch in northwest San Diego, which had been a sticking point during the preparation of the MSCP because proposed "development bubbles" threatened the integrity of the preserve. According to the county's subarea plan, Otay Ranch "comprise[d] the largest privately held ownership of coastal sage scrub vegetation in the U.S." (County of San Diego 1997, 3–13). Its size and undeveloped character, as well as the diversity of its terrain and strategic location, made it the single most biologically desirable tract in the planning area. In late 1994, however, the County Board of Supervisors approved an Otay Ranch General Plan Amendment that set aside some 11,375 acres (as well as another 1,166 acres of open space) while allowing the development of 15 villages that could accommodate 68,000–70,000 residents. After the area's owner, Baldwin Vista Associates, agreed to exchange rights to develop some sensitive areas for permission to build on other parts of the property, the wildlife agencies approved the county's subarea plan. Shortly after the deal was struck, however, Baldwin went into bankruptcy reorganization and sold some of the land it had promised for the preserve. After building more than 5,347 houses without setting aside any open space, Baldwin tried to cut a deal to substitute lower-quality habitat for the land it had originally pledged, but the county was not receptive, and negotiations resumed.

Management and Monitoring

Because so much of the MHPA consists of small habitat patches immediately adjacent to urban development, the wildlife agencies emphasized that intensive and adaptive management would be crucial if the plan was to have any chance of achieving its biological goals. In the ten years since the plan's approval, some monitoring has gotten under way: the city and county have been working with the U.S. Geological Survey to conduct rare plant monitoring; San Diego State has been developing a priority scheme for species and habitat types; and the FWS is preparing to initiate an animal monitoring component.

Both management and monitoring have been hampered by insufficient funding, however, and are heavily dependent on volunteers. Despite a three-year deadline in the plan, by 2006 a regional funding source for the MSCP had not yet materialized. In November 2004 the region got a reprieve when county voters approved a $14 billion sales tax renewal measure, TransNet, that includes about $850 million for habitat conservation and management, but it is not clear how much of that will benefit the MSCP. Meanwhile, although the county has implemented its program at minimal levels, using Local Assistance Grants from the Department of Fish and Game, plan-related spending by the financially troubled city has plummeted to zero, as a result of which, according to then-Deputy Planning Director Keith Greer, the city has "basically had to beg, borrow and plead for other staff to assist with the biology" (M. Lee 2005).

Furthermore, because there is no central authority overseeing management and monitoring, both vary considerably from one tract to another, and no entity gathers and synthesizes management and monitoring data; as a result, no one can judge the biological effectiveness of the preserve as a whole or prescribe changes in management. The original *Biological Monitoring Plan* (Ogden Environmental 1996) stated that the wildlife agencies would coordinate monitoring efforts throughout the MSCP study area to ensure consistency in data collection and analysis. But the agencies do not have the resources to attend to these tasks and are fully occupied permitting new plans (Wynn 2006). The plan also noted the importance of establishing a centralized repository for data, so that it would be accessible to researchers and managers. Although environmentalists had hoped for a regional conservancy to manage and monitor the MSCP preserves, neither development interests nor jurisdictions were interested in creating such an entity. Instead, the jurisdictions purportedly coordinate their activities through a Habitat Management Technical Committee, formed to address day-to-day open space management issues, and an Implementation Committee, which provides a forum for discussing regional funding, public outreach, and implementation issues. These committees meet annually, but as Keith Greer (2004, 237) points out, "The greatest hurdle has been the fragmented approach to the monitoring effort, with no one lead agency looking at the entire program."[19]

The wildlife agencies are working with the jurisdictions to address the deficiencies in management and monitoring. In 2006 the state of California unveiled the Biogeographic Information and Observation System (BIOS), which is designed to serve as a central repository for monitoring data that can then be synthesized and analyzed. The wildlife agencies also hope that TransNet money will fund personnel who can coordinate monitoring and improve the relationship between regional monitoring and management decisions. But resistance to centralizing management and monitoring remains potent.

Even if monitoring data become more readily available, adaptive management will be constrained by assurances in the plan that the preserve will not be expanded. The original 1996 *Biological Monitoring Plan* lists potential actions that could be taken if monitoring indicates that habitats or species are declining. These include erecting fences or signs, redirecting trails, removing invasive exotic plants, enhancing or restoring habitat, conducting prescribed burns, and reintroducing plants. The report adds (7–2) that adaptive management "may include reconfiguring preserve boundaries to include more or different habitat if a species is declining." The prospects for adding more land to the preserve are poor, however, given the dearth of additional land; the "no surprises" clause, which insulates landowners from responsibility for addressing unanticipated habitat requirements; and the shortage of money to buy more acreage.

Meanwhile, it is apparent that at least some preserve areas are suffering from edge effects that management is unable to control. It is also evident that jurisdictions lack the manpower and resources to control weeds, trash-dumping, and off-road vehicle use on many of the protected tracts (M. Lee 2005). For example, managers at the Crestridge Ecological Reserve—which the Department of Fish and Game describes as southern California's model preserve and one of the most significant pieces of land set aside in the region—is hemmed in on all sides by residential development and, as a result, frequently experiences vandalism and illegal motorcycle use (Hennessey 2003). Preserve managers rely heavily on volunteers to repair the damage.

An article in the *Los Angeles Times* (Jacobs 1997) elaborates on the problems of managing southern California's preserve lands. The author points out that in 1995 Bruce Babbitt flew in to inaugurate the state's first "conservation bank"—a 180-acre preserve in Carlsbad. But two years later the mesa, the Carlsbad Highlands, had become a playground for motorcycles and off-road vehicles, and a shooting range for hunters. The Department of Fish and Game had neither posted signs nor built barriers to keep intruders out, even though the acreage came with a substantial endowment. A 1997 report documented a lack of "critical site maintenance" on three-quarters of the department's burgeoning preserves. Among the most frequent problems were illegal use, serious vandalism, unauthorized dumping, and trespassing by off-road vehicles. The threats to preserves in San Diego and the rest of California are increasing as public authorities increasingly look to public lands as sites for gas pipelines, electricity transmission lines, train tracks, and other public amenities that would require condemnation if sited on private land.

Conclusions

Compared with the trajectory that San Diego was on, the MSCP is an improvement, and its formulation confirms the predictions of the optimistic model of EBM with respect to the benefits of a landscape-scale focus. Although the plan probably will not increase the total amount of land conserved over what the status quo approach would have yielded, it did shift local officials' attention to ensuring that developers set aside high-quality habitat in a more biologically sound configuration (Greer 2003; Oberbauer 2003). It has enhanced the city and county's ability to obtain state and federal dollars for land acquisition and prompted both state and federal agencies to acquire more land themselves. It has increased coordination among jurisdictions that formerly often worked at cross purposes—although the enhanced coordination is also attributable to new storm water regulations and transportation planning. And it has prompted efforts by the wildlife agencies to institute a comprehensive, region-wide monitoring strategy. More subtly, the language of conservation biology has become more pervasive in San Diego than it was prior to the mid-1990s: local officials and even developers routinely talk about species' requirements, and the need to protect core and linkage areas is a given (Greer 2003; Rolfe 2003; Spencer 2003). The region's planners have adopted the term "green infrastructure," and there is a greater focus among local officials on quality of life rather than simply on growth (Fairbanks 2003; Wynn 2006).

On the other hand, some aspects of the MSCP are more consistent with the pessimistic model of EBM. The preserve itself is only minimally protective of San Diego's biological diversity: it has substantially higher edge-to-area ratios, greater internal fragmentation, and less connectivity than scientists recommended, even in their minimally protective prescription. In addition, because the preserve lacks buffering and contains many steep and degraded canyons, it provides less habitat than its total acreage suggests. Furthermore, San Diego's endangered species bear the risk associated with the plan's shortcomings. Despite a paucity of data to support doing so, wildlife agencies agreed to treat as covered by the plan dozens of plants and animals, many of which stood to lose more than half of their habitat, and the wildlife agencies guaranteed that land-owners would be held harmless in the event that more protective action was required. (The idea was that these species would get more protection than they were receiving under the status quo approach because their habitat and populations would be managed and monitored; the result, however, has been that take of these species is allowed, but management and monitoring is not happening at the level envisioned.)

Consequently, managers have little flexibility to change their approach in the face of information suggesting the MSCP preserve is inadequate to conserve biological diversity. In any case, because there is so little money available for data collection, and no centralized entity that is synthesiz-ing, analyzing, and disseminating the data that are gathered, there is no way for managers to know whether such action is required. Moreover, although there was considerable momentum when the subarea plans were approved, some observers worry that turnover of leadership and staff—at both the wildlife agencies and the jurisdictions—may jeopardize implementation in the long run. Technically, the wildlife agencies could withdraw the permits if their holders are not complying with their imple-menting agreements. But neither the FWS nor the Department of Fish and Game regional field offices have the authority or political backing to take such a step, so in effect, with the issuance of take permits the wild-life agencies relinquished their oversight role, leaving enforcement to the jurisdictions.

The preserve's fragility was underscored in the fall of 2003, when almost 80 percent of it went up in flames during a ferocious series of wildfires. At the time, scientists pointed out that San Diego's species are adapted to fire, and predicted it would take at least seven years for the coastal sage scrub and other habitats to return. They acknowledged, however, that species' recovery would depend on the availability of refugia from which they can recolonize recovering landscapes. Adding to scientists' concern, jurisdictions continued to approve developments at the western ends of canyons, where they are highly vulnerable to the east–west winds that sweep wildfires across the landscape (Stallcup 2004). Another wildfire in 2007, which forced the evacuation of more than 500,000 residents from 250,000 homes, exposed the risks of this strategy.

The MSCP's biological weaknesses are a result of the development-friendly context within which collaboration and flexible implementation have occurred. At the inception of the planning process, species listings enhanced environmentalists' clout, and the recession of the early 1990s slowed development pressure, giving planners some breathing space. From the outset, however, a unified development community, often with the support of local officials, insisted that the MSCP process was about accommodating growth. The wildlife agencies' reluctance to hold localities' feet to the fire, particularly as the planning process wore on, exacerbated the power of development interests vis-à-vis environmentalists. As the FWS's Gail Kobetich explains, the service saw its role as advisory: "We were very careful not to be out front," he says. "It had to be a local effort. If it wasn't, there would be no local buy-in" (Cohn 1998, 52–53).

The development-friendly context, combined with environmentalists' inability to join forces and mobilize the public, allowed developers' framing of the issues to prevail during the planning process: although the wildlife agencies pressed for an emphasis on conserving the region's biodiversity, discussions both inside and outside the Working Group focused increasingly on averting species listings, streamlining the regulatory process, and "balancing" growth and development. Furthermore, preserve boundaries were largely the cumulative result of agreements between jurisdictions and large developers, not the result of a biologically based evaluation.

Despite its shortcomings, many who participated in the MSCP planning process firmly believe the plan was the best deal they could have gotten, given the region's conservative political culture and the extent of development in San Diego when planning got under way; in fact, less than ten years after the plan's completion, there were few lots left that were large enough for the massive planned communities that had dominated the city's buildout in the preceding decades (Showley 2005). Federal District Court Judge Rudi Brewster, a Reagan appointee, disagrees. In 2006 he ruled that the MSCP is not sufficiently protective to pass legal muster. He pointed out that the MSCP virtually guaranteed development but "would permit monumental destruction" of several protected species that live in vernal pools (M. Lee 2006). He added that the city's strategy for funding its portion of the MSCP was vague and speculative, and ordered the FWS to stop current or planned development on sites containing vernal pools.

5

Restoring South Florida's River of Grass

As regional habitat conservation planning took hold in the 1990s, there was a simultaneous burst of interest in landscape-scale aquatic-system planning and management. In December 2000 President Clinton signed a bill authorizing the Comprehensive Everglades Restoration Plan (CERP), an aquatic-system ecosystem-based management (EBM) experiment described by many observers as the largest ecosystem restoration project ever attempted. Preparing CERP brought together hundreds of scientists and engineers from more than a dozen state and federal agencies, as well as from academic and tribal institutions, in a collective endeavor of unprecedented magnitude. According to its proponents, the $8 billion plan—in combination with another $8 billion worth of projects already in the works—promised to revitalize the world's largest freshwater marsh, the South Florida Everglades. Since its adoption, CERP has attracted millions of dollars from federal, state, and local sources—money that has funded extensive scientific research and modeling, and enabled the state of Florida to acquire more than 200,000 acres of land. Furthermore, both the Army Corps of Engineers and the South Florida Water Management District, CERP's primary implementing agencies, have become more attuned to environmental concerns and more inclined to work cooperatively with other state and federal agencies.

As currently constituted, however, CERP is unlikely to restore a healthy, resilient Everglades ecosystem, because it perpetuates a heavily engineered and intensively managed approach, and imposes the risk of failure on the natural system. It focuses on manipulating the distribution, depth, and duration of flooding rather than on restoring the Everglades' north–south flow, even though many scientists have long recognized that the uninterrupted movement of water was an essential characteristic of the historic ecosystem. In theory, CERP's strong rhetorical commitment to adaptive management creates the possibility of a midcourse correction in response to concerns about flow. In practice, however, managers' flexibility is limited because the plan promises users additional future water supplies and guarantees they will retain their existing water allocation during the plan's 30-year implementation. Moreover, the plan contains only minimal provisions for buffers and, as yet, contains no contingency plan if its central technology—underground water storage—proves infeasible.

The main explanation for CERP's minimal protectiveness is that its conceptual basis was generated by a collaborative planning process within a context that heavily favored development interests. Although environmentalists had succeeded in raising the salience of the Everglades' declining health by the late 1990s, the state's environmental community was unified only briefly; more important, the state's agricultural and development lobbies were (and continue to be) enormously powerful at the state and federal levels. Rather than taking strong positions on behalf of environmental protection, political leaders abdicated the task of setting goals to a collaborative process. To gain consensus among stakeholders, planners promised to "expand the pie" and make ecological restoration a coequal goal with ensuring current and future water supplies and providing flood protection.

By late 2007 obstacles to implementing CERP's projects had dimmed its prospects for restoring the Everglades' health even further. As the Army Corps of Engineers and the South Florida Water Management District began fleshing out the plan's details, the tenuous coalition that brought it to fruition dissolved, delaying the start dates of key projects and causing the federal government to withhold its share of funding. Because of South Florida's explosive growth, those delays alone have undermined the plan's potential ecological effectiveness: development is encroaching on the Everglades from all sides, eliminating the buffer between the natural system and the man-made one, occupying some of the land that was slated for restoration, and raising the price of the remainder. Despite widespread recognition of the relationship between land-use decisions and the health of the natural system, Florida's political officials have been unwilling to regulate land use in ways that would protect CERP's footprint. As a result, the plan may slow but is unlikely to reverse the Everglades' demise.

The Evolution of Landscape-Scale Planning

Within several decades in the middle of the twentieth century, engineers replumbed the Everglades, compartmentalizing and dramatically reducing its spatial extent and causing ecological devastation. In the 1970s concerned scientists and newly mobilized environmentalists capitalized on a series of crises to draw public attention to the Everglades' deteriorating health. Their efforts prompted changes in state agencies' legal mandates, as well as some projects that aimed to repair elements of the natural system, but these piecemeal efforts were insufficient to halt the Everglades' decline. In the late 1980s, however, scientists and environmentalists converged on the idea of a comprehensive approach that would restore a semblance of the historic Everglades. Three themes emerged as a result of this landscape-scale focus: reconnecting the region's undeveloped land, reviving the flow of water across it, and buffering it from the urban east coast.

Early Environmental Initiatives

The evolution of South Florida's contemporary landscape began 5,000 years ago with the retreat of the polar ice caps. By the time Europeans arrived in the sixteenth century, the Everglades was a slow-moving sheet of water, 100 miles long and between 30 and 40 miles wide. This "river of grass" originated at Lake Okeechobee, which overspilled its banks during the summer rainy season and traveled south to Florida Bay. Because South Florida's slope is so gradual, the water moved slowly—at a maximum rate of only about two feet per minute (Lodge 1998). This "sheet flow" nourished a mosaic of marshes, wet prairies, sloughs, ponds, and creeks that was vast and—as a result of frequent floods, droughts, and fires—heterogeneous enough to support an enormous diversity of animal and plant life.

Native Americans had inhabited South Florida for centuries with only marginal effects on the landscape, but in the late 1800s white settlers began trying to drain the Everglades to make it more habitable. Their initial attempts largely failed, but in 1947, after a series of severe hurricane-caused floods, Congress approved the U.S. Army Corps of Engineers' request to undertake a massive hydrologic reconfiguration of the region. The Central & Southern Florida (C&SF) Project took two decades to complete and involved digging, widening, or deepening 978 miles of canals; erecting 990 miles of levees; installing 212 tide gates, floodgates, and other control structures; and building 30 pumping stations. The project compartmentalized what had been a fundamentally interconnected system. The Everglades Agricultural Area (EAA)—700,000 acres of rich peat soil immediately south of Lake Okeechobee—severed the link between the lake and the rest of the marsh. Levees created three separate wetland impoundment areas, called water conservation areas (WCAs), whose water levels are controlled by canals and structures, as well as Everglades National Park, which comprises less than one-quarter of the original Everglades and occupies its southern tip.

The ecological impacts of the C&SF Project were immediately apparent. In the early 1970s, shortly after the project was finished, the population of South Florida's famous wading birds began to decline precipitously. Because they rely on its quirky wet–dry cycle, wading birds are potent indicators of the Everglades' health: beginning in October or November, they follow the "drying front," finding prey concentrated in the pools that remain as the water recedes. Historically, the system was so huge the birds had many options for when and where to nest, so they almost invariably enjoyed a long breeding season, regardless of local fluctuations in habitat conditions. But the use of the water conservation areas as receiving waters for storm runoff from surrounding areas, combined with the unnatural timing and quantities of water released into Everglades National Park, reversed drying fronts, thereby eliminating prey concentrations and causing nesting failures (Lodge 1998).[1] Other impacts of the C&SF Project began to manifest themselves as well: after a severe drought in 1970–1971 muck fires raged, rich peat soil subsided at the alarming rate of an inch a year, and seawater intruded into freshwater supplies.[2]

These worrisome ecological symptoms coincided fortuitously with the emergence of a newly energized environmental community in South Florida. In the late 1960s, two major development proposals—for a jetport in Big Cypress Swamp and a cross-Florida barge canal in North Florida—had galvanized the state's environmentalists. Buoyed by their success in defeating both initiatives, environmentalists began lobbying elected officials to pay more attention to protecting South Florida's natural resources. In response to the increasing urgency and effectiveness of environmental activism, as well as the publicity surrounding the drought, in 1971 Governor Reuben Askew convened a conference of experts to formulate recommendations for managing South Florida's water. The conferees agreed that the region's ecology was deteriorating and advised serious policy changes, such as prohibiting further drainage of wetlands, purchasing or zoning land in water recharge areas, and limiting population growth. Five years later Harold Odum's Center for Wetlands at the University of Florida and the state's Bureau of Comprehensive Planning released the *South Florida Report*, which recommended even more draconian solutions: a no-growth agenda with a focus on improving the quality of life. Reasoning that it required large amounts of energy to move water around, the authors urged the state to rely more heavily on natural solutions to its drainage problems (Blake 1980).

In response to both public concern and experts' insistence, the state legislature passed a series of laws aimed at improving Florida's water management. The 1972 Water Resources Act reorganized the South Florida Flood Control District and made part of its mandate the protection of water quality throughout its 17,000-square-mile jurisdiction, which spans the Everglades ecosystem from its headwaters in the Chain of Lakes south of Orlando to Florida Bay.[3] (To capture this expansion in the agency's mission, the 1975 Environmental Reorganization Act changed its name to the South Florida Water Management District, or SFWMD.) The Florida State Comprehensive Plan, also mandated by legislation passed in 1972, called for managing the state's water resources to approximate the hydroperiod that existed prior to modification, albeit "within the constraints of existing development and planned land use" (Blake 1980, 269).

Despite these changes in Florida's legal and political context, the Corps and the SFWMD continued to operate the C&SF Project primarily to serve the state's historically powerful and well-connected agricultural and urban development interests (Blake 1980; Grunwald 2006). The effects of water managers' disregard for the environment continued to manifest themselves. In the late 1970s, extensive algal blooms on Lake Okeechobee prompted another brief surge in public alarm. Then, in the early 1980s, decisions made during a yearlong drought, followed by torrential rains, wrought ecological havoc on Everglades National Park. In response, noted park biologist William Robertson advocated a return to sheet flow to restore the park's health. Marine biologist Art Marshall, the state's foremost environmental advocate, put together an even more ambitious solution: a comprehensive Everglades restoration plan whose main purpose was to restore sheet flow throughout the historic system. Marshall (1980) emphasized that "An extremely important characteristic of sheet flow is that it involves moving—rather than standing—water."

Governor Bob Graham responded: in August 1983, the governor announced his "Save Our Everglades" campaign, a five-point plan based on Marshall's concept. According to the governor's spokesperson, Jill Chamberlin, "Wherever possible, the Governor wants to restore the natural ecology" (Gruson 1983a). In Graham's own words, "Success in this endeavor means turning back the clock 100 years. By the year 2000 the Everglades will look and function more like they did at the turn of the century than they do today" (Shabecoff 1986). The governor's initiative bore some fruit, but the SFWMD continued to see its primary responsibility as protecting agricultural areas and cities from floods, and delivering water and drainage for development (Hansen 1984). Moreover, South Florida was enjoying a real estate boom in the mid-1980s, and a Chamber of Commerce mentality dominated local land-use decision making.

Landscape-Scale Thinking Catches On

Thus, despite more than a decade's worth of environmental activism in South Florida, by the end of the 1980s the Everglades ecosystem was in a death spiral. Water depth and distribution patterns had been altered, often fundamentally, throughout the natural system, as a consequence of which three of the seven major landscape features in the predrainage Everglades had disappeared; thousands of acres of South Florida wetlands were infested with the thirsty melaleuca tree (*Melaleuca quinquenervia*) and other invasive exotic species; the alligator population, which had recovered after enforcement officials cracked down on poaching in the 1970s, was in decline; the number of wading birds nesting in the southern Everglades had fallen by over 90 percent since the 1940s; and 16 species or populations of Everglades vertebrates were listed by the state or federal government as endangered or threatened, and another 11 were listed by the state as "species of special concern" (J. Ogden 1999). Furthermore, in the late 1970s the Everglades Agricultural Area had begun diverting nutrient-laden agricultural runoff south into the Everglades (previously the EAA had backpumped its wastewater into Lake Okeechobee). By the late 1980s the injection of large quantities of phosphorus was stimulating an explosion of cattails, which were displacing the native sawgrass and changing the microscopic life of the marsh (Nordheimer 1987).

The first major reaction to mounting evidence of the Everglades' continuing deterioration came from an unexpected quarter. In late 1988, at the behest of environmentalists, acting U.S. Attorney for South Florida Dexter Lehtinen sued the state of Florida for failing to enforce its own water quality laws. A three-year legal battle that pitted the state and the sugar industry against the federal government ensued, ultimately resulting in a settlement that required a massive reduction in the phosphorus level of water flowing into the Everglades. Spurred by media coverage of the lawsuit, in 1994 the state legislature passed the Everglades Forever Act, which required the state to set phosphorus limits and codified solutions to the water quality problem, such as requiring farmers to adopt best management practices and converting 40,000 acres of private and state-owned land into filter marshes known as storm water treatment areas (STAs).

In hopes of broadening the scientific understanding of the Everglades' problems beyond water quality, in October 1989 Steven Davis, an ecologist with the SFWMD, and John Ogden, an ornithologist with the Everglades National Park, organized a scientific symposium to discuss what was known about the Everglades and what could be done to save it. That symposium was the catalyst for a new, more integrative understanding of the ecosystem. After viewing a natural system model developed by SFWMD engineer Tom MacVicar, participants recognized for the first time the deep connections among ecological elements and processes from Lake Okeechobee to Florida Bay (Boucher 1995). Symposium participants identified three features crucial to the evolution of the historic Everglades: its vast spatial extent; its continuously changing hydrology, dominated by dynamic storage and sheet flow; and its pattern of highly varied and patchy habitats.[4]

Simultaneously with the development of a more holistic view among scientists, in the political realm longtime Wilderness Society activist Jim Webb was reviving Art Marshall's ideas about a comprehensive Everglades restoration. Webb's notion was to capture the billions of gallons of water shunted to the Atlantic Ocean and Gulf Coast by C&SF Project canals, store it, and release it according to the rhythms of the historic cycle. Intrigued, environmentalists translated this general concept into a strategic plan, Everglades in the 21st Century, and began promoting a full-scale reevaluation of the C&SF Project (Grunwald 2006). In response to environmentalists' activism, in the fall of 1989 Congress took a first step by passing the Modified Water Deliveries to Everglades National Park Act, known as Mod Waters, which authorized the acquisition of 107,600 acres on the park's eastern border and directed the Army Corps of Engineers to institute a rainfall-driven water delivery regime that would mimic natural flows there. Three years later, as part of the 1992 Water Resources Development Act, Congress asked the Corps to undertake a comprehensive review of the C&SF Project.

To conduct this "Restudy," the Corps assembled a team comprising staff from 27 federal, state, and local agencies and asked them to determine

... Whether modifications to the existing project are advisable ... due to significantly changed physical, biological, demographic, or economic conditions, with particular reference to modifying the project or its operation for improvement of the quality of the environment, improving protection of the aquifer, and improving the integrity, capability, and conservation of urban water supplies affected by the project or its operation. (USACE 1994, 77)

The purpose of the Restudy's first stage, the Reconnaissance Study, was to define the Everglades system's ecological problems, formulate an array of conceptual plans for solving those problems, evaluate each of those plans, and recommend plans or components of plans deserving more detailed study during the Restudy's second phase, the Feasibility Study.

In an effort to influence the Restudy, the Everglades Coalition, a network of 32 environmental groups, prepared its own Greater Everglades Ecosystem Restoration Plan, which it released in July 1993. Environmentalists envisioned "re-creating a free flowing 'River of Grass' where water rises and falls in natural harmony with seasonal and annual variations in rainfall" (Everglades Coalition 1993, 4). More specifically, they recommended that the Army Corps of Engineers create a storage area of about 100,000 acres within the Everglades Agricultural Area, in addition to 70,000–80,000 acres of storm water treatment areas. They also proposed "reconnecting the Water Conservation Areas... and restoring historic natural volumes of water in the form of continuous sheet flow from the southern end of the Everglades Agricultural Area... to Florida Bay, distributed and delivered on a timetable matching historic natural conditions" (Everglades Coalition 1993, 17). And they advocated increasing the self-sufficiency of agricultural and urban water supplies, rather than continuing to mine water from the natural system.

The Interior Department agencies, particularly the National Park Service and the Fish and Wildlife Service, also had a particularly strong interest in promoting an environmentally protective redesign of the C&SF Project. In hopes of ensuring a prominent role for the department in the Restudy, in 1993 Interior Secretary Bruce Babbitt convened an 11-agency South Florida Ecosystem Restoration Task Force (Task Force) whose purpose was to coordinate the activities of the many federal agencies with disparate missions and statutory responsibilities in service of a common purpose. The Task Force articulated three broad and uncontroversial aims: "get the water right"; "restore, preserve, and protect natural habitats and species"; and "foster compatibility of the built and natural systems." To help it formulate science-based objectives consistent with those goals, the Task Force established a Science Subgroup composed of 29 agency scientists. Journalist Heather Dewar (1994) wrote:

Imagine what would happen if you put thirty top government scientists…in a room and told them not to come out until they'd figured out how to save the Everglades. Assume nothing is impossible, you'd tell them. Don't worry about money. Don't think about whose ox is being gored. Just find a way to get South Florida's liquid heart pumping again.

Not surprisingly, the Science Subgroup's recommendations were precautionary. In November 1993 the subgroup released its first report, which reiterated many of the ideas put forward at the 1989 symposium and recommended ambitious ecological and hydrological objectives for nine subregions in South Florida. The authors defined the conceptual target of restoration as predrainage South Florida, and laid out principles for restoring the elements and processes that characterized the historic Everglades to the degree possible, given its reduced spatial extent. They began by asserting that "Hydrologic restoration is a necessary starting point for ecological restoration" (Weaver and Brown 1993, 1), and strongly urged policymakers to decompartmentalize and reestablish sheet flow throughout the system. "Given the historic, predrainage role of massive sheet flows emanating from the upper reaches of the Everglades watershed in structuring the physical and biotic landscape of the South Florida ecosystem," the authors wrote, "it is imperative to establish sheet flow conveyance on the system's historic north–south gradient" (Weaver and Brown 1993, 22). To that end, the scientists recommended: acquiring as many square miles of land from farmers and developers as possible, and using that acreage to create flow ways stretching from Lake Okeechobee to Florida Bay; establishing a buffer zone between the natural system and the rapidly expanding east coast cities; filling in long stretches of canals and knocking holes in levees; and getting more water flowing in a wide, shallow sheet across marshes and leaving it there longer.[5] As Dewar (1994) explains, there was a growing scientific consensus that "The only way to save the Glades [was] to mimic nature's design, which sent wide, shallow sheets of water slowly south across South Florida's interior."

To some extent, the Corps's Reconnaissance Study, released in late 1994, echoed the Science Subgroup's message. According to journalist Robert McClure (1994), under the most ambitious option laid out by the Corps, engineers would convert nearly 50,000 acres of sugar and vegetable fields in western Palm Beach County to marsh, reestablishing natural water flow from Lake Okeechobee to the Everglades; knock down huge earthen berms inside the Everglades that impeded the natural flow of water; fill in large parts of the Miami Canal in western Broward and Palm Beach counties to restore the natural sheet flow; and raise parts of the Tamiami Trail to allow water to flow more naturally into Everglades National Park. The study report also endorsed as "critical to Everglades restoration" plans being developed by the SFWMD to set aside buffer areas of reservoirs and marshes to separate urban land from undeveloped Everglades.

Collaborative, Landscape-Scale Planning

With the momentum building for comprehensive restoration, over the next four years a multiagency team led by the Army Corps of Engineers worked interactively with a stakeholder group established by Governor Lawton Chiles to develop the underpinnings for a restoration plan based on the results of the Reconnaissance Study. Consistent with the optimistic model of EBM, a landscape-scale focus prompted interagency coordination and an improved understanding among managers and stakeholders of how the Everglades functions. In addition, the collaborative process forged some level of trust among participants and increased their mutual understanding of one another's interests. More consistent with the pessimistic model, however, as the process progressed, planners focused on configuring a system that would satisfy the region's many stakeholders rather than one whose primary goal was ecological restoration. As a result, instead of adopting an approach that would remove barriers to flow and allow natural processes to reestablish themselves, planners chose a package of elaborately engineered projects whose overall goal was to re-create historic depths, duration, and distribution of surface water through precise management. Planners pointed out that such an approach enabled managers to avoid the potentially adverse consequences of reestablishing sheet flow across a much smaller and significantly altered Everglades. Importantly, however, it also allowed them to retain control of the water, thereby assuring agricultural and urban users that their water supplies would be secure throughout the restoration process and into the future.

Building Consensus: The Governor's Commission for a Sustainable South Florida

The dual goal of protecting water users' interests while restoring the ecosystem emerged from the collaborative process established by Governor Lawton Chiles in early March 1994, as the Reconnaissance Study was nearing completion. Executive Order 94-54 established the Governor's

Commission for a Sustainable South Florida, whose purpose was to provide stakeholder input to the Interior Department's Task Force and establish a political consensus that could underpin the Corps's final Restoration
Plan.[6] With the help of the state's departments of Community Affairs and
Environmental Protection, the governor's office identified "statesman-
like" individuals who would provide balanced and complete representa-
tion of the region's various stakeholders (Pettigrew 2003). The governor
then appointed 37 voting members from the South Florida business and
economic communities, the public interest and environmental communities, and local, state, and tribal governments. He also chose five nonvot-
ing members to participate on behalf of federal agencies with an interest
in restoration. He asked the commission to "recommend actions for the
restoration, management, preservation and protection of [the Everglades
ecosystem] and to recommend strategies for ensuring the South Florida
economy is based on sustainable economic activities that can coexist with
a healthy Everglades ecosystem" (State of Florida, Office of the Governor
1994). Between 1994 and 1999, the commission met every month for
two or three days. A facilitator from the Conflict Resolution Consortium
helped to create processes for regulating the pace and tone of discussion.
But it was the charismatic and well-respected chair, Dick Pettigrew, a
former state legislator, who ensured that the group ultimately reached
consensus (Applebaum 2002; E. Barnett 2002; Collins 2002; Hurchalla
2003; Ring 2002).

Early on, it became clear to the commissioners that they would be able
to agree on a plan only if it promised gains for all stakeholders. At the
inception of the process, animosity among the region's stakeholders ran
deep: according to one observer, "Simply put, the Everglades is a war
zone where environmental interests clash with agricultural and urban
interests over water quality and quantity...[and] federal interests in the
form of national parks and wildlife refuges clash with the state's rights
in land use planning and water allocation" (C. Vogel n.d., 85). Despite
a history of bitter conflict, the commissioners came together around a
grim vision of South Florida—whose population was expected to double from 6 million in the mid-1990s to 12 million by 2050—besieged
by water shortages, poor water quality, dirty air, and social degenera-
tion. Although they agreed the region was on an unsustainable course,
the commissioners initially were skeptical about the prospects for agreement on a solution. But they were able to move forward by defining the
problem as insufficient and wasted water, rather than as a more systemic

failure to develop within environmental constraints. The logical solution, then, entailed storing as much water as possible; as Audubon's Stuart Strahl (2002) observed, "The more you store, the more you have, and the less conflict there is." Similarly, as the Corps's Stuart Applebaum (2002) concluded: "Enlarge the pie. Everybody can get a bigger slice and you avoid the conflict."

Having agreed to maximize storage, however, the group had to confront thorny questions about the actual mechanisms for storing and distributing water, as well as the amount that would be allocated for each purpose. Over the next six years, as they debated these questions, relationships among participants improved and trust developed. As Ernie Barnett (2002) of the Florida Department of Environmental Protection explains: "I think the ag guys were able to realize that the environmental groups weren't all nuts and looneys. The environmental groups were able to recognize that the ag guys weren't evil people out to do the environment in." Similarly, according to Dick Ring (2002), former superintendent of Everglades National Park, "People came to believe others would honestly represent their position and would debate in good faith."

Most of the participants also came to understand the complex Everglades ecosystem in a more holistic way (Applebaum 2002; E. Barnett 2002; Collins 2002). In addition, as the commission's technical teams patiently worked through numerous iterations of restoration alternatives in response to commissioners' questions and challenges, members of the group learned how changes in the C&SF Project would affect the natural system, as well as various stakeholders' water supplies. According to Pettigrew (1995), as a result of those discussions, "Diametrically opposed stakeholders gradually realigned their positions to reach sustainable solutions."

Not all participants' perceptions of the relationship between human activity and environmental health were transformed by the experience, however. In particular, representatives of the water utilities remained intransigent in their demands for guaranteed future water supplies, regardless of evidence suggesting that the natural system would suffer (Hurchalla 2003; Pettigrew 2003). Furthermore, some commissioners attended meetings only sporadically, and many were uncomfortable when the group's focus shifted to human, rather than natural, systems (Oyola-Yemaiel 1999). As a result, participants remained skeptical of some of their fellow commissioners' motives. As Maggy Hurchalla (2003), a former Martin County official appointed to represent local government,

explains, "I learned when I could and when I couldn't trust people." It was easier to develop working relationships, she adds, with people who "didn't have such a severe dog in the fight." Similarly, Malcolm (Bubba) Wade, vice president of U.S. Sugar (2001), notes that environmentalists were not always true to their word because their constituents did not feel bound by deals made at the negotiating table.

Moreover, because participants remained divided over their fundamental interests, the commission declined to tackle issues on which it could not reach consensus (Pettigrew 2003). Instead, the commission agreed on five general objectives: restore key ecosystems; achieve a healthier, cleaner environment; limit urban sprawl; protect wildlife and natural areas; and create quality communities and jobs (GCSSF 1995). It then took the long list of plan components assembled by the Army Corps of Engineers in its Reconnaissance Study as the starting point for developing its own Conceptual Plan. The commission rejected the Reconnaissance Study's "narrow focus," however, on the grounds that

The alternatives proposed may provide sufficient water for the natural resources of the South Florida ecosystem, but do not address people's water supply needs. There will not be sufficient support for proposals to spend billions of dollars on environmental restoration unless adequate attention is paid to meeting the needs of the public health, safety, and welfare. (GCSSF 1995)

The commission proceeded to task its Technical Advisory Committee with developing a water budget that would meet *both* restoration goals and future water supply needs for the region. Given those constraints, that committee concluded: "Paradoxically, achieving balance in the total South Florida system, including more natural Everglades and estuaries, will require additional structural facilities and increased operational flexibility" (GCSSF 1996). This observation became the premise for the commission's Conceptual Plan, which consisted of 112 structural and operational changes to the existing C&SF Project.

Devising CERP

With a mandate from the 1996 Water Resources Development Act to incorporate the commission's conceptual framework, the Corps's and SFWMD's multi-agency team of government and tribal biologists, ecologists, engineers, geographic information system specialists, hydrologists, public involvement experts, and real estate specialists set out to craft a restoration plan that could restore the ecological integrity and functionality of the Everglades while simultaneously enhancing supplies for water users and maintaining flood control.[7] From the outset, planners recognized that because of the drastically reduced spatial extent of the system, they could not attain full-fledged restoration; nevertheless, they hoped to "shift the substantially degraded system in the direction of a more natural one" (SFERTF 2000). At the same time, they had been repeatedly admonished by the Governor's Commission to recognize that "It is an important principle that has helped gain consensus for the restudy that human users will not suffer from the environmental restoration provided by the restudy" (GCSSF 1999, 5–6).

Working within these parameters—which dovetailed nicely with the engineering culture of the Corps and, to a slightly lesser extent, the SFWMD—the Restudy team began with the list of components approved by the Governor's Commission. Using a screening process to eliminate options that were either too costly or inconsistent with the SFWMD's projections of lower east coast water supply needs, between the summer of 1996 and the summer of 1997 the team assembled a subset of those components into a single plan called the Starting Point. Then, in the eight months from September 1997 to April 1998, they evaluated a series of Restoration Plan alternatives using an iterative process that involved assessing each alternative with respect to a set of hydrological criteria for the depth and duration of flooding, as well as a set of ecological attributes that characterized the health of the natural system. After formulating and evaluating ten major alternative comprehensive plans and running more than two dozen intermediate computer simulations, the team selected a preferred alternative. In mid-October 1998 the Corps released a 4,000-page draft report laying out its proposal.

The draft plan contained 68 projects that, taken together, would capture an average of 1.7 billion gallons per day of water normally shunted to tide, store it in massive lagoons and underground aquifers, and then release it according to a "natural" schedule. The plan ensured that urban water demands through 2050 would be met by 2010, and agricultural water supply demands would be met by 2015. Only by about 2017 or 2019 would the Everglades and coastal areas begin to see significant benefits. Despite the belated appearance of environmental improvements, most environmental groups were fairly pleased with the draft plan. Some were critical, however, and the Sierra Club immediately called for an independent scientific review.

Even more damaging than the Sierra Club's challenge was the scathing critique issued by Everglades National Park scientists, who—because of the park's longstanding distrust of the Corps and SFWMD—had participated only halfheartedly in the Restudy. Park scientists fired off a

missive to the Corps alleging the plan favored drinking water supplies over ecological restoration. According to the letter, there was "insufficient evidence to substantiate claims" that the plan would "result in the recovery of a healthy, sustainable ecosystem." Rather, the scientists said, "We find substantial, credible, and compelling evidence to the contrary" (Pittman 1999a). Park scientists pointed out that CERP barely increased flows to the southern Everglades, increasing them from about 60 percent of predrainage levels to 70 percent—and even that increment would not occur until 2036—though models suggested that at least 80 percent to 90 percent would be necessary to trigger conditions favorable to marl prairies, the southern estuaries, and wading bird nesting (Levin 2001; Ring 2002). To CERP planners' dismay, in late January six nationally prominent scientists buttressed park scientists' critique and echoed the Sierra Club's call for independent review of the plan. Stuart Pimm, an internationally recognized ecologist, was especially outspoken, saying the group was having difficulty finding a "thread of restoration" in the plan. The outside scientists were particularly appalled that CERP retained the altered system's fragmentation and compartmentalization and continued to rely on intensive management rather than trying to reestablish sheet flow.[8]

After some initial resistance, the Corps agreed to submit CERP to a detailed independent scientific review, although it still intended to hand the plan over to Congress by July 1, as scheduled. In hopes of quelling charges that CERP did too little, too late for the environment, hydrology modelers tested a scenario that sent more water to the Everglades through the Everglades Agricultural Area. They concluded that this approach would yield "a series of improvements to the ecosystem by 2010," including "vast improvements" to Everglades National Park. The scenario would have reduced water supply benefits by a modest amount, however, and was dropped from further consideration (Grunwald 2002b). Instead, Corps engineers proposed a more modest solution: collecting runoff from Palm Beach and Broward counties, pumping it into wells, reservoirs, and filter marshes, and ultimately releasing it into the park and Biscayne Bay. But critics pointed out that this solution, which the Corps said could pour 112 million additional gallons into natural areas each year, had a variety of defects, the most obvious of which was that urban runoff contains a host of pollutants, including fertilizers, pesticides, oil residue, and heavy metals (King 1999).

In early April 1999 the Corps released a revised CERP that doubled the pace of restoration outlined in earlier proposals, aiming to finish 44 of the 68 projects—including several key environmental projects—by

2010. In addition, the 30-page Chief's Report that accompanied the plan said the Corps's analyses had shown CERP would deliver 80 percent of the water stored in wells and reservoirs to the Everglades and estuaries such as Biscayne and Florida bays, giving the ecosystem half of the total expanded water supply (M. Davis 2002). (Under the status quo, cities and farmers were getting 70 percent of a smaller supply.) The plan retained assurances for water users, however, because, as Stu Applebaum reminded people, "We don't want to set up a competition for water because that will mean that the Everglades gets shortchanged again. We're trying to enlarge the pie—provide more water—for everyone" (Zaneski 1999).

Nevertheless, Florida's water users protested the last-minute commitments in the Chief's Report, saying the emphasis on restoring the natural system over supplying water to users jeopardized CERP's broad-based support and violated the "consensus-based, balanced" approach produced by the Governor's Commission (Anon. 2000). To hold the fragile consensus together Senate staffers working on the bill agreed to ignore the Chief's Report and authorize only the original plan (Grunwald 2006). To appease senators who were convinced by environmentalists' arguments, however, the final bill submitted to Congress on July 1, 1999, stated that its "overarching purpose" was restoration (though its provisions retained strong assurances for flood control and water supply) and required concurrence by the Interior Department with programmatic regulations. In addition, the Corps pledged to deliver 80 percent of the new water to the natural system—although that promise was not legally binding. Although some critics remained dubious about the plan's restoration prospects, they recognized the importance of unity in order to receive federal support, and in September 2000 a coalition of environmentalists, agricultural interests, utilities (representing urban interests), and the governor's office threw its weight behind the first phase of the plan, a $1.4 billion proposal to build ten major projects and four pilot projects (Balz 2000).[9] In late September the Senate passed the Water Resources Development Act of 2000, which authorized phase 1 of CERP. The House approved the bill in October, and in December President Clinton signed it into law.

The Comprehensive Everglades Restoration Plan

Consistent with the optimistic model of EBM, CERP is relatively comprehensive: it aims to restore a healthy Everglades ecosystem. Although they acknowledged the Everglades' greatly reduced spatial extent, the plan's authors nevertheless assert that "Implementation of the recommended Comprehensive Plan will result in the recovery of healthy, sustainable ecosystems throughout south Florida" (USACE and SFWMD 1999, x). The plan, they say, will restore the "essential defining features of the pre-drainage wetlands over large portions of the remaining system." As a result, "At all levels in the aquatic food chains, the number of such animals as crayfish, minnows, sunfish, frogs, alligators, herons, ibis, and otters will markedly increase. Equally important, animals will respond to the recovery of more natural water patterns by returning to their traditional distribution patterns" (xi). Although the restored Everglades will be different from what existed in the past, they add, "It will have recovered those hydrological and biological patterns which defined the original Everglades, and which made it unique among the world's wetland systems" (xii).

More consistent with the pessimistic model, however, the plan avoids imposing costs on water users at the expense of the environment. CERP articulates a threefold goal, to: "restore, preserve and protect the South Florida ecosystem, while providing for other water-related needs of the region, including water supply and flood protection" (WRDA 2000, P.L. 196–541). To this end, the plan adopts a heavily engineered approach: building 181,300 acres of surface storage capable of holding 1.5 million acre-feet of water, including one above-ground reservoir spanning 60,000 acres in the Everglades Agricultural Area and two underground reservoirs in former limestone quarries in the eastern Everglades; injecting as much as 1.6 billion gallons of water per day into underground wells, using a technology known as aquifer storage and recovery (ASR); building about 35,600 acres of storm water treatment areas; removing more than 240 miles of canals and levees within the remaining Everglades, while adding almost 500 miles of levees and canals on the periphery; designating multipurpose water management areas to serve as buffers between the eastern Everglades and rapidly urbanizing Palm Beach, Broward, and Miami–Dade counties; and modifying water deliveries to improve the timing and amount of freshwater flowing into estuaries, Everglades National Park, and Florida Bay to more closely mimic a rainfall-driven regime. The plan also includes provisions to reduce seepage by erecting impermeable barriers beneath levees, installing pumps to direct water back into the Everglades, and holding water higher in undeveloped areas than in adjacent developed areas. Finally, it prescribes two wastewater reuse plants for Miami–Dade County to clean water that can then be discharged to Biscayne Bay (see figure 5.1).

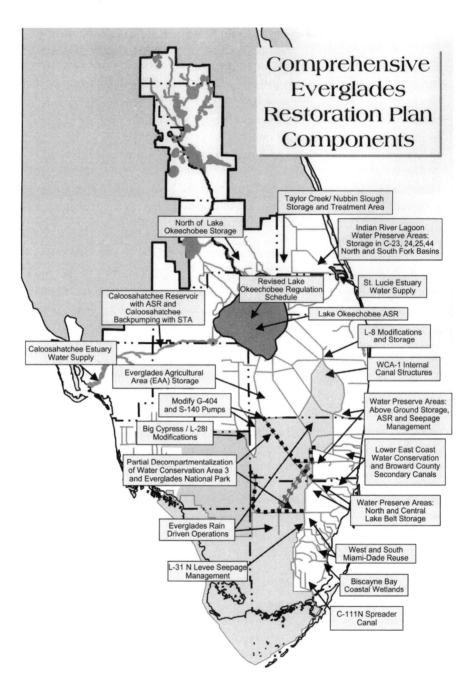

Comprehensive Everglades Restoration Plan Components

Taylor Creek/ Nubbin Slough Storage and Treatment Area

North of Lake Okeechobee Storage

Indian River Lagoon Water Preserve Areas: Storage in C-23, 24,25,44 North and South Fork Basins

Revised Lake Okeechobee Regulation Schedule

Caloosahatchee Reservoir with ASR and Caloosahatchee Backpumping with STA

St. Lucie Estuary Water Supply

Lake Okeechobee ASR

L-8 Modifications and Storage

Caloosahatchee Estuary Water Supply

WCA-1 Internal Canal Structures

Everglades Agricultural Area (EAA) Storage

Modify G-404 and S-140 Pumps

Water Preserve Areas: Above Ground Storage, ASR and Seepage Management

Big Cypress / L-28I Modifications

Lower East Coast Water Conservation and Broward County Secondary Canals

Partial Decompartmentalization of Water Conservation Area 3 and Everglades National Park

Water Preserve Areas: North and Central Lake Belt Storage

Everglades Rain Driven Operations

West and South Miami-Dade Reuse

L-31 N Levee Seepage Management

Biscayne Bay Coastal Wetlands

C-111N Spreader Canal

There are several reasons to wonder whether the approach delineated in CERP will yield environmental improvements sufficient to restore the Everglades' ecological health. The plan's overarching approach to ecological restoration is to re-create, where possible, historic depths and durations of flooding in the remaining natural areas by reengineering and intensively managing the system. Doing this will require precise knowledge as well as enormous skill and vigilance on the part of water managers: although planning was based on the "average" year, successful implementation will require devising and operating a system that can mimic South Florida's highly variable climate of wet and dry seasons and years. Moreover, the original plan paid scant attention to the movement of water that is such a critical aspect of sheet flow. Although scientists and some stakeholders asked the Corps to model the effects of enhanced flow, the scale of the hydrologic models it was using—primarily the South Florida Water Management Model and the Natural System Model—was too coarse (two-mile-square grids) to reliably predict local ecological effects, and its parameters were not chosen to accommodate flow (MacVicar 2001; Mazzotti 2006).

By contrast, journalist Michael Grunwald (2006, 326) observes:

The critics [of CERP] envisioned a more natural, less structural CERP that would provide more water to the park and faster environmental benefits to the entire ecosystem. Instead of trying to store water at the side of the Everglades and have water managers squirt it wherever and whenever they thought it was needed, the critics wanted to store more water at the top of the Everglades and let it flow south in an uninterrupted sheet.

Dissenters would have preferred to build larger reservoirs in the EAA and create a flow way to the remaining natural system, and to reconnect the central and southern Everglades by elevating the Tamiami Trail and removing as many other barriers as possible.[10]

Furthermore, despite CERP's rhetorical emphasis on ecological restoration, the risks associated with the plan fall heavily on the environment. For example, the plan allows rock mining companies to excavate up to 21,000 acres of Everglades wetlands in the so-called Lake Belt. Once the pits are depleted—some 35 years from the plan's inception—the Corps plans to spend an estimated $1 billion converting them into reservoirs. It is not known, however, whether water will seep out of the mines or

Figure 5.1
Comprehensive Everglades Restoration Plan Components

whether the remnants of mining will contaminate the well fields with deadly microbes; nor is it obvious that it makes sense to allow the destruction of thousands of acres of wetlands, even as scientists insist that the system's spatial extent is critical to its survival (Morgan 2004a).

Similarly, the performance of aquifer storage and recovery (ASR) technology, which accounts for much of the plan's new water storage capacity, is speculative, yet without ASR scientists say they will have an extremely difficult time restoring water to the Everglades (Pittman 1999b). Although there are more than 50 ASR projects in North America, and 36 in Florida alone, the proposed ASRs would be much bigger and more numerous than anything that currently exists, pumping as much as 20 times as much water as the biggest system in existence. Each well would reach down about 1,000 feet into the brackish Floridan Aquifer and create a bubble spreading 1,000 feet or more in every direction. But some experts warn that mixing surface water with underground sources may cause the constituents in the rock to dissolve. They suggest that mercury and arsenic in the Floridan could wind up in ASR water; and they worry that pumping so much water underground in so many locations will fracture the rock that separates the aquifers from areas where utilities have injected treated sewage (Morgan 2001a; NRC 2002).

Acknowledging that many of the storage technologies envisioned by the plan—in fact, projects accounting for nearly half the plan's cost—are highly uncertain, planners designed pilot projects to test wastewater reuse, seepage management, water storage in quarries in the Lake Belt, and ASR. As an additional means of dealing with uncertainty, the plan made provisions for "adaptive assessment"—a passive version of adaptive management that involves redesigning projects in response to improved modeling, rather than management experiments. To garner the information on which improved modeling is based, scientists planned to monitor the health of the natural system relative to a set of criteria derived from conceptual ecological models linking the major hydrologic stressors to attributes that characterize the health of the natural system.

Nortwithstanding CERP's provisional nature (WRDA 2000 explicitly recognized that CERP was a work in progress), planners face serious constraints on modifying projects in light of information suggesting they will not achieve environmental restoration. Chief among these is the "savings clause" in WRDA 2000's "Assurance of Project Benefits." This provision prevents any elimination or transfer of water from an existing use to another until a new source of supply of comparable quantity and quality as that available on the date of the law's enactment (December 13, 2000) is available to replace the water lost as a result of plan implementation. Nor can CERP reduce the level of flood control provided as of December 13, 2000. Similarly, Florida state law (Sec. 373.1501(5)(e), F.S.) mandates that the SFWMD must "provide reasonable assurances that the quantity available to existing users will not be diminished by implementation of project components so as to adversely impact existing legal users." These requirements circumscribe the extent of experimentation and adaptation because as each project moves forward, the existing balance must be maintained; they also contribute to delay in projects with environmental benefits, which cannot be initiated until water supply demands are met.

Implementation: The Consensus Unravels

Consistent with the optimistic model of EBM, implementation of CERP has yielded some policies and practices likely to benefit the environment: it has prompted the state of Florida to acquire more than 200,000 acres of land; it has also resulted in dramatic advances in the scientific understanding of the South Florida ecosystem, better coordination among the many agencies that operate in the region, and efforts to engage in more environmentally sound practices by the Corps and SFWMD. Consistent with the pessimistic model, however, CERP's implementation has been mired in delays. The ascension of leaders who are only weakly committed to environmental protection has made the political context for ecological restoration even less hospitable than was the planning context, undermining trust and support for earlier agreements; efforts to design specific projects have exposed conflicts among environmental, water supply, and flood control priorities that were disguised during the planning process by agreement on vague objectives; and some stakeholders have seized opportunities created by CERP's laborious project planning requirements to challenge projects they dislike. The delays have cut two ways: on the one hand, they have allowed scientists to improve the quality of information and models on which CERP is based; on the other hand, because neither the plan itself nor any complementary initiative contains mechanisms for ensuring that land-use decisions are consistent with restoration goals, South Florida's sprawl is rapidly encroaching on the Everglades, increasing demands for water and flood control while raising the price of or paving over land that was intended for restoration projects.

The Benefits of CERP

Like the other planning processes described in this book, CERP has succeeded in attracting money: between 1999 and 2006 the federal government spent $341 million on CERP-related activities, and state and local governments contributed another $2 billion (USGAO 2007). Most of that money has been used to buy land: the state of Florida has acquired 207,000 acres, just over half the estimated 406,000 acres needed to complete CERP; the remaining 199,000 acres are projected to cost at least $1.34 billion (USACE and SFWMD 2006).

In addition, CERP implementation has institutionalized more integrative and holistic science in South Florida, and many expect that better scientific understanding will result in improvements to critical restoration projects, such as the Decompartmentalization and Sheet Flow Enhancement Project, known as Decomp.[11] In 2000 the Corps and SFWMD established Restoration, Coordination, and Verification (RECOVER), a multidisciplinary, multiagency team of scientists charged with providing support to CERP. RECOVER has devised a set of 83 performance measures based on conceptual ecological models for 11 physiographic regions, as well as for the system as a whole. It has incorporated these performance measures into a Monitoring and Assessment Plan (MAP) to support CERP's cutting-edge adaptive management approach, which replaces the originally envisioned adaptive assessment. In September 2005 RECOVER issued the Initial CERP Update, laying out proposed modifications to CERP projects based on improved modeling. Beginning in 2007 RECOVER will issue System Status Reports that will evaluate the status of the Everglades relative to CERP's interim goals.

Beyond integrating scientific research across disciplines, agencies, and jurisdictions, CERP has increased coordination among federal agencies as well as among state and federal entities and tribes. The Task Force and its Working Group continue to provide arenas for debating and potentially resolving interagency and interjurisdictional differences. (The Task Force also established a Science Coordination Team, later renamed the Science Coordination Group, to improve communication, coordination, and cooperation in applying science to South Florida's ecological and socioeconomic problems.) Since undertaking the Restudy, the Corps has adopted a more consultative approach as well, working more cooperatively with managers and technicians in other agencies—an effort that is reflected in gradually improving relationships with the Everglades National Park staff and other longtime critics. One manifestation of this shift is the 2006 joint Corps/Interior Department announcement of their "bold plan" to adopt an active adaptive management approach to the Decomp project.

Finally, CERP has reinforced efforts by the Corps and SFWMD that began in the 1980s to manage water in a more environmentally sound fashion. Many observers remark on a genuine evolution in the Corps's Jacksonville office, which has become progressively more committed to restoring the natural system (Best 2006; Kraus 2006). The SFWMD has adopted some more environmentally protective practices as well—although to some extent these changes have come in response to legal mandates rather than increased awareness created by CERP. For example, in March 2001, despite bitter opposition from utilities, the district proposed to regulate "minimum flows and levels," thereby enabling it to reject water-use permits on the grounds that they would cause "significant harm" to the total supply and natural systems (Morgan 2001b).[12] In 2005, after the state passed a new Growth Management Act that tied growth proposals to the availability of water, the district announced that it was rejecting a request by Miami–Dade County for a 100-million-gallon-per-day increase in Everglades water; instead, the agency agreed to increase the county's supply only enough for the next 18 months (P. King 2006). The SFWMD subsequently opposed Hendry County's bid to amend its comprehensive plan to allow development on 18,400 acres of land adjacent to the Caloosahatchee River, urging the Department of Community Affairs to reject the plan because it did not identify a source of the estimated 15 million to 20 million gallons of water per day the county's additional 64,000 residents would demand (Burnham 2007). Broward County residents faced limits on water use as well, prompting County Commissioner Kristin Jacobs to remark: "There has been this laissez-faire attitude about water, but the district is doing business differently now. The utilities and cities aren't hearing the message. This is a freight train coming" (Wyman 2006).

These benefits, though genuine, are qualified. Although there is some evidence the SFWMD is trying to manage the water supply in a more environmentally conscious way, it is difficult to detect a genuine culture change in the agency; because its board and executive director are political appointees, the district's overall direction strongly reflects the values of the state's administration. Moreover, despite improved interagency cooperation, tensions remain between the water supply agencies and those—such as the Park Service and the FWS—with a mandate to protect the environment. Finally, substantial advances in the breadth and quality of South Florida's ecosystem science may not translate into better planning and management. An independent scientific review by the National Academy of Sciences warns that both staffing and funding levels are insufficient to implement CERP's MAP fully (NRC 2006). In any case, it remains to be seen whether decision makers will adjust their operations in the face of new information generated by scientists, since the historic disconnect between scientists and engineers in South Florida continues to be a source of frustration for both groups (Best 2006; L. Brandt 2002; R. Johnson 2003; Salt 2003).

Disagreements Delay Implementation

Furthermore, although it has yielded some outputs that are likely to benefit the environment in the long run, because of delays it is impossible to attribute actual environmental improvements to CERP. The first project—the Southern Golden Gates Hydrologic Restoration Project, known as Picayune Strand—did not break ground until October 2003, and others have been substantially delayed: as of 2006, the ten components authorized in WRDA 2000 and scheduled for completion by 2005, as well as six pilot projects originally slated to be finished in 2004, were delayed by eight years, on average (NRC 2006). Implementation of CERP began with a burst of enthusiasm and funding; as a result, by the end of 2003 the SFWMD had already acquired three-quarters of the land needed for the first ten projects approved in WRDA 2000 and nearly half the total acreage slated for acquisition. But shortly thereafter land acquisition virtually ground to a halt: because Congress repeatedly deferred WRDA 2002, federal funding was not forthcoming for the three projects—Indian River Lagoon, Picayune Strand, and Water Preserve Areas—that contained more than half the land in the restoration and are among CERP's most environmentally beneficial elements (Franz 2002). Particularly serious has been the delay in starting the Decomp project, which most environmentalists regard as the "heart of the restoration."[13] Also vexing to environmentalists, the Corps has shifted the Indian River Lagoon project timetable, delaying the major environmental component until 2020 and thereby all but ruling out acquiring much of the land, which is likely to become too expensive or simply unavailable (Anon. 2004).

Changes in the political context—particularly the 1999 ascension of Jeb Bush to the governorship of Florida and the 2001 inauguration of President George W. Bush—have contributed to the delays in implementation by undermining the fragile consensus among environmentalists that under-pinned CERP's passage. Although both leaders were rhetorically commit-ted to Everglades restoration—in 2002 they cosigned the congressionally mandated agreement that 80 percent of the water captured by CERP would go to the ecosystem—their administrations have retreated from the strongly protective positions of their immediate predecessors. For exam-ple, in 1999 Governor Bush appointed a new Governor's Commission con-sisting almost entirely of agriculture, development, and business interests; he also appointed six pro-development members to the nine-member SFWMD board. In 2003, in response to a white paper documenting the importance of sheet flow, the White House weakened the role of the fed-eral Science Coordination Team by requiring it to address only questions specifically put to it by the Task Force, rather than using its discretion to scrutinize scientific research. And at both the federal and state levels, there have been several highly publicized instances of agencies firing, demot-ing, or transferring scientists who have dissented from their institutions' pro-development policies.[14]

In addition to shifts in the political context, interagency conflicts have arisen, reflecting unresolved tensions between water supply guarantees and environmental restoration that were masked by CERP's vague lan-guage. The Corps's development of programmatic regulations—the legally mandated guide to CERP implementation—exposed ongoing turf battles between the Corps and the Department of Interior, which in turn reflect philosophical differences over the plan's approach. In devising the 2001 draft of the rules, the Corps declined to provide assurances for the Everglades, suggesting instead that trust between federal and state offi-cials would be sufficient to ensure environmentally protective outcomes. Eventually, after nearly two years of wrangling, the Corps released a final set of regulations prescribing that 80 percent of new water supplies would be set aside for ecological restoration—although the Corps retained flexibility to change that allocation during implementation—and chang-ing the Interior Department's role from "consulting" to "concurrence" (or signing off) on the restoration's individual projects and scientific stud-ies (Morgan 2003). (Interior was elevated to full partnership in approving the three-party agreement on CERP interim goals.)

As with interagency disputes, stakeholder tensions have emerged dur-ing implementation that had been papered over or avoided during the planning process. For example, environmentalists have challenged the plan to build reservoirs in abandoned lime rock mines in the Lake Belt region of the eastern Everglades. In late August 2002 the Sierra Club, the Natural Resources Defense Council, and the National Parks Conservation Association sued the Corps in U.S. District Court, claiming the ten-year mining permits it approved—which allowed ten companies to mine limestone in a 22,000-acre region of the eastern Everglades—violated several environmental laws, would contaminate the Miami–Dade water supply, and threatened CERP. In May 2006 U.S. District Court Judge William Hoeveler issued a scathing ruling that held the permits were issued in blatant violation of many existing laws and regulations. He ordered an 18-month reassessment before new permits could be issued (Morgan 2006).

Meanwhile, urban development interests have been working assiduously to ensure that environmental protection measures mandated by CERP do not threaten their water supplies. For example, in March 2004, after two years of work, the Department of Environmental Protection (DEP) released a proposed "water reservations" rule to guide water management districts' decisions on how to allocate water for developers and municipalities, as well as how much to set aside for the environment to meet the baseline requirements for CERP.[15] The Association of Florida Community Developers immediately asked an administrative law judge to block the proposed rule. Reprising the logic that underpinned CERP, Cathy Vogel, a lobbyist for the association, explained, "If all uses are going to be met, they're going to have to find a way to make the pie bigger" (Caputo 2004). The two sides eventually worked out a compromise that would allow environmental set-asides, but only on the condition that a regional plan spelled out how much water would be available for development. At that point, however, farming groups and the state Agriculture Department objected. According to spokesman Terence McElroy, the Agriculture Department had no objection to considering environmental needs, but "don't do it at the expense of everybody else" (R. King 2004). The issue was finally resolved in early 2006, when the judge rejected developers' legal challenge to the DEP's statutory authority to adopt a water reservations rule.

The fate of the Mod Waters project exemplifies how changes in the political context, interagency disagreements, and stakeholder differences—all of which are exacerbated by a cumbersome project planning process—have impeded progress on Everglades restoration. Although not part of CERP, Mod Waters is a foundation project for the restoration, and Congress has forbidden the Corps to spend money on Decomp and other CERP projects until it is complete. Authorized by Congress in 1989, Mod

Waters aims to restore more natural timing, distribution, and quantity of water flowing into Everglades National Park by steering water under the Tamiami Trail and into the Northeast Shark River Slough. Originally projected to cost $89 million, by 2007 the project's cost had ballooned to more than $400 million; its target completion date is 2009, but agency officials do not expect to meet that deadline (USGAO 2007).

One of the main obstacles to proceeding with Mod Waters has been disagreement over a rural neighborhood in the eastern Everglades known as the 8.5-Square-Mile Area. Congress initially required the Interior Department to protect the 8.5-Square-Mile Area from floods that would be caused by enhancing water flows to the park, but in 1998 the SFWMD board voted unanimously to raze the entire area. A year later, however, a new board appointed by Governor Bush reversed that decision, throwing the project into limbo. In 2000 the Corps, the park, and the state finally agreed on a compromise: to buy 44 percent of the land and 12 percent of the homes in the disputed neighborhood, and build flood protection for the remainder. Several residents sued to stop the buyout, however, and a judge ruled it was not authorized by Congress. After an extended battle with property rights advocates, in February 2003 Congress approved the new plan. Finally, in early March 2004 the Corps began demolishing homes, and nearly two years later engineers broke ground on a new canal, pump station, and levee that would protect the remaining homes from higher water flows in the Northeast Shark River Slough.

A second impediment to Mod Waters' completion remained, however: what to do about the Tamiami Trail. The Corps's original plan had been modest, calling for raising a small section of the highway and expanding culverts to allow more water to flow into the park (Cusick 2006a). But after several redesigns, in June 2001 the Science Coordination Team recommended ripping out nearly 11 miles of Tamiami Trail and building an elevated skyway in order to restore sheet flow to Everglades National Park. As U.S. Geological Survey ecologist Ronnie Best explained, "It's not simply the flow of water itself. We're talking about all things that flow. It's the transport of sediments. It's the movement of nutrients through the system. It's the flow of biological species" (Morgan 2001c). But the Corps remained skeptical of the more ambitious plan, which would cost three times the amount of the cheapest alternative. After offering a series of more modest proposals, in late August 2005 the Corps proposed to build two bridges—a one-mile-long span west of Krome Avenue and a two-mile bridge near the Everglades Safari attraction (Morgan 2005).

Demands for proof of environmental benefits have dogged other aspects of Mod Waters as well: for instance, the proposal to backfill some of the L-67 canals, which stretch from central Broward County to the Tamiami Trail, has provoked opposition from anglers. Along with raising the Tamiami Trail, the overhaul of the L-67 levees was supposed to open up water flow through the Shark River Slough. But the plan would also cut off as much as 17 miles of one L-67 canal to boaters, who want to retain access to the rich largemouth bass fishery that has developed in the canals. Before taking such a dramatic step, anglers argued, the Corps should have to demonstrate that it was absolutely necessary. "We're all for restoration," said Al Olvies, president of South Florida Anglers for Everglades Restoration. "What we're not in favor of [is] backfilling the canals, especially with the lack of scientific evidence for the need" (Morgan 2004b).[16]

Paralysis accompanied by rising costs describes the cumulative impact of changes in the political context, interagency disputes, stakeholder disagreements, and a planning process that puts the onus on proponents of change to demonstrate environmental benefits. In March 2005 these problems gained the spotlight when someone leaked a highly critical Corps memo. The memo, written by Gary Hardesty, the Corps's top Everglades manager in Washington, suggested that the CERP process had degenerated into endless meetings where little was accomplished. Hardesty noted that, largely as a result of the Corps's failure to build any of the originally authorized CERP projects, the cost of restoration had risen by more than $2 billion (Cusick 2005). Impatient with the federal government's unwillingness to provide funding for CERP, in October 2004 the state of Florida announced an initiative dubbed Acceler8, which promised to raise $1.5 billion in bond revenue to jump-start eight major restoration projects—comprising 11 CERP project components and three non–CERP project components—proposing to complete them by 2011.[17] Acceler8 focuses primarily on increasing surface-water storage, however, and although it includes three environmentally important projects, it does nothing to restore sheet flow to the southern Everglades (Hain 2004).

The Consequences of Delay
Delays and funding shortfalls impair the state's ability to acquire land, which in turn jeopardizes two critical and related aspects of CERP: the ability to increase, or even maintain, the spatial extent of the remaining ecosystem, and the flexibility to adjust in response to information

suggesting more acreage is needed for ecological restoration. Originally, planners had envisioned an extensive buffer between the natural system and the urbanized east coast that would serve as water treatment areas, habitat, and water storage (the Water Preserve Areas); restoration of 90,000 acres of habitat—virtually all the remaining undeveloped land around the St. Lucie Inlet (the Indian River Lagoon); and restoration of 13,600 acres of Biscayne Bay wetlands. But while the state struggles to raise money to buy land for these projects, real estate values are rising by 20 to 40 percent a year, and South Florida's growth machine is gobbling up property in the restoration's footprint. Thus, Kim Dryden, a FWS biologist observed, "We're tearing down the ecosystem a lot faster than we'll ever be able to fix it" (Grunwald 2002a).

One prime target of developers is the wetlands around Biscayne Bay, which planners had hoped to acquire as part of an effort to restore the struggling sea grass beds, mangroves, and wetlands in Biscayne National Park. In the spring of 2002, the Lennar Corporation proposed expanding a development called Lakes by the Bay by building up to 3,000 town houses on a 516-acre site in the middle of the Biscayne Bay Coastal Wetlands. The environmental group 1000 Friends of Florida challenged the initial SFWMD permit, but the district and the Corps ultimately approved the project, saying Lennar would minimize damage by establishing a 145-acre wetland preserve on the site.

Sprawl development threatens the restoration most acutely, however, in the agricultural lands of western Palm Beach County. There commissioners have faced an onslaught of proposals to increase residential density. For example, in the summer of 2004 Florida Crystals Corp. unveiled a plan to build a 16,000-acre "smart growth green community" on a swath of sugarcane fields between the Loxahatchee National Wildlife Refuge and the J. W. Corbett Wildlife Management Area that the company was going to have to retire from farming anyway because it had so badly depleted the soil. Aware that the development would undermine any possibility of restoring sheet flow through the EAA, the company sweetened its offer by offering to donate up to one-quarter of the property for Everglades restoration. Although the Palm Beach County commissioners were not persuaded by this proposal, they did give Palm Beach Aggregates, Inc. the right to build 2,000 homes and 50,000 square feet of commercial space on a 1,219-acre plot of agricultural land close to the proposed Florida Crystals community. In 2007 other proposals were on the table as well. The owners of Callery–Judge Groves wanted to put 2,999 homes and

235,000 square feet of commercial development (down from 10,000 homes and 3.8 million square feet of commercial development, which the county rejected) on their 4,000-acre orange grove. Another large landowner, GL Homes, was pitching 12,000 houses on 5,000 acres of the former Indian Trail Groves (Florin 2007). In addition, the county itself had requisitioned a piece of the Corbett Wildlife Management Area for an electrical substation and was considering extending PGA Boulevard through the Loxahatchee Slough—using land it had purchased with money set aside for conservation.

Proponents of restraint have offered alternatives to sprawling development in Palm Beach County, but local officials have been inclined to support growth. For example, in November 2004 Audubon, the Florida Wildlife Federation, and 1000 Friends of Florida presented a proposal to keep most of western Palm Beach County's sugar fields in agriculture, put one-fifth of them underwater, and steer development to existing western cities such as Belle Glade. But that plan was never seriously considered; instead, Palm Beach County commissioners approved a sector plan covering some 50,000 acres that opened up the cane fields to development. The plan required large landowners to set aside only 50 percent of their property in exchange for higher densities on the remainder; it counted as open space shrubs, medians, and landscaping. In November 2007—after a two-year battle with the state's Department of Community Affairs, which rejected the sector plan in 2005—the county scrapped it and began work on new guidelines.

Nor has the Corps, the SFWMD, or the Jeb Bush administration supported environmentalists' efforts to conserve land in CERP's footprint.[18] For example, in 2004 Governor Bush and Palm Beach County officials tried to entice the Scripps Research Institute to build a research institute, as well as a surrounding community, in a remote 1,900-acre site known as Mecca Farms, at the headwaters of the Loxahatchee River. Both the SFWMD and the Corps approved the site. In April 2005, after Scripps and the county signed a contract on Mecca Farms, environmentalists filed suit challenging the project's environmental impact statement, which had considered only the impacts of the 500-acre campus, not the surrounding development. Two months later, a judge ordered Scripps to suspend construction until the Corps completed a full environmental review. Eventually, Scripps decided to locate elsewhere, but the county appeared likely to pursue housing on the remote Mecca Farms site anyway, since it had already spent $122 million there (Hafenbrack 2006; D. Poole and

Florin 2006). In June 2006 Jeb Bush, whose administration had abandoned efforts to promote increasing urban densities and redevelopment, signed a bill that would prevent local governments from using concerns about urban sprawl as a reason to bar development on small, undeveloped agricultural parcels.

External reviewers have highlighted the failure to protect land in CERP's footprint as a major obstacle to the restoration. In January 2005 the National Academy of Science's Committee on the Restoration of the Greater Everglades Ecosystem (CROGEE) released a report saying land acquisition, particularly in the EAA, should be the top priority. "The worst from the point of view of Everglades restoration," the report said, "would be commercial, residential and industrial development of the [EAA]" (NRC 2005, 9). According to the report's lead author, Jean Bahr, "The whole plan as it is designed is predicated on being able to have land that can be restored. The more land that gets taken out of the potential restoration pot, the less likely the success" (Anon. 2005).

But as development continues and CERP remains mired in delay, the South Florida ecosystem continues to decline. Scholars have documented extensive development of wetlands at the urban fringe, including disturbed, marginal areas that were intended to serve as a natural buffer between the natural system and the urbanized Palm Beach and Broward counties. The number and acreage of tree islands have declined, wading birds have colonized the northern Everglades but continue to vanish from the south, and invasive exotics are spreading. The region's human population grew 23.5 percent between 1995 and 2005, and water withdrawals from the Everglades have been increasing. As the National Academy's Committee on Independent Review of Everglades Restoration Progress (NRC 2006, 36) observes, "Many components of the Everglades ecosystem have moved away from historical conditions rather than toward them."

Conclusions

CERP has indisputably prompted some environmentally beneficial policies and practices, advancing a trajectory toward comprehensive Everglades restoration that began in the early 1980s. Consistent with the optimistic model of EBM, several improvements have accrued as a result of trying to address South Florida's environmental problems at a landscape scale. Treating the Everglades as a single ecosystem prompted the acquisition of thousands of acres of ecologically valuable land from Lake Okeechobee to Florida Bay. Landscape-scale planning has also yielded integrative assessments of the natural system, as well as sophisticated hydrological and ecological models that facilitate learning by scientists and managers. And it has prompted the agencies responsible for managing South Florida's water supply to consider the interrelationships among the system's components and to coordinate with other state and federal agencies when making decisions.

By contrast, stakeholder collaboration has yielded few tangible gains. As predicted by the pessimistic model, the consensus-based approach adopted by the Governor's Commission enabled South Florida's water users to avoid confronting the root cause of the ecosystem's decline: excessive and wasteful consumption of land and water. Instead, constrained by the need to meet the demands of all stakeholders, the commission put its imprimatur on an elaborately engineered and intensively managed plan that imposes the risk of failure squarely on the natural system. CERP designers hope that adaptive management will, over time, improve the environmental benefits of the restoration; however, such an outcome rests heavily on the ability of highly constrained managers to hew closely to the recommendations of scientists—a rare occurrence in the history of U.S. environmental policymaking.

An alternative approach would have aimed to reduce urban and agricultural users' dependency on Everglades water and restore as much flow as possible within a large, protected area from Lake Okeechobee south to Florida Bay. Such an approach would have entailed requiring strict conservation measures by urban utilities—currently South Florida's per capita water consumption is among the highest in the nation, and the state has yet to adopt the kinds of policies that have spurred conservation elsewhere (C. Barnett 2007)—as well as creating incentives to reuse wastewater and develop alternative sources, acquiring a flow way through the EAA, gradually expanding the protected area as cane fields are retired, and removing the myriad barriers within the remaining Everglades. Although some additional engineering would be necessary early on, in the long run human inputs would be greatly reduced. As the Committee on Restoration of the Greater Everglades Ecosystem (CROGEE) observes, storage components with fewer requirements for active controls, frequent equipment maintenance, and fossil fuels for operation are likely to be less vulnerable to failure, and hence more sustainable over time (NRC 2005). But a plan that relies more heavily on restoring natural processes and less on human control would have emerged only from a process that began with the constraint of restoring the health of the natural system and tried to meet water supply needs within that limit. It also would have entailed severely restricting development of South Florida's remaining wetlands during planning and implementation, particularly those adjacent to existing protected areas.

Defenders of the consensus-based approach employed by the Governor's Commission contend it was necessary to make the plan politically legitimate and durable. But the agreement on general principles attained by the commission has proven to be fragile: during implementation environmentalists and other stakeholders have seized opportunities provided by changes in the political context to pursue advantages, and development interests have resisted efforts to manage growth or designate water supplies for the natural system. Nor is it likely that a commitment to flexible, adaptive implementation will be sufficient to overcome the obstacles that have arisen. Despite the development by RECOVER of a sophisticated adaptive management strategy, it remains unclear whether managers can and will adjust in response to new knowledge, such as the importance of flow.[19] In any case, strong assurances for water users, combined with the rapid disappearance of wetlands, have greatly reduced the options.

Ironically, the settlement of the 1988 water quality lawsuit by the much-maligned 1994 Everglades Forever Act has yielded more immediate and tangible environmental benefits than CERP by prompting substantial reductions in the amount of phosphorus being pumped into the Everglades. Of course, the cleanup has experienced delays of its own: in 2003 the state legislature extended the deadline established in 1994 for meeting the 10 ppb phosphorus standards from 2006 to 2012. The cleanup has also been the target of criticism; for example, environmentalists argue that the approach to measuring attainment—averaging phosphorus levels from different locations over a five-year period—is insufficiently protective. Nevertheless, the court-ordered cleanup has prompted relatively quick and dramatic results: 35,000 acres of newly constructed storm water treatment areas have dramatically reduced phosphorus concentrations coming off agricultural land from an average of 147 ppb to an average of 41 ppb (NRC 2006); best management practices have also proven effective, reducing by more than 1,300 metric tons the amount of phosphorous flowing into the Everglades since 1996 (SFWMD 2006).

Nevertheless, defenders of CERP's approach offer two rebuttals to the critique in this chapter. The first is technical: that removing barriers to flow in the Everglades would have either flooded residential areas or made some areas too wet and others too dry. But this conclusion may well be an artifact of the models used to analyze impacts. Arguably, planners' resistance to a flow-based approach resulted from an excessive reliance on hydrological models that were poorly suited to the task. It is at least arguable that if planners had invested as much time and effort into figuring out how to restore flow as they did on making their compartmentalized approach work, they might have come up with a viable plan. In fact, in 1999, when hydrology modelers tested a scenario that closely mimicked the original north-south flow, they concluded that although it created some problems, it also produced "a series of improvements" to the northern part of the ecosystem and "vast improvements" to the park (Grunwald 2006, 327).

A second, and more compelling, rebuttal is that CERP is the best deal planners could get, given the political constraints, and that without a consensus among stakeholders, forward movement would have been impossible. That, of course, is tantamount to saying that genuine restoration simply may not be politically palatable. It is worth considering, though, whether determined and sustained pro-environmental leadership, backed by a cohesive environmental community and a mobilized public, could have enabled planners to consider a more environmentally protective approach and forced users to make concessions. Instead, although the Everglades restoration plan has benefited from the abiding interest of several prominent and extremely persuasive advocates and the extraordinary dedication of some of the region's most prominent scientists, it continues to suffer from a dearth of political leaders willing to challenge Floridians to grapple with difficult tradeoffs between human consumption and the health of the natural system.

6

Averting Ecological Collapse in California's Bay–Delta

As planners worked on the Comprehensive Everglades Restoration Plan in Florida, a second major aquatic ecosystem-based management (EBM) initiative got under way on the West Coast. The process formally began in December 1994, when California Governor Pete Wilson, Interior Secretary Bruce Babbitt, and EPA Administrator Carol Browner signed a historic agreement that became known as the Bay–Delta Accord. The agreement did two things: it established an interim water quality standard for northern California's Sacramento–San Joaquin Delta, and it committed the state and federal governments to developing a long-term plan to restore the health of the Delta and San Francisco Bay ecosystem (the Bay–Delta). By August 2000, negotiations among state and federal policymakers, in consultation with stakeholders, had yielded a second agreement, the CALFED Record of Decision (ROD). The ROD constituted a commitment to "develop and implement a long-term comprehensive plan that will restore ecological health and improve water management for beneficial uses of the Bay–Delta system" (CALFED Bay–Delta Program 2000a, ES-3).

Like the other EBM programs described in this book, CALFED succeeded—at least initially—in attracting large sums of money: between 1995 and 2005, the program garnered about $3 billion, mostly in state bonds. It spent about one-third of that on ecosystem restoration projects, such as installing fish screens and ladders and acquiring and restoring wetlands; a portion of the remainder funded CALFED's science program. In addition, the program enhanced interagency coordination; in particular, under the auspices of CALFED, water managers have worked with fisheries scientists to devise and institute an environmental water account that aims to facilitate real-time adjustments in water project operations to prevent the deaths of migrating and estuarine fish species at state and federal water export pumps.

Despite these achievements, there is overwhelming evidence that the Bay–Delta has continued to deteriorate. Most notably, in 2003 the delta smelt—a prime indicator of the ecosystem's biological vitality—went into a precipitous downward spiral, and by 2005 scientists were warning of its imminent extinction. The blame for the ecosystem's collapse cannot be laid at CALFED's door, but neither can the program be credited with halting the decline, much less reversing it. CALFED's management-intensive approach put the risk associated with failure directly on the natural system, rather than on water users: instead of trying to restore natural functions and processes by reducing exports of freshwater from the system, the program relied on a precisely manipulated water delivery system that aimed to ensure that urban and agricultural water supplies were not disrupted, while providing just enough water to satisfy legal requirements for the region's endangered fish. Nor did a commitment to adaptive management compensate for the program's weaknesses. Policymakers were unwilling to experiment with reducing water diversions and unable to agree on performance measures, making it impossible to assess whether the program's ecological restoration activities were achieving their objectives and, if not, what to do differently. The EWA, CALFED's most frequently cited instance of adaptive management, was severely constrained in its ability to respond to information about its effectiveness by assurances to water users and by consistent underfunding for environmental water.

CALFED yielded an approach that had little prospect of restoring the ecological health of the Bay–Delta ecosystem because the ROD rested on a promise to meet the demands of all stakeholders—simultaneously ensuring flood control; reliable, high-quality water supplies; and environmental improvement. This requirement, deemed essential to ongoing collaboration, made it extremely difficult for planners to recommend options that would impose costs on powerful stakeholders. Similarly, CALFED's flexible governance—in which each participating agency chose whether to comply with the program's mandates—undermined the environmental protectiveness of its implementation: when CALFED's goals conflicted with their institutional missions, water project operators simply circumvented the interagency process.

In response to the ecological collapse in the Delta and in recognition of CALFED's shortcomings, in 2006 the state of California undertook a massive reassessment of its Bay–Delta policy. That process is occur-

ring within a context in which power is shifting in favor of environmentalists and fishing interests, thanks to a series of pro-environment
court decisions. In the absence of leaders committed to an overarching
goal of restoring ecosystem health and willing to employ their regulatory authority to ensure compliance, however, powerful water users may
nevertheless succeed in blocking major changes to the status quo.

The Evolution of Landscape-Scale Planning

The CALFED planning process began not in response to a change in the
Bay-Delta's environmental conditions, which had been deteriorating for
decades, but after legal action by environmentalists precipitated a regulatory "crisis" that threatened the reliability of water supplies to urban and
agricultural users and jeopardized the state's autonomy. Galvanized by
the impasse that ensued, high-level state and federal policymakers initiated a process to address the system's problems in a comprehensive fashion. CALFED created a forum in which all the state and federal agencies
involved in regulating and distributing water could devise a long-term
solution cooperatively, in collaboration with stakeholders and based on
knowledge gathered by reputable scientists.

The Origins of Environmental Problems in the Bay–Delta
The Bay–Delta watershed originates in the Sierra Nevada and part of the
Cascade range, where each spring the mountain snowpack melts into
hundreds of streams that merge into the Sacramento and San Joaquin
rivers (see figure 6.1). For centuries those two rivers flowed west out of
the foothills, through the Central Valley, to the Sacramento–San Joaquin
Delta, a vast expanse of marsh and grassland where freshwater and seawater mixed, and finally to San Francisco Bay. Because of its brackish,
estuarine conditions, the 1,315-square-mile Delta supported a rich diversity of plants, fish, birds, and wildlife.

Serious human modification of the Bay–Delta ecosystem dates back
more than 150 years. In the 1800s settlers began to transform the watershed by draining marshes, logging forests, and building dikes to control flooding. During the Gold Rush, miners dumped millions of tons of
sludge and debris into the Sacramento and San Joaquin rivers and their
tributaries. In the early 1900s, as mineral exploration slowed, miners-
turned-farmers built canals and levees to control the region's unruly
water so they could exploit its fertile soil. The most significant alterations

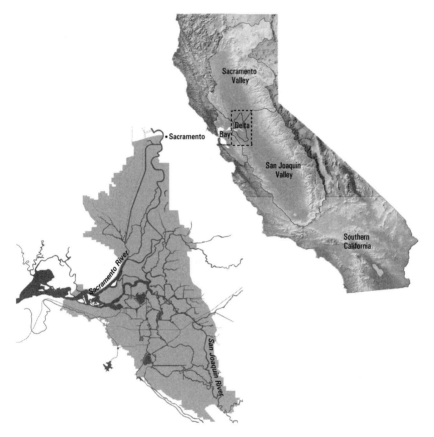

Figure 6.1
California Bay–Delta Watershed

to the system, however, began in the 1930s, when the federal government initiated the Central Valley Project (CVP), a massive system of reservoirs, dams, canals, aqueducts, and tunnels whose purpose was to provide water for farmers in the Sacramento and San Joaquin valleys. In the early 1970s the state of California completed its own water management infrastructure, the State Water Project (SWP), to provide additional irrigation to San Joaquin Valley farms and facilitate growth in California's south coastal basin.

Undertaken with little concern for their environmental consequences, the CVP and SWP had a variety of potentially damaging effects: they established a freshwater flow regime that was antithetical to the one within which native species evolved, dramatically reduced the areal extent and function of floodplains and other habitats, imposed numerous barriers to fish movement, and encouraged the invasion of exotic species into the system. Not surprisingly, shortly after portions of the SWP came on line in the 1960s, signs of an ecological breakdown appeared (Zakin 2002). For instance, the striped bass index—a measure of the number and distribution of striped bass in the estuary and considered an indicator of the Bay–Delta's ecological health—began to decline (Nolte 1990).

Nevertheless, water exports increased steadily over the next three decades, as both state and federal managers strove to meet the growing demands of water users. By the early 1990s the Delta was supplying drinking water to 22 million Californians—furnishing as much as 40 percent of the state's supply in some years—and providing irrigation water for about 4 million acres of farmland. To accomplish this, project operators diverted anywhere from 30 to 60 percent of the freshwater flowing down the Sacramento and San Joaquin rivers each year. In an "average" year, and counting both upstream and in-Delta withdrawals, the CVP and SWP together diverted about 6 million acre-feet of the 23 million acre-feet flowing through the system.[1] Further stressing the system, 7,000 small upstream and in-Delta water operators diverted another 5 million acre-feet per year (Lund et al. 2007; G. Martin 1999b).

The Impetus for Landscape-Scale Planning

As the environmental impacts of these diversions became evident, scientists raised concerns about the state's water policy. Critics recognized that water diversions were not the only cause of problems in the ecosystem: industrial pollution, the loss of more than 90 percent of the region's historic wetlands, overfishing, toxic pesticides in agricultural runoff, and the proliferation of invasive species also had compromised the integrity of the ecosystem. Many scientists suspected, however, that changes in the timing and amount of freshwater flows were critical causes of the ecological decline. For example, some believed that reductions in freshwater entering the Bay had forced the zone where freshwater and seawater mix to compress and shift upstream, facilitating the proliferation of invasive species while making the system less hospitable to natives, particularly the pelagic species that spend part of their lives in the Delta. Others hypothesized that changes in the temperature and timing of freshwater releases, as well as the presence of physical barriers in the forms of dams and diversions, hampered the migration of salmon and other anadromous fish.[2]

While scientists in various settings refined their understanding of the Bay–Delta ecosystem, northern California's environmentalists and fishermen tried to bring about changes in the state's approach to water management through lobbying and public education campaigns. They faced a formidable array of adversaries, however; urban and agricultural users who rely on the Bay–Delta's water mounted a formidable defense of the status quo. For example, the Central Valley's massive Westlands Water District, which represents 600 farm conglomerates spanning 600,000 acres, is a beneficiary of CVP water and lobbies aggressively to protect its (junior) water rights. The entity with the largest entitlements to the SWP is the Metropolitan Water District (MWD) of Southern California, which is "a giant among public water agencies" (Gottlieb and Fitzsimmons 1991, 5) that serves an area of 5,135 square miles and a population of 15 million. Robert Gottlieb and Margaret Fitzsimmons (1991, xvi) explain, "Water politics in Southern California have always been politics of growth, of heating up the local economy by finding strategies to subsidize an increased and reallocated supply of a necessary natural resource so that, no matter how rainfall might fluctuate from year to year, economic growth would anticipate no checks and limits." Historically, California's urban and agricultural users made common cause and dominated water policymaking at the expense of the natural system (Hundley 2001).

By the early 1990s, however, the balance of power was beginning to shift, as the catastrophic effects of a multiyear drought that began in 1986 gave environmentalists and fishermen some legal leverage over state policymakers. State and federal endangered species laws provided one legal hook. During the drought, the CVP and SWP increased exports to record levels, causing the populations of several key fish species to plummet. In August 1990 the California Department of Fish and Game reported that the striped bass index had reached a record low, and state biologists advised the state's Fish and Game Commission to list the once-plentiful delta smelt as threatened because of its precipitous decline (Nolte 1990). In November 1990 the National Marine Fisheries Service (NMFS, pronounced "nymphs") listed the Sacramento River winter-run Chinook salmon as threatened after the population dropped to 533 fish (Zakin 2002). By the spring of 1992, the U.S. Fish and Wildlife Service (FWS) had listed the delta smelt as threatened, and environmentalists had filed petitions to list other fish species as well. Biologic al opinions issued by the FWS and NMFS had caused intermittent, brief pumping shutdowns, and water users feared the Endangered Species Act consultation process would ultimately result in permanent restrictions on CVP and SWP operations (Rieke 1996).

During the early 1990s, state and federal water managers also ran afoul of the Clean Water Act. Responding to a 1986 appellate court mandate, the State Water Resources Control Board (SWRCB) undertook extensive Bay–Delta water quality and water rights adjudicatory proceedings, which resulted in a 1988 draft order that contemplated a new "water ethic."[3] The order contained water quality standards that would have limited water exports to the south and given the Bay an additional 1.5 million acre-feet in the spring, when fish need water to carry them through the mazelike Delta (Diringer 1991a). Although the new draft standards were only moderately protective, irate southern California water interests lobbied ferociously against them, and an intimidated SWRCB promptly withdrew the proposal. Then, in January 1989 the board eliminated the plan's two most controversial provisions—freezing the amount of water diverted from the Delta and increasing flows into the Bay—and in August 1990 it published a revised order that skirted the issue of freshwater flows altogether (Ingram 1990).[4] Despite environmentalists' protests that the plan's protective measures were virtually indistinguishable from those rejected by the court years earlier, in the spring of 1991 the SWRCB voted unanimously to adopt the revised plan.

Environmentalists immediately threatened to sue both the EPA and the Water Board, noting: "The water interests won't take us seriously as long as they have a state plan that allows them to continue to divert most of the freshwater flow from the delta. The environmental community won't have the leverage it needs to have an equal place at the bargaining table" (Diringer 1991b). In early September 1991 the EPA reinforced the environmentalists' position by rejecting the board's salinity and water temperature standards, describing them as "not adequate to protect the health of the estuary" (Petit 1991). Shortly thereafter—again at the urging of environmentalists—the EPA issued an ultimatum: if the state did not adopt more protective standards expeditiously, the agency would substitute its own.[5]

In the spring of 1992, in hopes of reestablishing state control over the allocation of water, Governor Pete Wilson proposed to "solve" the Bay–Delta's problems through a combination of conservation, marketing, and new water storage facilities, based on the recommendations of a state water policy task force. The task force report established the principle, popular among water users, that crisis could be averted only if "assurance is given that no user group will prevail at the expense of another" (Diringer 1992a). Environmentalists were not appeased by the governor's proposal, however, and in late July 1992 a coalition led by the Sierra Club Legal Defense Fund filed notice of intent to sue the federal government if it did not follow through on its threat to enact new protections for the Bay and Delta within 60 days (Diringer 1992b). The EPA responded by presenting a set of restrictions on freshwater exports and pressing the state to adopt them. State water managers balked, however, again warning that the EPA's prescription would commit so much freshwater that users up and down the state could face cuts as high as 30 to 60 percent of current deliveries, causing "tremendous economic damages" (Diringer 1992c).

As California and the EPA were facing off over Bay–Delta water quality standards, in October 1992 Congress approved (and President George H. W. Bush signed) the Central Valley Project Improvement Act (CVPIA). The CVPIA, which was the result of lobbying by an unusual coalition of environmental and urban interests, declared environmental protection an official purpose of the CVP and set a goal of doubling anadromous fish populations in rivers affected by the project by 2002. To this end, the legislation allocated 800,000 acre-feet of water to fish and wildlife (along with numerous other environmental provisions). Although environmentalists regarded the CVPIA as a triumph, it was nevertheless clear that addressing the Bay–Delta's problems in piecemeal fashion would not suffice in the long run.

At the same time, the federal government's increasing intervention in California water policy provoked state officials to try to regain their autonomy by taking a more comprehensive approach. On December 9, Governor Wilson created two entities: the 21-member Bay–Delta Oversight Council and the California Water Policy Council. He directed the Oversight Council, which included both water managers and prominent stakeholders, to develop a comprehensive plan to reverse the environmental decline in the Delta while simultaneously enhancing the quality and reliability of water supplies. He asked the Water Policy Council, composed exclusively of state officials involved in water management, to coordinate with federal officials in formulating a long-term water policy for the state.

On the same day, the SWRCB announced a plan to impose a series of interim measures that could deprive water users of well over 1 million acre-feet of water each year, to be put in place while the Oversight Council devised its long-term plan. Environmentalists pointed out that although the board's proposal was an improvement over its 1991 plan, it nevertheless assured the Bay and Delta of only half as much additional freshwater as would have been provided under the board's 1988 draft decision. The EPA was similarly unimpressed, and in mid-January it declared that the state's latest plan was still too weak, so it would move ahead with its own scheme. Despite this threat, the SWRCB—which faced simultaneous pressure from users—proceeded to release a plan that was even weaker than its predecessors, although insufficiently lax to satisfy agricultural interests. Kern County Supervisor Ben Austin reprimanded the board, saying, "Water means life, not only for the species this plan proposes to protect, but also for the 3 million people who live and work in the San Joaquin Valley. You will effectively write off an entire region. Certainly people are at least as important as fish and wildlife" (Diringer 1993a). In a gesture of support for agricultural interests, Governor Wilson asked the SWRCB to shelve its latest proposal (Olszewski and Kershner 1993).

Fed up, in mid-April 1993 the Sierra Club Legal Defense Fund—on behalf of 18 fishing and environmental groups—sued the EPA for failing to set standards for the Bay in accordance with federal law. In hopes of settling the suit, in mid-September the EPA agreed to issue more stringent water quality standards by the end of the year. At the same time, the four key federal agencies involved in California water management— the EPA, the Bureau of Reclamation, the FWS, and NMFS—signed a memorandum of understanding that created the Federal Ecosystem Directorate (Club Fed), whose role was to coordinate federal resource protection and management decisions.[6] In mid-December, Club Fed announced a sweeping set of measures for restoring freshwater to the Bay and Delta. The group estimated the restrictions could result in a 9 percent cut in exports from the system during "average" years, and as much as a 21 percent cut (1.1 million acre-feet) in dry years (Diringer 1993b). The centerpiece of the package was a set of water quality standards based on the X2 salinity criterion developed in the early 1990s by a technical team that had been convened as part of the San Francisco Estuary Project.[7] In April 1994, the EPA settled the environmental lawsuit against it by agreeing to adopt and enforce Club Fed's water quality standards by December 1994.

The Bay–Delta Accord: A Decisive Step toward Landscape-Scale Planning
The federal government's aggressive moves to list endangered species and institute water quality standards unilaterally made clear to Governor

Wilson, as well as to urban and agricultural users and much of the business community, that the state was no longer in control of water policy, and prompted tentative steps toward cooperation (Rieke 1996). In June 1994 the governor's Water Policy Council signed a memorandum of understanding with Club Fed that committed all parties to working cooperatively on a resolution of California's water wars. The primary goal of this endeavor was not ecological restoration but rather to "minimize the overall costs in water and dollars for achieving environmental protection and provide meaningful regulatory stability for users of the Bay–Delta's resources." More specifically, under the Framework Agreement state and federal agencies agreed to (1) work together to develop new state standards that would satisfy federal Endangered Species Act and Clean Water Act requirements, (2) coordinate operations of state and federal water projects, and (3) develop a long-term planning process (Wright 2001).

The agreement prompted an intense round of negotiations with Bay–Delta stakeholders. With the settlement deadline looming, Betsy Rieke, the Assistant Interior Secretary for Water and Science, got the disputants to agree on a plan, and on December 15, 1994, state and federal officials, joined by ten "interested parties," signed the Bay–Delta Accord. The Accord retained the EPA's water quality standards, which regulated the amount and timing of freshwater flows on behalf of fish,[8] created some operational flexibility in complying with the Endangered Species Act to prevent disruptions to users' water supply, and aimed to improve conditions for fish by taking steps unrelated to freshwater flow, such as installing fish screens and restoring habitat. The Accord also set in motion the CALFED Bay–Delta Program, an effort to generate a comprehensive, long-term solution to the Bay–Delta's problems. Although CALFED followed in the footsteps of a series of collaborative initiatives—including the San Francisco Estuary Project and a "three-way process" among environmentalists, agricultural interests, and urban users—it devised its own collaborative problem-solving structure. The Accord established a Policy Group, comprising high-level officials from ten state and federal agencies and presided over by Executive Director Lester Snow, as the program's decision-making body. It also created the Bay–Delta Advisory Council (BDAC), a 32-member panel made up of stakeholders whose purpose was to solicit public input and advise the Policy Group.

Formerly warring stakeholders agreed to participate in the CALFED process for a variety of reasons. For environmentalists, many of whom had been skeptical about earlier collaborations, CALFED promised to generate results that might actually be implemented. By contrast, previous processes, such as the San Francisco Estuary Project, involved enormous commitments of time and resources, but had no mandate. Furthermore, rather than continuing to address problems as they arose in piecemeal fashion, CALFED had the potential to yield a sustainable, holistic regime for managing the Bay–Delta's water, and therefore to improve the overall health of the regional ecosystem. Some environmentalists welcomed the program's collaborative approach out of concern that continuing to use regulatory clubs—although they had changed the balance of power somewhat—would end up weakening support for the Endangered Species Act in the long run.[9] Agricultural and urban interests hoped the process would both alleviate the uncertainty caused by enforcement of environmental laws and pave the way for building additional storage and conveyance facilities that could improve drinking water quality and enhance overall supply reliability. Everyone wanted to move away from year-to-year, crisis-driven decision making; as CALFED's executive director Lester Snow (2005) put it, "People understood that as bad as things [were then], they could get a lot worse."

Collaborating with Stakeholders on a Comprehensive Plan

From the outset, CALFED planners focused on gaining consensus among stakeholders on the goals and elements of a comprehensive plan.[10] That focus yielded outputs that are consistent with the predictions of the pessimistic model of EBM. Reaching agreement on broad goals was relatively straightforward; planners were able to move forward by promising to improve conditions for everyone. On specifics, however, consensus was elusive. Spending money to acquire and improve wildlife habitat provoked little conflict because restoration projects neither were funded by users nor threatened any change in the amount of water delivered. By contrast, figuring out how to provide adequate amounts of freshwater for fish without reducing exports from the system was extremely contentious, and intensive negotiations among stakeholders and policymakers failed to produce agreement. A final round of discussions among high-level officials yielded an innovative solution that broke the impasse: an environmental water account that was designed to ensure that water users faced no disruption in their supplies as a result of fish protection measures.

Establishing CALFED's Mission, Principles, and Goals

During its initial phase, the summer of 1995 through the summer of 1996, CALFED staff, in consultation with stakeholders, established as the program's dual mission "to develop and implement a long-term comprehensive plan that will restore the ecological health and improve water management for beneficial uses of the Bay–Delta system." That mission was tightly linked to and qualified by the program's six solution principles, according to which a long-term solution had to be affordable, equitable, implementable, and durable; in addition, it had to reduce conflicts in the system and could not solve problems in the Bay–Delta by redirecting them elsewhere. Finally, planners delineated four primary objectives: (1) to provide good water quality for all beneficial uses; (2) to improve and increase aquatic and terrestrial habitats and improve ecological functions in the Bay–Delta to support sustainable populations of diverse and valuable plant and animal species; (3) to reduce the mismatch between Bay–Delta water supplies and current and projected beneficial uses dependent on the Bay–Delta system; and (4) to reduce the risk to land use and associated economic activities, water supply, infrastructure, and the ecosystem from catastrophic breaching of Delta levees (http://calwater.ca.gov).

Although these goals and principles were acceptable to all, there was neither a recognition that CALFED's four main objectives might be in conflict nor an effort to establish priorities among them; in fact, the CALFED mantra—"getting better together"—implicitly assumed the existence of a win–win solution and the sustainability of the existing export regime. The program's unwillingness to make trade-offs was reflected in the three alternatives that emerged from the first stage of the planning process, all of which promised something for everyone: each combined an ecosystem restoration program with projects that would enhance the water supply by increasing storage, widening and deepening channels, or building a canal to circumvent the Delta. None departed substantially from the status quo.

Collaborating with Stakeholders

For the next three years, policymakers worked extensively with stakeholders in an effort to gain consensus on the specific elements of a comprehensive plan. Although the Bay–Delta Accord established a formal stakeholder advisory committee, participants made it clear that the most productive collaboration occurred in subcommittees, interagency teams, and ad hoc work groups, rather than in large public meetings. In the more informal settings they were able to explore ideas without having to defend them or posture for their constituents. Some of these groups broke down in conflict, but others built trust and engaged in joint learning and creativity (Bobker 2005; Innes et al. 2006; Innes, Connick, and Booher 2007; Wright 2006). Some observers firmly believe these negotiations wrought a genuine transformation among water users, and hence a concurrence among all stakeholders that the Bay–Delta's environmental problems were real and needed to be addressed (S. Johnson 2005). Public statements by some stakeholders suggest that attitudes and positions did, in fact, change. For example, Tim Quinn, then-deputy general manager of the MWD, said that historically, Delta water users thought only about how to transport water out of the system reliably. By the late 1990s, however, they were being forced to think about the health of the Bay–Delta. "If we are not taking care of that Bay–Delta watershed," said Quinn, "we are not taking care of California's economic future" (Curtius 1998).

But other urban and agricultural interests endorsed ecosystem restoration projects because they were funded with bond money, rather than user fees, and they anticipated that restoration would create tangible evidence of the program's benefits, thereby easing opposition to new storage and conveyance projects. Moreover, deep divisions among stakeholders persisted over conservation (known within CALFED as water-use efficiency) and the appropriate allocation of water, and those differences surfaced as the discussion moved from general principles to specifics. Lester Snow explained (Anon. 1997): "Early on, when we said there were no preferred approaches, everyone agreed intellectually. As we move to more detail, we see people move back to 'our way is right, your way is wrong.' "

Reflecting an ongoing interest in retaining their existing water allocations, water users employed a variety of tools to shape the context for CALFED's deliberations and, in particular, to influence the baseline for future water allocation decisions. In the fall of 1995 San Joaquin Valley water agencies sued the Interior Department, saying that implementation of the 1992 CVPIA would illegally divest districts of their water. The department survived this legal challenge, but in 1997 a Central Valley water consortium filed suit when the department tried to implement the CVPIA's 800,000-acre-foot environmental water set-aside from the CVP's "unsold" capacity, and in that litigation the users prevailed. Similarly, in 1995 and again in 1996, Rep. John Doolittle (R, Rocklin) introduced bills on behalf of agricultural interests that aimed to overturn the CVPIA, and in the summer of 1998 Governor Wilson lobbied President Clinton on behalf of farmers, saying that implementation of the law could imperil CALFED.

In addition to contesting the allocation of water for fish, agricultural interests joined forces with urban water districts to press their representatives to circumvent CALFED. In late 1996, representatives of the urban and agricultural water agencies formed a "working group" separate from CALFED to "develop a solution package to present to CALFED and other stakeholders as a foundation for discussion" (Anon. 1997). Around the same time, the Bay Area Council, a group representing regional business interests including the California Farm Bureau and the Los Angeles Chamber of Commerce, sought to persuade Governor Wilson, Secretary Babbitt, and Governor-elect Gray Davis that the best way to ensure the water supply needed to accommodate the state's anticipated population growth was to expand the system of water storage and accelerate the pace of its construction (Clifford 1998). In the fall of 1998, at the urging of agricultural interests, Rep. George Radanovich (R, Mariposa) attached a rider to the omnibus spending bill in the House that authorized a study of raising the Shasta Dam near the headwaters of the Sacramento River (G. Martin 1998). Governor Wilson continued to advocate on behalf of agricultural interests as well, and in mid-November 1998 he infuriated environmentalists by appointing a 33-member panel of representatives of big agriculture to advise him on agriculture and water policy issues (Gledhill 1998).

While development interests emphasized new construction, environmentalists—who were organized in a loose coalition called the Environmental Water Caucus—continued to advocate publicly for their preferred approach. They argued that CALFED was relying on inflated demand forecasts while overlooking water savings achievable through conservation, recycling, new irrigation technology, groundwater banking, and water marketing. Peter Gleick, of the Pacific Institute, a Bay Area think tank, claimed that the state had grossly underestimated the potential for reducing waste, and proceeded to generate a series of studies showing the potential for conservation in California. The Bay Institute's Gary Bobker pointed out that CALFED had adopted an engineering logic, even though it had not shown that it was possible to take more water out of the system and bank it without harming the environment. In response to these arguments, both Secretary Babbitt and Governor Wilson characterized environmentalists as "intransigent" and "obstructionist" (T. Perry and

Clifford 1998). According to Wilson, environmentalists were motivated by "the unrealistic belief they [could] stifle growth" (Ritter 1998).

State and Federal Officials Cooperate to Draft a Final Plan

Given the fractiousness of the stakeholders, it is unsurprising that CALFED's release in March 1998 of a draft programmatic Environmental Impact Statement/Environmental Impact Review (EIS/EIR) provoked intense debate at public hearings around the state and drew more than 1,800 comments, even though it did not specify a preferred alternative. Undaunted by the ongoing posturing and dissension, as the Wilson administration drew to a close Secretary Babbitt and George Dunn, Governor Wilson's chief of staff, engaged in a "monumental effort to reach consensus among various stakeholder representatives on an appropriate preferred alternative" for CALFED (Wright 2001, 339). During the summer and fall of 1998, the two leaders sponsored weekly meetings to tackle thorny issues. As the deadline neared, however, divisions among stakeholders remained, and "The agencies were forced to weaken or qualify many of the key recommendations as they sought to gain universal support" (Wright 2001, 339)

On December 18, 1998, Wilson and Babbitt presented a $4.4 billion draft plan that attempted to appease all the key parties while acknowledging it failed to resolve the toughest issues (Barnum 1998). The plan contemplated raising several dams and building new off-stream reservoirs, but made construction of new projects contingent on progress in conservation. To proponents' relief, upon taking office in early January, Governor Gray Davis expressed support for CALFED and "insisted upon a balanced plan that would address the key short- and long-term needs of each of the major stakeholder groups" (Wright 2001, 339).

Six months later, after holding extensive public hearings and soliciting comments on the draft plan, CALFED unveiled the details of its preferred alternative in a revised EIS/EIR. Rather than increasing freshwater flows through the Bay–Delta, as environmentalists had hoped, the plan relied on elaborate water delivery techniques to save fish and wildlife, such as timing water diversions more strategically and conducting those diversions more carefully. To solve the problem of maintaining a reliable water supply for users, the plan also recommended engineering solutions such as recharging subterranean aquifers, building off-stream reservoirs (supplied by pipes or canals), and augmenting existing reservoirs. (It eschewed the most controversial approach, construction of a canal around the Delta,

because program staff feared that proposal would provoke a massive public backlash.[11]) The basic idea underpinning the plan, according to Lester
Snow, was that by rebuilding the channels and pumping when threatened
fish were not present, managers could run the pumps at maximum levels
and store water elsewhere for later use (G. Martin 1999b).

Despite the extensive consultation that preceded its release, public reaction to the draft plan was tepid, and hearings held during the comment
period offered little indication that stakeholders' views had changed as
a result of the collaborative process: CALFED officials were bombarded
with complaints, dire predictions, and uncompromising rhetoric from all
sides. Farmers charged the plan was tilted too heavily in favor of fish and
birds and would be ruinous to agriculture. Environmentalists countered
that the plan did nothing to curb farmers' wasteful ways. The MWD complained the proposal would not improve the quality or reliability of water
flowing through the California Aqueduct. Observers warned that "The
regionalism and tribalism that [had] blocked previous efforts at restoring
the delta [were] once again reasserting themselves" (T. Perry 1999).
Stakeholders' intractable differences manifested themselves internally as
well: it was proving impossible to craft a final plan with the stakeholders
in the room because negotiators could not get beyond lowest-common-
denominator language. According to one CALFED staff person, "The
phrase 'CALFED-speak' entered the lexicon in a derogatory manner,
meaning that to get everyone to agree you'd end up with mush" (Innes
et al. 2006, 23).

With efforts to build consensus among stakeholders bogging down,
Secretary Babbitt convened a small group of high-level public officials
representing a diversity of interests to hammer out the details of a final
plan in advance of the impending presidential election. Critical to reaching agreement was the creation of a novel mechanism for allocating water:
the Environmental Water Account (EWA). Because curtailing water
exports appeared unthinkable, Secretary Babbitt directed agency officials
and stakeholders to find a mechanism for applying export reductions on
a real-time basis, rather than a fixed schedule, in hopes that such a mechanism could meet the legal requirements of the Endangered Species Act
without jeopardizing the total quantity of water available for export. The
development of this mechanism brought together fishery scientists and
project operators in months of gaming exercises that simulated how real-
time changes to project operations might be made in response to monitoring data—in the process creating mutual understanding of one another's

constraints. Final agreement on the EWA, however, hinged on the establishment of a baseline allocation of environmental water. According to Alf Brandt (2002), the stakeholders battled fiercely over how much water fish had a right to, and the SWP contractors prevailed. At the Davis administration's insistence, Secretary Babbitt agreed to reduce the amount of water allocated for the natural system under the CVPIA in exchange for a promise of additional water purchases under the EWA. That decision facilitated agreement but put the risk of tight water supplies on the ecosystem by further squeezing the baseline, which is the only "guaranteed" environmental water supply.

The CALFED Record of Decision

On August 28, 2000, Secretary Babbitt and Governor Davis signed the record of decision (ROD) that laid out CALFED's Preferred Program Alternative, a 30-year plan that sought to achieve several goals simultaneously: restoring the Bay–Delta's ecological health while improving water-supply quality and reliability and enhancing flood control. Although the agreement had no legal force, it was backed by a commitment to joint implementation among the 18 state and federal agencies with management and regulatory responsibility in the Bay–Delta. The ROD did recommend creating a new, 12-member, joint federal–state commission. In the meantime, however, it relied on an interim governance structure similar to the one that crafted the ROD: a federal–state Policy Committee to advise participating agencies, which retained final decision making authority, and a Bay–Delta Advisory Committee to furnish input from stakeholders. To ensure that decisions were based on the best available information, the ROD established a formal science program whose purpose was to furnish policymakers with an "objective" understanding of the Bay–Delta system and thereby reduce conflict over technical issues.

The primary means by which CALFED aimed to achieve environmental benefits was by devoting substantial resources to ecosystem restoration. The goal of the Ecosystem Restoration Program (ERP) was to "improve and increase aquatic and terrestrial habitat and improve ecological functions in the Bay–Delta system to support sustainable populations of diverse and valuable plant and animal species." To that end, the ERP prescribed hundreds of actions to conserve and restore critical ecosystem elements and processes, such as: restoring, protecting, and managing diverse habitat types representative of the Bay–Delta ecosystem and its watershed, acquiring water from sources throughout the watershed to provide flows and habitat conditions for fishery protection and recovery, restoring in-stream flows, improving Delta outflow during key springtime periods, reconnecting Bay–Delta tributaries with their floodplains, developing prevention and control programs for invasive species, restoring sediment, and reducing or eliminating fish passage barriers. In addition, the ERP conducted research to help decision makers define problems and establish priorities for action. Implementation of the program relied heavily on voluntarism; for example, the ROD emphasized that the ERP would preserve as much agricultural land as possible through partnerships with willing landowners, and would acquire land from willing sellers only as a last resort.

Another environmentally oriented CALFED element, the Water Use Efficiency Program, aimed to reduce the strain on the Bay–Delta ecosystem by accelerating the implementation of water conservation and recycling practices throughout the state. Planners estimated that if such practices were widely adopted, the urban sector could save between 520,000 and 688,000 acre-feet of water; the agricultural sector could save from 260,000 to 350,000 acre-feet; and water reclamation projects could save between 225,000 and 310,000 acre-feet (CALFED 2000b, 59). Like the ERP, the water-use efficiency program relied on incentives and voluntary mechanisms, such as a competitive grant/loan program and public relations efforts to encourage better local groundwater management.

To ensure that an adequate portion of the existing water supply benefited fish and wildlife, the ROD established three "tiers" of environmental water supply protection. The first tier included regulatory requirements already in place: the state's Water Quality Control Program, the federal CVPIA, and rules for project operations under the Endangered Species Act. The third tier committed state and federal agencies to make water available if the combined protections of tiers 1 and 2 proved insufficient to protect endangered species. The element that had clinched the deal on the ROD, however, was the second tier, the Environmental Water Account, which facilitated the acquisition of an average of 380,000 acre-feet for the environment through a combination of setting aside water during system operation (195,000 acre-feet) and purchasing water from willing sellers (185,000 acre-feet). In exchange for adjusting their pumping schedules, the EWA absolved operators of liability for fish that were "taken" at the pumps (Rosenkrans and Hayden 2005).

In some respects the CALFED ROD improved on historic practices. Previously, single-purpose environmental protection efforts ignored, and often exacerbated, other problems within the watershed, and year-to-year water management decisions were crisis-driven and reactive. By contrast, consistent with the optimistic model of EBM, CALFED was relatively comprehensive: it defined the geographic scope of the problem as the legally defined Delta, Suisun Bay, and Suisun Marsh, while encouraging solutions from a much broader area—from southern California north to the Oregon border and from the Central Valley west to the Farallon Islands. Furthermore, CALFED explicitly recognized the interrelationships among its eight program elements: ecosystem restoration, water quality, levee system integrity, water-use efficiency, water transfer, watershed management, storage, and conveyance. And it provided a forum in which agencies formerly working at cross purposes could coordinate their permitting and other decisions (Freeman and Farber 2005).

According to the ROD (CALFED 2000b, 24), "Compared to the [No Action Alternative] and existing conditions, the [Preferred Program Alternative] provide[d] significant improvements in terms of ecosystem quality, water quality, water supply reliability, and levee system integrity effects. Under the [No Action Alternative], each of these four areas of critical concern would [have continued] to deteriorate." As predicted by the pessimistic model, however, the ROD departed only marginally from the status quo, and its finely tuned, intensive management approach imposed the risk of drought on the natural system while assuring that urban and agricultural users would experience " … no reductions, beyond existing regulatory levels, in CVP or SWP Delta exports resulting from measures to protect fish under FESA and CESA" (CALFED 2000b, 57). Relatedly, the ROD retained firm control over the Bay–Delta, seeking to maintain it as a static, homogeneous system rather than one whose salinity fluctuates over time and across locations. The ROD also perpetuated the status quo by adopting the premise that there was enough water in the system to meet all demands, but not enough storage capacity. Reflecting this emphasis, it called for serious consideration of a variety of water storage initiatives: raising Shasta Dam by nearly six feet, increasing the storage capacity of the Los Vaqueros Reservoir, building an off-stream storage facility north of the Delta, or building a reservoir south of the Delta. (According to the ROD, construction of new storage projects would begin only after careful review and public comment, and investment in local groundwater management and conservation programs.)

Moreover, despite the program's apparent comprehensiveness, planners skirted the question of trade-offs among its elements and declined to address the question of "restore to what?" They also focused exclusively on the two large water projects while ignoring the 7,000 permitted diverters who get their water from the Bay-Delta watershed, even though the latter divert an amount nearly equivalent to the water that historically flowed through the system. Finally, although CALFED planners paid some attention to terrestrial habitat, they made no attempt to address land-use decision making within the Delta, despite its obvious implications for water consumption and destruction of wetlands.

Implementing CALFED

Although they had concerns about CALFED's approach, many environmentalists initially held out hope for the program—and, in at least some respects, it delivered. Like the other initiatives described in this book, one of CALFED's main accomplishment was to attract money: over a ten-year period the program garnered $3 billion in funding, nearly one-third of which it spent on habitat acquisition and restoration. In addition, as predicted by the optimistic model of EBM, CALFED improved coordination among water managers, allowed for real-time responses to newly acquired monitoring information, and enhanced scientists' understanding of fish movements. Furthermore, according to Judith Innes and her coauthors (2007), the collaborative processes undertaken under the auspices of CALFED built social and political capital among some former adversaries. Consistent with the pessimistic model, however, the program fell short in other respects: the promise of adaptive management was largely unrealized because agency officials refused to experiment with reducing water diversions and were unwilling to adopt performance standards that, in turn, would have facilitated testing hypotheses about the impacts of management. Also consistent with the pessimistic model, CALFED neither inspired managers to go beyond legally required environmental protection measures nor proved durable. Both federal and state water operators continued to manage at the edge of regulatory standards and evaded CALFED oversight and coordination when making decisions about pumping volume and water delivery contracts. The tenuous consensus among stakeholders that allowed for the ROD did not endure, and by 2007 the program was in disarray: environmentalists and fishing interests had filed several lawsuits, prevailing in most of them, and the state was in the midst of a two-year "visioning" process for the Bay–Delta.

Building a Scientific Foundation for Decision Making

A central aspect of CALFED implementation was the formation of a science program to furnish decision makers with neutral, policy-relevant advice. To spearhead the construction of CALFED's science base, the program hired Sam Luoma—a well-respected hydrologist with the U.S. Geological Survey who had extensive experience in multidisciplinary investigations—and asked him to establish a science board that would be independent of, but could furnish policy-relevant advice to, the multi-agency planning process. With Luoma's guidance, the science program instituted a variety of activities: it held annual conferences at which experts interested in the Bay–Delta could interact and present new findings; it sponsored an electronic journal, *San Francisco Watershed and Estuary Science*, which disseminated peer-reviewed science relevant to the Bay–Delta and its watershed; and it held public forums to air new findings on the two fish species of greatest political concern, salmon and delta smelt. With its abundant funding and inclusive, transparent, peer-reviewed system for distributing money, the CALFED science program succeeded in attracting scientists from academia and nongovernmental organizations, thereby providing an important corrective to formerly near-exclusive reliance on science generated by mission-driven agencies. It also defused some of the conflict that traditionally had surrounded agency-by-agency water management decisions.

Unlike most other EBM initiatives, CALFED's science program declined to construct a holistic model of the Bay–Delta system to use as a basis for decision making. According to Luoma (2005), such models are of limited utility and run the risk of being dramatically wrong, particularly for systems that have been as radically modified as the Bay–Delta. Instead, Luoma's approach was incremental and reflected his political sensitivity: he wanted to prevent negotiations from collapsing as a result of disagreements over science. The program did invest in several important scientific ventures. Program-sponsored scientists devised a preliminary set of performance measures by which to gauge ecosystem health. In addition, they conducted research on the delta smelt and on the impacts of climate change, yielding work that subsequently proved invaluable when environmentalists and others petitioned the state and federal governments for additional protections.

In response to policymakers' concerns, however, most of CALFED's initial research aimed to reduce uncertainties in managing flows, diversions, and fish populations in the Delta. The advantage of this approach was that it enabled the CALFED agencies to begin taking actions, particularly to restore anadromous fish populations. On the other hand, it also allowed policymakers to skirt the central question of what the Delta—which has been thoroughly transformed from a vast, shallow-water tidal marsh to a complex system of levees and deep-water channels—should look like in the future, and what sorts of policies and practices would bring about such a vision. It also meant that when the Delta's pelagic fish species, whose health reflects a complex interaction among multiple factors, began to crash in 2003, the science program had to scramble for a diagnosis.

Habitat Restoration

Another aspect of CALFED's implementation, and one that yielded tangible environmental benefits, was its financial support of ecosystem restoration projects, at least some of which otherwise might not have garnered funding. From the outset—even before the ROD was signed—CALFED identified and began allocating money to a variety of "no-brainer" actions its agencies could take without conducting a lot of preliminary research, such as providing cold water upstream for salmon. These actions were "targets of opportunity," in the sense that many scientists had been urging their adoption for years (Kier 2006). In pursuing restoration so aggressively, CALFED officials were banking on the notion that improving habitat would result in healthier fish populations, so that fewer, if any, cuts in water supplies would be needed. Moreover, they hoped restoration projects would create goodwill and credibility, making believers out of skeptics.[12] By the time the ROD was signed, the Ecosystem Restoration Program had already spent about $250 million to fund 271 projects (CALFED 2000b); between 2000 and 2005, the program dispensed another $550 million for habitat restoration. With that money, the CALFED agencies protected or restored 100,000 acres of habitat and built or improved 68 fish screens (CALFED 2005).

Among the most highly touted ecosystem restoration initiatives supported by CALFED was the restoration of anadromous fish habitat, a task for which there is no mechanism under the Endangered Species Act. For example, CALFED backed the removal of four small barriers along Butte Creek, a tributary of the Sacramento, to facilitate the spring-run salmon migration. CALFED also promoted the dismantlement of diversions on Cold Creek and Clear Creek that impeded the movement of fall-run salmon. On Battle Creek, officials negotiated the removal of five Pacific Gas & Electric power-generating dams, retrofitted two other dams with fish ladders, and increased the volume of downriver flows tenfold (G. Martin 1999c). In addition to subsidizing in-stream modifications, CALFED funded a variety of watershed restoration projects to improve habitat for fish and wildlife. For example, the program gave a $980,000 grant to the Feather River Coordinated Resource Management Committee to restore 2,000 acres along Last Chance Creek, a major tributary of the Feather River. The program also subsidized efforts to buy land along a 200-mile stretch of the Sacramento River in hopes of doubling the amount of streamside vegetation from 10,000 acres to 20,000 acres.

By 2005, when the state reviewed CALFED's progress, the Central Valley's salmon populations had rebounded in several streams. For example, in 2002 Butte Creek's spring-run salmon shot up to 6,000 from a low of ten fish a few years earlier (Zakin 2002), and the Sacramento River's winter-run salmon appeared to be recovering below the Shasta Dam. Some other species were also faring well: the Department of Fish and Game reported an increase in the Swainson's hawk population in 2004, increases in sandhill cranes, and stability in waterfowl populations over a 16-year period (LHC 2005). It is difficult to attribute these improvements directly to actions supported by CALFED, however. Bird populations began stabilizing prior to the program's inception, and salmon populations increased partly as a result of the Pacific Decadal Oscillation, not just changes in spawning habitat or the removal of migration barriers (Kier 2006; Luoma 2005).[13] Cutbacks in the salmon harvest, which are also independent of CALFED, have likely contributed to the trend as well (Kier 2006). Furthermore, some important habitat modifications for which CALFED got credit—such as the installation of an $8.5 million variable-level intake device on the massive Shasta Dam, which enabled operators to maintain cold-water habitat below the dam—were funded by the CVPIA, not CALFED, and so likely would have happened anyway.

Moreover, it is difficult to assess progress toward ecological health and would have been impossible to manage adaptively because the program neither adopted performance measures nor adequately funded monitoring. CALFED's Ecosystem Restoration Program tried to establish an adaptive approach: following an independent review, the ERP devised a systems model that integrated all the components of the ecosystem, which it anticipated would allow it to develop hypotheses about the impact of management decisions. But the agencies that would have had to implement the strategy proposed by the ERP found it too flexible, and could not, in any case, agree on performance measures against which to gauge progress. So the program resorted instead to a long list of fixed milestones that measured completed tasks, or outputs, not outcomes (Bobker 2005; S. Johnson 2005). An additional impediment to adaptive management was the profound unwillingness by both the wildlife and water-management agencies to experiment with the factor many suspect is the most important determinant of ecosystem health: the overall amount of water that is diverted from the system.

Finally, although CALFED deserves credit for improvements in anadromous fish habitat, a substantial fraction of the wetland acquisition and restoration projects undertaken since 1995 might well have happened even without the program. Efforts to restore wetlands around San Francisco Bay date back to the 1980s. The impetus for a more coordinated effort came in 1993, when the San Francisco Estuary Project released a report—prepared by the Association of Bay Area Governments, San Francisco State University, the FWS, and the EPA—that recommended devising a plan to restore the Bay's tidal wetlands. In 1994 the state completed its purchase of 10,000 acres of Cargill Salt Co. land on the west side of the Napa River, using money from a settlement with Shell Oil Co. to mitigate for its 1988 oil spill (Barnum 1996). By the mid-1990s, when CALFED came on the scene, cooperative ventures were already acquiring acreage around the Bay as soon as it became available—from retiring farmers, ranches, the Army, the Catholic Church, private developers— and restoring it to tidal marsh (Kay 2001). In addition, in 1974 Congress established the Don Edwards San Francisco Bay National Wildlife Refuge, which by 2004 comprised 30,000 acres. All told, with or without CALFED, environmentalists aimed to restore about 60,000 acres of tidal marshes around the Bay, thereby bringing the total to 100,000 out of a historic 190,000 acres.

The Environmental Water Account

A third aspect of implementation, and CALFED's other highly acclaimed environmental achievement, the environmental water account, made the region's water management more responsive to new knowledge and therefore, according to some, constituted the program's most notable example of adaptive management (Luoma 2005; L. Snow 2005).[14] Certainly the EWA—with its emphasis on flexible, real-time decision making—changed the way state and federal agencies managed water. Biologists from the wildlife agencies and the operators from the CVP and SWP began sitting down together and making decisions about pumping based on when fish were near the pumps, and therefore most vulnerable to "take." As a result of this process, Lester Snow (2005) points out, a whole generation of biologists and pump operators came to understand one another's perspectives. The change in practice was striking: under a traditional Endangered Species Act-based approach, each year the wildlife agencies established seasonal pumping restrictions according to a biological opinion on the status of the endangered fish species. Once project operations hit the take limits set in those permits, they had to reconsult with the FWS or NMFS—a process that could result in a requirement to cut back or shut down pumping, potentially disrupting water supplies to users. By contrast, with the EWA as collateral, fishery agencies could call for more moderate and precisely timed pumping reductions—thereby simultaneously helping fish and water users (Innes et al. 2006).

On the other hand, between 2001 and 2007 the EWA was sharply constrained in its ability to respond to information suggesting it was delivering insufficient water for fish because it acquired far less water than originally anticipated. The EWA got off to a rocky start: in late March 2001 the SWP reported that more than 18,000 young salmon migrating from the Delta to the ocean had been sucked into its pumps and killed— far exceeding the "red light" limit of 7,000 smolts specified by CALFED (Brazil 2001). Then, in February 2002 the EWA's baseline (tier 1) lost 200,000–300,000 acre-feet of environmental water after a judge struck down the Interior Department's rule for allocating the CVPIA's 800,000 acre-feet of water for anadromous fish. The ruling forced EWA managers to redo their calculations, models, and spreadsheets; more important, it seriously eroded the foundation of water availability on which the EWA was built. As a consequence, tier 2 water (the EWA), which was supposed to *supplement* tier 1 water, instead was used to compensate for shortfalls in water that regulators had expected would be available for fish protection (Swanson 2006).[15]

Despite these problems, a technical assessment by a CALFED review panel released in early 2005 concluded that the EWA had yielded several benefits. Reviewers commented that managers had developed complex criteria based on the dynamics of fish populations, rather than a single indicator (fish taken at the pump), and that communication among agencies, as well as scientific knowledge, had improved during the account's first four years of operation. The panel speculated that fish probably had gained more protection than they would have under a traditional approach (Innes et al. 2006). Another review, sponsored by Environmental Defense, was more critical, however. Its authors argued that funding constraints limited managers' ability to adjust their pumping schedules in response to information suggesting problems in the Delta fisheries; therefore, the EWA provided reliable water supplies for users at the expense of the fish. The report pointed out that, although it met its target in 2000 and 2001, in subsequent years tiers 1 and 2 combined were underendowed by about 420,000–460,000 acre-feet annually (Rosenkrans and Hayden 2005). These shortfalls were attributable only partly to the accounting changes in the CVPIA allotment made in response to the court ruling. In addition, only about 29 percent of the expected 195,000 acre-feet of projected operational assets materialized, on average, and—because state and federal funding had dwindled—the EWA was not able to compensate by purchasing water from willing sellers. "As a result," said the authors, "fishery agencies [were] significantly constrained in their ability to dedicate water at key times of the year to protect fisheries…as promised in the CALFED plan" (Rosenkrans and Hayden 2005, v). Making matters worse, CALFED's backstop measure—tier 3—had no assets to make up the shortfall, even though "The health of the estuary largely depends on a reliable set of environmental safeguards, including dedicated water supplies" (Rosenkrans and Hayden 2005, v).

Intensive Management and Environmental Stewardship
In any case, a growing body of evidence suggested that manipulating the pumping schedules of the CVP and SWP was at best insufficient to remedy the Delta's problems, and at worst exacerbated them. In the early 2000s freshwater exports from the Delta increased markedly, reaching a record high of 6.3 million acre-feet in Water Year 2005 (Nelson et al. 2006). At the same time, the Delta's pelagic fish species hit new lows: in 2005, after three years of decline, surveys of the delta smelt detected the smallest population recorded in nearly four decades of counting.[16] Although 2006 was a wet year—the snowpack in the High Sierra was 170 percent of normal—the spring 2006 survey detected no recovery in smelt numbers (Boxall 2006a, 2006b). Some scientists suspected that increased winter pumping was partially to blame for the decline in delta

smelt and other pelagic species, several of which were dropping simultaneously. Pumping during winter had increased markedly, ostensibly to compensate for reduced spring pumping (Bay Institute et al. 2007); during the same period scientists documented an increase in the proportion of fish being killed by the pumps in winter (Swanson 2006; Thompson 2006). (A similar pattern emerged during the 1980s after winter pumping was increased.) "You can't really deny that smelt have gone down while pumping has gone up, and the big crash took place when they changed the timing of the pumping," said Peter Moyle, a fisheries biologist and delta smelt expert at the University of California at Davis (Boxall 2006b).

After conducting additional research, scientists were more confident about other export-related causes of the downturn in the Delta's pelagic species. Some studies suggested that water releases from upstream reservoirs, which slow significantly in late fall, allowed saltwater from the Bay to intrude farther into the Delta, providing a "highway" for Asian overbite clams (*Corbula amurensis*), an invasive species that has infested the estuary. The clams, which arrived in the late 1980s, reproduce rapidly; they also voraciously consume the particular species of zooplankton that is the smelt's main source of nutrients and whose population has declined sharply. A shortage of prey, along with the higher salinity in the lower Delta, forces smelt to search for food farther upstream, where they get sucked into the giant pumps and killed before their eggs are fertilized.[17]

In short, the CALFED solution of adjusting the timing of water releases may actually have harmed the Delta's native species. Moreover, the program's focus on manipulating pumping schedules allowed policymakers to elide the more fundamental questions of whether the Bay–Delta ecosystem could actually sustain such a high level of freshwater withdrawals. Some scientists doubted the ecosystem could recover as long as diverters continued to remove about half the freshwater from the system. Veteran Bay Area fisheries specialist Bill Kier pointed out that a study conducted in the 1980s of several Gulf of Mexico river deltas that examined biodiversity and fish abundance found that key species started disappearing when more than 40 percent of a river's water was diverted (G. Martin 1999a). If increasing the flow of water through the Delta was essential to restoring endangered fisheries, then no amount of tinkering with the timing and location of withdrawals was going to stem the decline in the system's biological diversity.

CALFED's emphasis on marginal adjustments to the status quo also begged the question of whether native species that evolved in a variable system could thrive in a static one. Bruce Herbold, a fisheries biologist with the EPA, explains: "[The Delta] used to be salty in the summer and really fresh all the way down to Suisun Bay in the springtime every year. And now it's fresh all the time. It's just stable" (Boxall 2006b). Tina Swanson, senior scientist at the nonprofit Bay Institute, points out that the overbite clam thrives in the Delta because export pumping has turned it into a freshwater system, and contends that if normal tidal forces were allowed to restore the system's historically variable salinity levels, the invaders that displace native species would perish (Weiser 2005c).

Overwhelming evidence that exports were implicated in the Delta's ecological decline did not spur support for precautionary measures, however; instead, those with strong interests in maintaining the status quo rejected the possibility that exports from the Delta would have to be reduced, and pointed out that scientists had not yet established a definitive causal link between pumping and fish declines. For example, in response to reports of the delta smelt's demise, Tim Quinn of the MWD equivocated, saying, "There's no evidence the pumping has had all that much effect. There's no doubt there is something going on out there in the Delta, and we need to figure it out. It probably has something to do with the food chain, but nobody's sure" (Weiser 2005a).[18] Maintaining the status quo level of exports was the best option, many water users argued, because of the Delta's complexity and uncertainty about the relative contribution of other factors, such as toxic pesticides and invasive species, to the pelagics' decline. Others, such as B. J. Miller, an engineering consultant to water contractors, asserted that factors beyond human control were largely responsible: "It's possible that fall salinity is affected by outflows that are not manageable," said Miller. "It's not water project operations. It's the weather" (Taugher 2006c). And stakeholders' elected representatives continued to emphasize the risk to users posed by reducing exports. For instance, at a February 2006 hearing on the issue, Rep. Richard Pombo (R, Tracy) pointed out: "Whatever we decide to do will have a big impact on the delta, but it will also have a big economic impact on California" (G. Martin 2006a).

A Durable Plan?

Furthermore, although CALFED "made progress in moving a highly polarized system toward a model of policy-making that is coordinated, communicative, and informed by a diversity of interests and options" (Innes et al. 2006, 33), stakeholders and agencies defected when shifts in the political context seemed to create better options for them. Within two months of the ROD's issuance, some stakeholders began bringing serious political and legal challenges. In late September 2000, the Municipal Water District of Orange County, the California Farm Bureau Federation, and a coalition of rural northern California counties all filed lawsuits against CALFED, alleging that plans to acquire or flood farmland violated farmers' property rights and that delays in building water projects favored the environment at the expense of people (T. Perry 2000). Then, in the spring of 2002 the Westlands Water District initiated a series of lawsuits and petitions aimed at unraveling the CALFED agreement (Martin 2002).

Dissension among stakeholders was severe enough to threaten the program's funding: Congress balked at reauthorizing it, citing disputes within California over the program's direction. At water users' urging, in 2001 Sen. Dianne Feinstein introduced a bill that would provide money to enlarge two reservoirs and create two new ones, and in the House, Rep. Ken Calvert (R, Riverside) introduced an even more aggressive bill that would have bypassed CALFED altogether by preapproving a handful of water projects. Over the next two years, as Feinstein struggled to work out a package that the state's own congressional delegation could agree on, CALFED's federal authorization languished.[19]

Dismayed by congressional resistance, state officials tried to breathe new life into the struggling program. On September 23, 2002, Governor Gray Davis signed SB 1653, creating a new governing body: the California Bay–Delta Authority. The Authority's board included public members from major regions appointed by the governor, two at-large members appointed by the legislature, a member of the Bay–Delta Public Advisory Committee (which replaced the Bay–Delta Advisory Committee), the directors of six of the most important federal agencies, and the directors of six key state agencies. The hope was that the new entity could rein in recalcitrant stakeholders, such as the Westlands Water District.

Like its predecessor, however, the Bay–Delta Authority had no means of enforcing its will, and agencies continued to pursue their own interests when cooperation with CALFED would have impeded their ability to fulfill their traditional missions. Under the Bush administration, for example, the Bureau of Reclamation began operating the CVP without consulting other agencies, unilaterally sending more water to farms and less to the environment. According to Mary Nichols, secretary of the State Resources Agency, "We're finding that our partners at Reclamation are not as interested in trying to coordinate...[T]hese agencies are more inclined to want to go it alone" (Robitaille 2003).

Even as evidence of the Delta's collapse accrued, state and federal water managers continued to seek ways to increase exports. In July 2003 the Bureau of Reclamation and the state Department of Water Resources (DWR) met secretly with the MWD, the Westlands Water District, and the Kern County Water Agency to forge the so-called Napa Agreement— a deal to raise the amount of water sent south from the Delta by the SWP pumps by 27 percent, from 6,680 cfs to 8,500 cfs, which they anticipated would result in an average increase in annual exports of about 200,000 acre-feet (Weiser 2005a). Beyond additional pumping, the agencies proposed an Operations Criteria and Plan (OCAP) that would weaken temperature standards for salmon on the Sacramento River, even though just months earlier the Department of Fish and Game and NOAA Fisheries (formerly NMFS) had rejected a similar proposal after their analysis revealed a potential for serious impacts on the endangered winter-run Chinook salmon.[20] Disregarding its own biologists' judgment, in late 2004 NOAA Fisheries issued a biological opinion on salmon and steelhead that allowed the OCAP to take effect (Weiser 2005b), and the bureau promptly began renewing long-term licenses with its 240 water users around the state. The DWR was similarly inclined: in July 2005 the *Contra Costa Times* disclosed that in two instances when biologists recommended temporary curtailments of water deliveries to protect the smelt, state water managers overrode that advice in deciding pumping levels (Taugher 2006b).

Unilateral moves by water managers aimed at increasing exports in turn triggered recourse by environmentalists to the adversarial practices of the past. In February 2005 a coalition of environmental and fishing groups sued the FWS, which in 2004 had issued a biological opinion saying that pumping increases would not harm the delta smelt (S. Young 2006). In August a similar coalition sued NOAA Fisheries over the biological opinion for salmon that formed the basis for the OCAP. In October 2006 the California Sportfishing Protection Alliance sued the DWR, charging it with lacking permits from the Department of Fish and Game required by the California Endangered Species Act (CESA) to operate state pumps. The suit followed a state Senate hearing in August 2005, which revealed the SWP had no permits or other formal documentation required by CESA. (The DWR claimed it had a set of agreements with state and federal regulators that comprised a "patchwork" of compliance.)

A Shifting Balance of Power Prompts New Thinking

In response to the resurgence in conflict around water management in northern California, in the spring of 2005 Governor Arnold Schwarzenegger ordered two reviews of CALFED: a financial and a management audit as well as a governance review by a blue-ribbon panel, the so-called Little Hoover Commission. The Commission's report noted that CALFED enjoyed a lot of support when it provided money that stakeholders believed otherwise would not have been available, but once the program faced difficult policy choices and its funding dwindled, stakeholders began to doubt its value. "Process and structure," the commissioners commented dryly, "cannot substitute for leadership and authority" (LHC 2005, iii). In April 2006 the Schwarzenegger administration released a plan to reorganize CALFED again, this time along the lines suggested by the Commission (Boxall 2006c). The governor also initiated a two-year "Delta Vision" process, the aim of which is to articulate "a view of future conditions to which decision makers must aspire" (BRTF 2007). In February 2007 he announced his appointments to a blue-ribbon panel that would spend the year developing management recommendations for that process.

As the Delta Vision process got under way, a series of legal and regulatory decisions eroded the legal foundation on which California's water delivery system is based, added weight to the contention that CALFED was insufficiently protective, and shifted the balance of power substantially in favor of environmentalists and fishing interests. First, in the fall of 2005 California's Third District Court of Appeals delivered a stunning blow by rejecting the premise on which CALFED rested—that water exports would have to be increased to accommodate the state's forecast population growth. Instead, a three-judge panel unanimously proclaimed: "Population growth is not an immutable fact of life," adding that "Smaller water exports from the Bay–Delta region [could], in turn, lead to smaller population growth due to the unavailability of water to support such growth" (Taugher 2005a). In January 2006 a panel of six independent scientists assembled by CALFED concluded that NOAA Fisheries' October 2004 biological opinion for salmon was not based on the best available scientific information and failed to err on the side of caution in the face of uncertainty (Taugher 2006a). The Bureau of Reclamation quickly moved to request a reevaluation of the biological opinion. In late January 2007 NOAA Fisheries retracted the permits needed to build new tide gates in the South Delta that would have made it possible to increase pumping under the state's proposed South Delta Improvements Package (Taugher 2007b).

For water managers, things went from bad to worse as the spring of 2007 wore on. In March 2007 Alameda County Superior Court Judge Frank Roesch ruled that, as the environmentalist–fishing coalition had alleged, the DWR was violating CESA by operating state pumps without a permit. He gave the state 60 days to comply with the law, or else he would require the state pumps to shut down (Taugher 2007d). In April the judge rejected pleas by state water officials and finalized his order, a decision the department promptly appealed. In May 2007 the environmentalist–fishing coalition again prevailed in court when federal Eastern (California) District Court Judge Oliver Wanger ruled that the FWS's delta smelt biological opinion was illegally lax.

In late May, shortly after Wanger announced his ruling, the DWR took the unprecedented step of shutting down its pumps for ten days. On June 9, however, water officials began ramping up deliveries again, ignoring the recommendations of the Delta Smelt Working Group, a team of biologists convened by the FWS. After hundreds of fish turned up dead at the pumps, environmentalists sued to cut off water deliveries altogether. Judge Wanger rejected their request, citing the immense economic damage that would ensue (Taugher 2007e, 2007f). In late August, however, he ordered a series of pumping cutbacks and other measures to protect the delta smelt. Judge Wanger's orders, which were expected to stay in place until the FWS issued a revised biological opinion, were less draconian than those proposed by environmentalists or the FWS, but far more severe than the DWR had hoped for: they threatened to curtail exports of as much as one-third of the 6 million acre-feet that is withdrawn from the Delta in a normal year (D. Walters 2007). According to Tim Quinn, now the executive director of the Association of California Water Agencies, "These reductions represent the single largest court-ordered redirection of water in state history" (Taugher 2007g).

Conclusions

CALFED's landscape-scale focus produced effects consistent with the optimistic model of EBM. First, it brought scientists, managers, and high-level policymakers from state and federal agencies to a forum where they generated shared language and concepts. Second, it facilitated a more cooperative approach to water management than the one that existed previously, in which water managers consulted only perfunctorily with the agencies charged with protecting fish and habitat. And third, it allowed for investment in restoration projects throughout the watershed, such as removing antiquated infrastructure, that may enable some ecological processes to reestablish themselves.

At the same time, efforts to garner consensus among stakeholders produced many impacts more consistent with the pessimistic model. To ensure stakeholder buy-in, planners simultaneously pursued several equivalent goals. An emphasis on "getting better together" limited, rather than broadened, their options, ultimately resulting in an even more intensively managed water system rather than one that would reduce the human impact and become, at least to some extent, self-sustaining. There was no serious discussion of limiting, much less reducing, withdrawals from the watershed; in fact, under CALFED the amount of water pumped out of the Bay-Delta system each year reached record highs. The results of CALFED's flexible, adaptive implementation were also consistent with the pessimistic model. Water managers, the main beneficiaries of CALFED's flexibility, did not exhibit stewardship but instead reverted to maximizing benefits for users when the political context allowed. Similarly, the most adaptive CALFED element, the EWA, consistently provided water for users, often at the expense of fish.

CALFED's intensive management approach imposed substantial risk on the Bay–Delta ecosystem, even as it provided a reliable water supply for water users. The most obvious consequence of relying on fine-tuning pumping operations is that natural variability, such as prolonged drought, may render the entire system unworkable (Boxall 2006c).[21] Experts have already documented changes in rain and snow patterns in northern California that are likely to threaten the state's water deliveries: tide gauges have recorded a sea-level rise of about seven inches at the Golden Gate during the past 100 years; snowmelt in the Sierra Nevada is starting a week earlier than it did before World War II; and more precipitation now falls as rain than as snow (Taugher 2007a). Under a worst-case global warming scenario devised by the California Climate Change Center at the University of California at Berkeley, the Sierra snowpack could be reduced by 90 percent from 2070 to 2090, and the average temperature may rise by 8 degrees Fahrenheit (Mooney 2006). Yet CALFED perpetuated an approach that strips the ecosystem of the resilience it would need to persist in the face of such dramatic change.

The outputs of CALFED's collaborative planning and flexible implementation made manifest the continuing dominance of water users in decision making regarding the allocation of water. In the 1980s environmentalists gained considerable legal leverage over urban and agricultural interests by invoking the Endangered Species and Clean Water acts, as well as the state's public trust doctrine (see chapter 9). Voters demonstrated their support through their willingness to approve bonds to finance conservation and restoration of the state's aquatic ecosystems. Environmentalists have been less successful at making the case among the public for a fundamental change in how water is allocated, however. Thus there has been little incentive for elected officials to espouse environmental improvements that will impose costs on users and to employ the regulatory leverage necessary to bring about such a shift. In fact, in 2006 Governor Schwarzenegger was promoting a huge bond package that featured major new storage projects, as well as some version of a peripheral canal. Moreover, although experts repeatedly have pointed out that the Delta's levee system is extremely vulnerable to earthquakes and climate change, and that new development is eliminating long-range management options for the region, the state has declined to restrain development in the Delta. Instead, in late 2005 Governor Schwarzenegger fired all the members of the state Reclamation Board after they raised concerns about development in flood-prone areas, and replaced them with more pro-development members (Taugher 2006d).

Defenders of CALFED will object that, given the system's physical and political complexity, the program did about as well as it could have, and that only a promise to meet all stakeholders' demands allowed political officials to move forward. But the San Joaquin River settlement, announced in the midst of the two-year Delta Visioning process, makes clear that an outcome in which everyone gets all they want is not inevitable. The agreement among parties to an 18-year-old lawsuit over damage done to the San Joaquin River by the operation of the CVP's Friant Dam requires farmers to relinquish about 15 percent of their historic water deliveries—about 170,000 acre-feet. It also requires the Bureau of Reclamation to double water releases from the dam from an average of 116,741 acre-feet each year to about 247,000 acre-feet in dry years and 555,000 acre-feet in wet years. The goal of the agreement is to rewater two river segments, totaling 60 miles, that were dried up after the construction of the dam in the 1940s and, by 2012, to reintroduce a spring Chinook salmon run to those stretches (Grossi and Schultz 2006).

The settlement did make accommodations for farmers: they got limits on water losses and guaranteed price breaks, and they are allowed to recapture water sent down the main channel as long as doing so does not harm fish. Furthermore, to quell the objections of the Modesto and Merced irrigation districts, which did not participate in the lawsuit, the introduced salmon will be treated as an experimental population, which frees property owners from concerns about having their land designated as critical habitat, and irrigation districts from liability for accidentally killing salmon. Nevertheless, the settlement retains an overarching goal of restoring the health of the San Joaquin's salmon runs, and water users who have become accustomed to the status quo will have to make do with less. Ironically, according to journalist Glen Martin (2006b), even though it stemmed from discussions pursuant to a lawsuit, "The agreement seems to have ushered in an era of good feeling in the San Joaquin Valley, a marked difference from the bitterness that characterized the past two decades."

7

Conserving the Sonoran Desert in Pima County, Arizona

Although the results of the four ecosystem-based management (EBM) initiatives described in earlier chapters are discouraging, some landscape-scale projects have achieved more environmentally beneficial results. For example, between 1998 and 2001 Pima County, Arizona engaged in a planning process that culminated in a relatively effective conservation program, the Sonoran Desert Conservation Plan (SDCP). The SDCP consists of five elements: conservation of biological corridors and critical habitat, riparian area protection, expansion of mountain parks and natural preserves, ranch conservation, and cultural resource preservation. The county anticipates that implementation of the plan's biological conservation and restoration features will facilitate the permanent protection of nearly 600,000 acres of high-value habitat in a biologically sound configuration as well as the rehabilitation of riparian areas, and will encourage environmentally sensitive land-use practices in the matrix of private land surrounding the preserve.

Like Austin's Balcones Canyonlands Conservation Plan (BCCP) and San Diego's Multiple Species Conservation Program (MSCP), the SDCP marks a sharp departure from Pima County's historic approach to land-use decision making, in which county officials routinely approved sprawl-inducing development. But the SDCP appears more likely than either Austin's or San Diego's plans to conserve the region's biological diversity, despite Pima County's similarly rapid population growth. In December 2001 the Board of Supervisors adopted, as part of its comprehensive land-use plan, the SDCP's Conservation Lands System (CLS), the overarching goal of which is to conserve Pima County's biological diversity; to this end, the CLS includes a map devised by the SDCP science team that identifies various categories of environmentally sensitive land, riparian area restoration programs, and a set of development mitigation standards and conservation practices. The county has applied the CLS's stringent development restrictions to nearly every rezoning approved since 2002, to maximize the likelihood that its biological objectives will be met. Moreover, in 2004 county voters approved a $174 million bond issue to purchase land designated as biologically valuable. Since then, the county has been assiduously acquiring property, including working ranches at the urban periphery that will buffer the preserve from encroaching development.

The SDCP is more environmentally protective than either the BCCP or the MSCP because Pima County's planning process, unlike those in Austin and San Diego, did not require stakeholders to reach consensus on the plan's goals. Instead, backed by a cohesive and effectively mobilized environmental community, the county's political leaders took a resolute stand from the outset that the plan would, above all, conserve native biological diversity. To increase the likelihood of achieving this goal, county officials used their regulatory leverage to sharply limit development during the plan's formulation. The combination of leaders' pro-environmental rhetoric and their willingness to employ regulatory tools shifted the balance of power between development and environmental interests in favor of the latter, ensuring that the stakeholder negotiations that did occur did not whittle away at the plan's environmentally protective standards. That said, although the county has established a pro-environmental trajectory, the plan's long-term benefits remain tenuous because of a variety of threats that may undermine its largely voluntary implementation.

Origins of the Sonoran Desert Conservation Plan

In the decades prior to the inception of the SDCP planning process, development in Pima County followed the traditional imperatives of economic growth: county supervisors routinely approved developers' requests to upzone parcels on the outskirts of Tucson and the fringes of the emerging suburbs of Marana and Oro Valley. By the 1990s, however, opposition to unregulated growth was increasing, and as the decade wore on, a series of events set the stage for more comprehensive efforts to manage the region's growth and protect its biological diversity. Despite fierce opposition from builders and property-rights activists, county leaders decided to pursue a pro-environmental agenda, and in October 1998 they initiated a planning process that aimed to conserve large swaths of undeveloped land and redirect growth toward less environmentally sensitive areas.

Tucson Sprawls

Pima County, which comprises more than 9,000 square miles, lies at the confluence of two ecoregions. Ringing the 1,000-square-mile Tucson Valley in eastern Pima County are several mountain ranges, whose "sky islands" rise more than a mile above the desert and—with their moister, cooler climate—support a wide variety of plants and animals (Tobin 2002). At the foot of the mountains, the Sonoran Desert stretches 120,000 square miles from Arizona into southern California and Mexico. Long favored by naturalists and biologists, the Sonoran Desert provides refuge for 500 bird species; 130 different kinds of mammals, including jaguars; 5,000 plants, among which are 3,500 native varieties; and 20 amphibian and 50 fish species (Jaffe 2001). In areas that have not been developed, the desert's signature resident, the majestic saguaro cactus, dots the landscape. Although unusually lush, the semiarid Sonoran Desert is also fragile: thanks largely to dry prevailing winds, it gets only 12.5 inches of rain each year, most of it during the extremely hot summer months when evaporation outpaces rainfall by a factor of ten (Jaffe 2001).

Despite the region's arid climate and scorching summers, humans have occupied it for centuries. Native Americans inhabited the Tucson area for more than 12,000 years prior to the arrival of Spanish settlers in the late 1500s. Tucson remained a small city until after World War II, when its population more than doubled, from 54,000 in 1940 to 120,000 in 1950. To accommodate the area's burgeoning population, planners proposed zoning the land outside the city limits for three to five homes per acre and the land in the mountain foothills for low-density (one home per acre) development—a plan that was subsequently described as "a blueprint for sprawl" (T. Davis 1999b).

In the early 1970s construction spreading outward in all directions provoked a burst of anti-development sentiment, and environmentalists managed to kill a series of highway proposals, as well as a plan to build a 17,000-home subdivision on the northwest side of the Catalina Mountains. After the Board of Supervisors voted unanimously to deny the rezoning for the latter project, County Supervisor Conrad Joyner remarked, "No one on this board who hopes to be re-elected as a supervisor or elected to some higher office could afford to vote in favor of Rancho Romero" (T. Davis 1999b). Skepticism about the merits of unbridled growth remained strong for the next couple of years, and a responsive Board of Supervisors seized the opportunity to institute a set of environmental ordinances.

Developers mobilized quickly, however, and began targeting politicians who favored limiting growth. In 1975 the development community orchestrated a full-fledged backlash in response to the county's proposed new Comprehensive Plan. (Although conceived during a construction boom, the plan—which suggested that Tucson should rely less on tourism and construction, that developers should pay the cost of new development, and that the boundaries of the city should remain the same despite anticipated population growth—came out in the midst of a recession and was poorly received.) The public seemed to embrace developers' position: in November 1976 voters replaced the pro-environmental county supervisor, Ron Asta, with the ardently pro-growth Katie Dusenberry. Then, in January 1977 Tucson voters recalled the pro-environmental City Council, presumably to punish it for approving a large water-rate increase the previous summer. Shortly thereafter, according to journalist Tony Davis, the words "controlled growth" and "sprawl" vanished from public discourse, much as they did elsewhere in the United States (T. Davis 1999a).

For the next 20 years, Pima County's growth machine—which consists of builders, realtors, construction workers, engineers, and others who derive their income from new development—maintained firm control over the county's land-use decision-making apparatus. Between the late 1970s and mid-1980s, the Board of Supervisors routinely approved zoning variances and development permits; between 1990 and 1998, landowners secured rezonings in nearly 78 percent of the 451 requests filed with the county (T. Davis 1998a). Residents' objections were muted and easily parried by developers, who argued that leaving land zoned for low-density development would preclude building houses for lower-and middle-income buyers. In the mid-1990s, builders were grading 12 acres of desert each day in Pima County; every month 580 new houses sprang up, and 1,400 newcomers arrived (Chesnick and Morlock 1999a).

As new development crept up mountainsides and spread across the desert, residents of the city's Northwest Side endured bumper-to-bumper rush-hour traffic and overcrowded schools. Meanwhile, downtown was deteriorating despite revitalization efforts. Tony Davis (1999b) observed that the area wore "a disheveled look." These changes galvanized residents and prompted the emergence of a potent alliance between neighborhood activists and environmentalists concerned about the rate, scale, and nature of development in Pima County.

The Impetus for Landscape-Scale Planning
In the latter half of the 1990s, the combination of growing public opposition to sprawling development, an endangered species listing, and changes in the composition and orientation of the county's political leadership set in motion a new feedback: abetted by tenacious local media coverage of development issues, political leaders reinforced public anti-sprawl sentiment, and vice versa. Crucial to the development of this political dynamic was turnover in the Board of Supervisors, which went from being staunchly pro-development to being strongly pro-environment. The process began in 1996, when Sharon Bronson, a longtime Tucson neighborhood activist, narrowly won election to the Board of Supervisors after the candidacy of Republican-turned-Independent Ed Moore split Republicans. Bronson joined Supervisor Raul Grijalva, who for years had been the most outspoken (and often the lone) environmentalist on the board. Then, in 1997, after the unexpected death of newly elected conservative Republican John Even, Grijalva engineered the appointment of green Republican Ray Carroll to replace him. (In September 1998 Pima County's most conservative district affirmed Carroll's appointment by a landslide.) As a result of these changes in its composition, within two years the board went from voting consistently 4–1 in favor of development proposals to regularly voting 4–1 against them.

During the same period, trouble was brewing over a small raptor, the cactus ferruginous pygmy owl (*Glaucidium brasilianum cactorum*), which makes its home primarily in Mexico but has established a small population in southern Arizona. In early 1997, after a five-year legal battle, the Fish and Wildlife Service (FWS) announced its intent to list the owl as endangered under the Endangered Species Act. The following year, a pygmy owl sighting stopped bulldozers at the massive Dove Mountain subdivision 20 miles north of downtown Tucson in the foothills of the Tortolita Mountains, an injunction blocked construction of the Amphitheater High School, and Pima County Community College announced it was having difficulty finding a suitable site to build a new campus on the Northwest Side, which contained most of the owl habitat. Alerted by the FWS about the option of formulating a habitat conservation plan (HCP) under the Endangered Species Act, County Administrator Chuck Huckelberry's office began work on such a plan.[1]

Fearing the county would propose what they regarded as the kind of permissive, development-oriented approach taken in San Diego, local environmentalists decided to be proactive. They formed the Coalition for Sonoran Desert Protection, which represented more than 30 groups and included all of the most contentious local environmental activists, to speak on behalf of the environmental community.[2] The coalition, led by the extraordinarily effective veteran environmental activist Carolyn Campbell, proceeded to devise its own HCP—one that, unlike its predecessors elsewhere in the United States, would not allow the "take" of *any* pygmy owls. In doing so, it established an "extreme" pro-environment position that it hoped would become the starting point for the county's deliberations.

The urgency created by the pygmy owl listing, combined with growing public animus toward new development, prompted the county's Board of Supervisors to establish an environmentally protective stance. Apparently sensing a change in the political climate, in early 1998 even longtime pro-growth Republican Mike Boyd announced that he was deeply concerned about the county's sprawling development. "We all recognize that continued sprawl will ruin what makes this area unique and will bankrupt the county with higher sewer fees, water rates and property taxes," he said (T. Davis 1998a). Boyd attributed his conversion to three things: public opinion polls, freedom from reelection concerns (he had already decided not to run in 2000), and a desire to fend off environmentalists' proposals for an urban growth boundary. At Chairman Boyd's request, the board convened a public study session on the issue of growth in late February 1998. At that meeting, the board surprised many by unanimously resolving to spend the spring and summer tightening the county's environmental ordinances; in addition, the board called on the county administrator to come up with a "comprehensive" approach to Sonoran Desert protection.

To developers' dismay, in May the board approved a request that the county proceed with the HCP concept devised by the Sonoran Desert Coalition, not the more conservative one proffered by the county administrator's office. The coalition's plan called for a biological survey of Pima County, after which the most biologically important land would be set aside in a preserve that would be surrounded by buffer zones. According to Chair Raul Grijalva (2006), he encouraged the board to choose the coalition's approach because it made sense to establish a position of strength from which to negotiate, rather than compromise before the dialogue even began. Although Huckelberry initially was reluctant to embrace the coalition's blueprint as a starting point, county staff soon began working intensively with environmentalists on a viable plan, while developers remained largely on the sidelines.[3]

For the rest of the summer, while Huckelberry's assistant county administrator, Maeveen Behan, fleshed out the fundamentals of the new Sonoran Desert Conservation Plan, the board debated tightening the county's environmental ordinances.[4] The ordinance proposals galvanized property-rights activists, however, and—led by realtor Bill Arnold—they stormed public meetings between June and August to express their dissent. Alan Lurie, executive vice president of the Southern Arizona Homebuilders Association (SAHBA), argued passionately that ordinance changes and comprehensive desert protection would drive up home prices and were being proposed without sufficient consultation with property owners, builders, and developers. Property-rights activists protested that county government was intruding on landowners' prerogatives. Nevertheless, the board adopted a new Native Plant Preservation Ordinance, as well as more stringent versions of existing rules—although it did agree to weaken some of the original proposals.[5] According to the *Tucson Weekly* (Nintzel 1998), "This was the first board in county history that had the political will to tackle these issues."

Landscape-Scale Planning

During the fall of 1998 and into 1999, Pima County's decision makers created an environmentally protective baseline for the SDCP: they established from the outset that the plan's preeminent goal would be to conserve the county's remaining biological diversity, and they began reining in rezoning approvals to ensure that development approved during the planning process did not undermine that goal. The planning effort that ensued was landscape-scale in the sense that it began with an integrative scientific assessment of the entire 6-million-acre county's biological diversity. On the other hand, the county declined to coordinate its planning effort with municipal jurisdictions or the state, which controlled substantial tracts of land within the SDCP's study area. Furthermore, although the county solicited public input, the process of involving stakeholders was hardly collaborative: the county administrator's office kept a tight grip on the newly formed Steering Committee and took primary responsibility for devising the plan framework. Throughout the process, county leaders consistently refused to accede to the demands of an increasingly disgruntled development community to make the plan less ambitious.

The County Approves a Draft Plan
In October 1998, County Administrator Chuck Huckelberry released an outline of what he described as a comprehensive approach to natural resource protection in the county, thereby establishing an environmentally protective baseline. The draft established the goal of creating a preserve system that would conserve the region's biological resources and set out a participatory, science-based process by which such a system would be developed. Huckelberry also called on the Board of Supervisors to limit rezonings of environmentally sensitive land and pass an ordinance allowing for the transfer of development rights from more to less environmentally sensitive areas—measures that would instantly give officials the kind of regulatory leverage that had been lacking in both Austin and San Diego, where routine upzonings and development approvals undermined incentives for developers to make concessions during the planning process.

Huckelberry infuriated many development interests by declining to attach a specific price tag to his proposal, estimating it would cost between $300 million and $500 million but downplaying the magnitude of the expense by comparing it to what the county regularly spent on building roads, sewers, and other infrastructure.[6] When the Chamber of Commerce insisted that the county perform a cost–benefit analysis on the plan, Huckelberry responded by forcefully redefining the problem. He encouraged people to consider the costs of *not* undertaking the plan—of losing pristine areas and of providing services for sprawling development. Supervisor Grijalva reinforced Huckelberry's position by commenting that "money is not an insurmountable obstacle," unless the board allowed it to be one (T. Davis 1998b).

Environmentalists described the desert plan as visionary and precedent-setting, and Tucson's main newspaper, the *Arizona Daily Star*, immediately came out in support of the concept, describing Huckelberry's proposal as "bold," "courageous," and "inspiring" (Anon. 1998b). Touting Tucson's strong sense of place, the editorial went on to say: "What is best is the evident activism and good faith of this farsighted document. Huckelberry's plan is gratifying because it shows a bureaucracy responding to a greening of public and supervisor sentiment." The paper's unrelenting coverage of the issue, including a seven-day series in

November 1998 on the landscapes the SDCP would conserve, kept public attention riveted on the plan's evolution.

As 1998 drew to a close, political momentum for a desert protection plan that could also serve as the basis for an Endangered Species Act section 10 permit continued to build. Interior Secretary Bruce Babbitt declared in December 1998 that he intended to do everything he possibly could to facilitate the project's success. Jim Kolbe, Republican congressman for southern Arizona, vowed to advocate for federal money to support the plan. Then, on December 30, 1998, the FWS—under court order—proposed designating more than 260,000 acres of critical habitat for the pygmy owl in Pima County, much of which was in the fast-growing Northwest Side. The designation, although it affected only projects that needed federal permits, increased pressure on the county to address the endangered species question in a more holistic fashion.

Although there had been several signs of a change in county leaders' approach to development, the first tangible confirmation came in January 1999, when the Board of Supervisors voted 4–1 to deny Fairfield Homes' application to rezone the Canoa Ranch, about 30 miles south of downtown Tucson. Fairfield had asked the supervisors to permit the construction of 6,111 homes (down from 6,573 homes in May 1988) on 3,000 acres, as well as two golf courses, a 750-acre commercial area, and an airstrip. This was the third largest rezoning proposal in Pima County's history, and in rejecting it the board reversed a 1995 decision to amend the county's Comprehensive Plan and recommend up to 37,000 houses on the property. In fact, the vote marked the first time in 25 years that the supervisors had turned down a large rezoning and, according to Supervisor Ray Carroll, symbolized a major transformation of public attitudes toward development (T. Davis 1999a). Three months later, the board approved Huckelberry's request that it crack down on upzoning in environmentally sensitive areas and allow developers to transfer development rights from more to less sensitive areas. (The board declined to approve Huckelberry's proposal to charge fees for building in the desert in order to raise funds for land acquisition, concerned that doing so would violate state law.) The board also committed to establishing a Steering Committee for the SDCP and indicated it would limit or end rezonings of more than 120,000 acres of privately owned desert land for two years while the plan was being developed. Then, in May 1999, the county's Design Review Committee—which heretofore had routinely granted exemptions to the county's Hillside Development Overlay Zone ordinance—quashed four variance proposals and delayed a fifth (T. Davis 1999c).[7]

Mapping an Environmentally Protective Preserve System

Like his counterparts in Austin and San Diego, Huckelberry began the process of fleshing out the conservation plan by establishing a ten-person Science and Technical Advisory Team (STAT) to generate an integrative assessment of the region's biological diversity. (He also convened Ranch Conservation and Cultural Resources teams, but the STAT received the bulk of the funding and attention.) Unlike policymakers in Austin and San Diego, however, Huckelberry repeatedly insisted that Pima County's plan would be "based on science and fact," rather than emphasizing the need to accommodate political and economic considerations. He pursued this pro-environmental course despite the increasing restiveness of developers, many of whom originally had cautiously supported the idea of an HCP, and the uncertain legal status of the pygmy owl. (SAHBA had filed lawsuits challenging both the FWS's listing of the owl and its designation of critical habitat.)

In an effort to ensure that the SDCP would be unassailable on scientific grounds, the county recruited academic and agency scientists and insulated the team from political pressures by allowing it to develop its own mission and methods without input from stakeholders. Huckelberry selected Bill Shaw, a University of Arizona wildlife biologist with nearly three decades of interest and expertise in urban wildlife conservation, to head the team. Unencumbered by the need to avoid antagonizing powerful stakeholders, the STAT began by formulating a holistic and protective mission: "to ensure the long-term survival of the full spectrum of plants and animals that are indigenous to Pima County through maintaining or improving the ecosystem structures and functions necessary for their survival." Next, the team laid out six specific objectives consistent with that single, overarching goal: (1) to promote recovery of federally listed and candidate species, to the point where their continued existence is no longer at risk; (2) to reintroduce and recover species that have been extirpated from the region, where feasible and appropriate; (3) to maintain or improve the status of unlisted species whose existence in Pima County is vulnerable; (4) to identify biological threats to the region's biodiversity posed by exotic and native species of plants and animals, and develop strategies to reduce those threats and avoid additional invasive exotics in the future; (5) to identify compromises to ecosystem functions within target plant communities selected for their biological significance, and develop strategies to mitigate them; and (6) to promote long-term viability for species, environments, and biotic communities that have special significance to people in this region because of their aesthetic or cultural values, regional uniqueness, or economic significance (STAT n.d.).

After articulating its aims, the STAT set out to construct a biological map of the entire county, without regard to political boundaries, that could serve as a foundation for a preserve design.[8] The team began by compiling a list of about 200 species of concern, based on a survey of biologists and local experts, ultimately arriving at a list of 107 imperiled species (T. Davis 2000a). Then they filtered out the animals that live primarily in areas outside the county's control, such as the endangered Mexican spotted owl, whose high-elevation-forest habitat is managed by the Forest Service and the Park Service. They also excluded species that, although declining in Pima County, were not at risk elsewhere and species for which conservation could best be done elsewhere.[9]

After arriving at a list of 55 priority vulnerable species, the team used an iterative process to construct the boundary of what became known as the Conservation Lands System (CLS), focusing exclusively on the land's biological value and development status without regard for ownership or jurisdiction. First, they established a set of criteria for identifying biologically valuable tracts using measures of species richness, the spatial distribution of vegetation communities, and landscape features, not just information about the habitat potential for individual plants or animals. Then, going back and forth between models, observation records, and the judgment of local naturalists and scientific experts, they came up with a biological preserve map that included every parcel with habitat suitable for at least three of the 55 species—an area that covered 1.16 million of the county's 5.9 million acres. They designated any parcel containing five or more species as part of the biological core, which comprised 635,000 acres of state and private land. Rivers and streams received the highest priority for protection. The remaining acreage—about 180,000 acres of private land and 300,000 acres of state land—was considered "sensitive" but acceptable for multiple uses.

In addition to this "coarse filter" focus on conserving species-rich habitat and ecological processes, the STAT employed a "fine filter" approach to ensure that the preserve system would protect individual species of concern. They also used "special elements"—particular highly valued plant communities and landscape features—to determine the exterior and interior preserve boundaries. In addition, the STAT noted that, although its basic preserve map would conserve most vegetation types, the county should restore some native vegetation types that have undergone or continue to experience heavy losses. In particular, the team recommended restoration of riparian processes and vegetation.

A County-Led Planning Process

In addition to the STAT and other technical teams, Huckelberry created a Steering Committee to provide stakeholder input, in hopes of fostering broad-based support for the SDCP. The county administrator's office retained tight control over the planning process, however, exerting consistent leadership and rebuffing efforts by development interests to derail it. The county released a flurry of technical reports that buttressed its conservation efforts; as criticism by ranchers, property rights activists, developers, and municipalities mounted, the county responded by reasserting and explaining the rationale for its position rather than simply accommodating challengers. By the time the county released the first draft of the SDCP in September 2000, it was abundantly clear to most observers that environmentalists' views were dominating the planning process and that neither Huckelberry's office nor the Board of Supervisors was receptive to the concerns of those who sought to dilute their conservation agenda.

The County Establishes a Stakeholder Group   Intentionally or not, county officials established a planning process that ensured they would retain control, particularly during the crucial early days. Unlike organizers elsewhere, Pima County did not select stakeholders to serve on its Steering Committee as representatives of key constituents, but instead sent out a blanket invitation and granted a spot to anyone who wanted to participate. The resulting 84-person group was too large and unstructured to have a meaningful impact on the desert plan in its formative stages, but county officials made no move to set up a smaller committee.

Well into its second year, while the STAT was developing its science-based preserve system, the Steering Committee endured "boot camp"—a series of presentations by county staff and outside experts on various aspects of land use, conservation biology, and endangered species law. The committee then spent a substantial portion of its second year establishing procedural ground rules, such as the requirement to have a super-majority, rather than consensus (which would have been impractical in a group that size), on recommendations to the county. Eventually the group progressed to discussing substantive issues, such as mitigation require-ments and financing, but there were few opportunities for members to have meaningful interactions—many did not even know other members' names—so trust did not develop among them. (The county hired a pro-fessional to facilitate the process, but he did little more than supply mate-rials to the group and take notes at meetings.) Instead of deliberation, the meetings featured posturing by various interests. Carolyn Campbell (2006), leader of the Coalition for Sonoran Desert Protection, describes it thus: "Nothing was getting done at these meetings because people just got up and talked for as long as they could get away with about what was important to them and why. And they'd repeat it over and over..."

Some members of the Steering Committee, impatient with their lack of input into the planning process, began referring to themselves as the "Steered Committee." Bill Arnold, a realtor and disaffected committee member, noted: "The whole premise of winning a debate is defining the terms. That's exactly what the county has done. They defined the terms, and all the discussion will now occur on their terms" (T. Davis 2000b). Despite their frustration, many participants stuck with the process—although envi-ronmentalists and neighborhood activists were far more tenacious than developers, most of whom stopped attending meetings altogether.

The County Rebuffs Critics   Ignoring stakeholders' complaints, in September 2000—after 72 formal public meetings and 200 community meetings—the county released, and the Board of Supervisors voted to accept, a first draft of the SDCP. The draft plan envisioned instituting a host of measures, such as the following:

- Creating more than 200,000 acres of new mountain parks
- Buying, protecting, and restoring riparian areas
- Combining the county's environmental rules into a single Environmentally Sensitive Lands Ordinance in order to streamline the regulatory process
- Increasing protection for ironwood trees and adding protection for other rare riparian vegetation to the county's Native Plant Protection Ordinance
- Requiring all new golf courses to use recyled water
- Buying development rights from ranchers on the urban fringe, to allow them to continue ranching
- Providing long-term lease assurances for ranchers on state and federal land, and compensating them if a federal agency reduces cattle numbers

• Changing property-tax laws to give ranchers incentives to preserve open space
• Requiring ranchers to pay assurance bonds guaranteeing they will carry out archaeological surveys, tests, and mitigation plans
• Requiring new development to limit water-intensive landscaping to 20 percent of the area for single-family homes and 30 percent for apartment buildings.

Huckelberry also recommended that the county adopt measures to protect the integrity of the planning process, such as delaying the rezoning or issuance of permits for development in pygmy owl habitat, ranch conservation areas, and riparian areas. In addition, he suggested that the county should lobby for legislative changes that would strengthen its ability to conserve natural resources, such as repeal of the ban on downzoning passed by the legislature in 1998, as well as grants of more county authority to regulate wildcat subdivisions, charge impact fees for parks and sheriff's facilities, and create tax incentives for those who preserve private land.

Opposition, which had been muted during the plan's early stages, began to mobilize in earnest between the release of the draft SDCP and the end of the first comment period in February 2001. Representatives of the city of Tucson as well as the suburbs of Marana, Oro Valley, and Sahuarita—who had rarely attended Steering Committee meetings but had repeatedly expressed annoyance at not being allowed to influence the content of the county's technical reports—publicly criticized the county for failing to establish a genuinely cooperative process that took their concerns seriously. Similarly, Michael Anable, director of the State Land Department, insisted he should be included in—not just consulted on—the county's decision-making process. In response to these charges, Huckelberry reiterated the importance of isolating the county's scientific process from politics, saying: "We haven't been wanting to debate a lot of issues or alternatives until the whole set of facts come before people." He added that once the final preserve plan was released in early 2001, it would be "wide open for public review" (T. Davis 2000c).

Not surprisingly, the most strident complaints about the plan came from ranchers, property-rights advocates, and developers. Ranchers and property-rights activists expressed skepticism about its scientific basis, pointing out that the STAT's maps were based on models that concealed enormous uncertainty and judgment. Developers disparaged the plan's comprehensiveness, saying the county ought to focus on compliance with the Endangered

Species Act, not growth management. SAHBA's president, Terry Klinger, explained that the STAT was "proposing an enormous biological preserve based on an arbitrary selection of additional species that aren't endangered or threatened." He added: "What they've done is come up with a computer model, and a computer model isn't exact science" (Jensen 2001).

County leaders forcefully rejected detractors' claims, however, adhering to their assertion that the plan was based on "science" and "facts." Complaints about the plan's scientific basis were further neutralized in October 2001, when two well-known HCP experts reviewed the SDCP preserve design process and certified its approach as scientifically defensible. Reed Noss, one of the country's foremost conservation biologists, and Laura Hood Watchman, who had done an extensive evaluation of HCPs for Defenders of Wildlife, deemed the SDCP a "credible, science-based process designed to achieve clear and laudable goals for the long-term conservation of biodiversity in Pima County" (Noss and Watchman 2001). Watchman described the SDCP as "at the cutting edge of conservation planning" (T. Davis 2001g).

County leaders also parried developers' efforts to shift attention to the cost of the plan by redefining the problem in a way that was pivotal. Huckelberry argued that existing patterns of development imposed economic burdens on county taxpayers, and he urged the public to consider the costs of development and of *failing* to conserve biologically valuable land. Maeveen Behan, Huckelberry's assistant, furnished technical backing for this position. Working with county staff, she generated maps depicting how sprawling development strained the tax base, whereas compact development enhanced the county's long-term fiscal stability. With respect to refusing to upzone property, Huckelberry made clear that he regarded upzoning not as an entitlement but as a discretionary privilege; the county's position, he said, was that landowners were entitled to the zoning of their property based on the zoning code established in 1958, but government was not obligated to enhance the value of individuals' land by upzoning (Huckelberry 2006). When developers suggested that acquiring ranchland for preserves would take a huge bite out of the county's tax base—a charge that had proven effective in Austin—Supervisor Grijalva countered that many of the owners of large tracts already contributed little to the tax base because they got large property-tax breaks for running cattle (T. Davis 1998b).[10]

As their allegations that the county was engaging in "full-blown growth management" (until recently, a lethal charge in Pima County) fell on deaf ears, developers turned to another claim that previously had been effective: that restricting the amount of developable land would preclude construction of affordable housing and would therefore hurt low-income and minority residents (Anon. 2001; Lurie 2001). Even though he had initially endorsed the plan, Don Diamond, the county's most prominent developer, came out in opposition in early 2001, saying that restricting development would drive up the price of land, which would be good for him personally but bad for the community. This claim had the potential to divide the plan's supporters, or at least mobilize residents who might not otherwise get involved, but environmentalists had already established strong relationships with neighborhood and labor groups (Grijalva 2006). Moreover, Supervisor Grijalva was a far more credible spokesman for the Latino community and for inner-city residents in general than developers were, and he ridiculed developers' newfound interest in Tucson's minority and low-income residents. Echoing Grijalva's arguments, Latino advocates pointed out that their constituents were not buying homes in pygmy owl territory (T. Davis 2001g). To bolster the county's position, Huckelberry presented evidence compiled by his staff showing that "Housing is as unaffordable for two thirds of the population as it was in 1991, seven years before the conservation planning process started." Local real estate and home-building interests, he observed, historically had shown little interest in affordable housing, but had instead "tailored their product to the high end income earners" (Huckelberry 2001).

The County Pursues an Environmentally Protective SDCP

Despite ongoing efforts by developers and their allies to undermine the SDCP, the county remained impervious as it worked on finalizing the plan. Moreover, although the self-imposed deadline for completion of the SDCP was not until December 2002, county supervisors expressed their intent to incorporate its biological elements into the Comprehensive Plan, which had to be updated by December 31, 2001, in accordance with the state's 2000 Growing Smarter Plus law.[11] Development interests objected fiercely to this idea, but to no avail.

Approving a Comprehensive Plan   In March 2001, the county's consultant, RECON of San Diego, unveiled the STAT's biologically preferred preserve map for the SDCP, triggering another round of protests from detractors of the plan. The map designated a 1.16-million-acre preserve that, if implemented, would conserve 55 species, as well as every major habitat in eastern Pima County (see figure 7.1); in hopes of defusing some

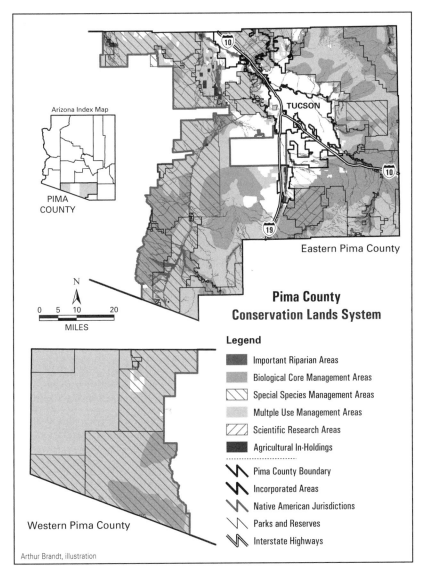

Figure 7.1
Pima County, Arizona, Biological Corridors and Critical Habitat Map

of the opposition, RECON's Paul Fromer characterized it not as a final prescription but as "a wish list for biological resources" (T. Davis 2001c). State Lands Department Director Anable immediately asked the county to remove all state lands from the plan, on the grounds that his department had not been given a meaningful role in the planning process. Undaunted, Huckelberry agreed to label state lands but not to remove them from the map. Members of the development community also reacted with dismay, partly to the stringent land-use controls and fees on development that accompanied the map, but also to its broad coverage. Their argument was the same as it had been from the outset: Pima County was using the STAT's map as a cover for its real goal of creating a growth management tool that could raise housing prices, cost taxpayers millions of dollars, and deprive people of the use of their land (Tobin 2002).

Over developers' continuing resistance, in April 2001 the Board of Supervisors agreed unanimously to tie the update of its comprehensive land-use plan to the SDCP. Then, at its June meeting, the board imposed a sweeping set of eight land-use restrictions proposed by the Coalition for Sonoran Desert Protection that would stay in place until 2002, when the board was scheduled to formally adopt the SDCP. Among those measures was a requirement that developers leave 80 percent of their properties unbladed if they were within the SDCP's proposed "draft biological reserve." Landowners outside the proposed preserve could buy or set aside between two acres (for unoccupied habitat) and four acres (for occupied habitat) of sensitive land within the preserve for every acre they bladed. In July the board approved two more ordinances: one required developers to conduct a site analysis and submit a biological impact report; the other required them to inform both the county and the FWS about how proposed projects could affect threatened and endangered species.

The board's conservation drive continued through the fall; in late October 2001 Pima County officials proposed adopting new rules—many of which were included in the draft SDCP—to protect the region's water supply from rising demand. For example, Huckelberry urged the board to prohibit new golf courses in unincorporated Pima County unless they used treated effluent for irrigation rather than groundwater. He also proposed limiting water-hungry landscaping to 20 percent of new developments, forcing those developments to use recycled wastewater, and tightening rules for drip and spray irrigation. He justified his recommendations based on a new water report prepared for the SDCP, which he said established "the scientific and factual foundation for further conservation" (Tobin 2001b).

2007 existed as a vision rather than a legal document, is remarkable for its comprehensiveness: it covers nearly 6 million acres and focuses on protecting critical habitat and biological corridors, expanding mountain parks, restoring riparian areas, preserving historical and cultural resources, and conserving ranches. Its biological element is holistic, in the sense that it aims to protect and restore areas of particularly rich habitat *and* the interconnectedness among them, as well as the region's ecological processes, particularly its riparian processes. And its emphasis is protective—its overarching purpose is to "ensure the long-term survival of the full spectrum of plants and animals that are indigenous to Pima County through maintaining and improving the habitat conditions and ecosystem functions necessary for their survival."

The plan's biological "backbone" is implemented primarily through the Conservation Lands System, the basis for which is the Priority Biological Resources Map that designates more than 1 million acres of biologically sensitive land—525,000 acres of biological core and 500,000 acres of multiple-use areas. In addition to the map, the CLS includes a set of voluntary guidelines to be used in permitting new development. According to the original CLS guidelines (which were later revised), development in (1) important riparian areas must set aside 95 percent of their existing biological resources; (2) areas that provide high potential habitat for five or more priority vulnerable species (biological core areas) must set aside 80 percent; (3) species management areas, defined as crucial for the conservation of particular native plant or animal species of concern, must set aside 80 percent; and (4) scientific research, multiple use, and recovery and agricultural recovery management areas, which generally support three or four priority vulnerable species, must set aside between 60 and 75 percent. The CLS also designates critical landscape connections and, although it provides no specific set-aside percentages, recommends both protecting them and removing barriers to the movement of fauna and pollinators. In addition, it suggests that development of agricultural holdings within the CLS proceed in ways that do not compromise the effectiveness of conservation efforts on adjacent land. Finally, the guidelines recommend that existing developments within the CLS retain 60 percent of their biological resources, and that urbanizing areas retain 30 percent of their biological resources.[15]

In addition to conserving land through the development approval process, the SDCP aims to acquire thousands of acres to expand mountain parks and establish new conservation areas through fee-simple purchase or the establishment of conservation easements. The plan also recommends restoring some of the county's riparian areas, which have been badly stressed by development, groundwater pumping, and the invasion of exotic species. Although it recognizes that many of the region's riparian areas have been irreparably damaged, the plan prescribes protecting the dynamics of those systems that remain and, where possible, restoring connections among their components, such as channels, overbanks, floodplains, vegetation, and shallow groundwater. Finally, the plan promotes ranch conservation, which provides essential buffering for the biological preserve: ranches occupy about 240,000 acres in eastern Pima County, and ranchers hold grazing leases on another million acres of state and federal land (Chesnick and Morlock 1999b). Under the plan, ranch conservation is accomplished in part through management agreements in which the county purchases and extinguishes ranchers' development rights. In exchange, ranchers agree to manage their lands in ways that maintain its biological integrity—for instance, by fencing off riparian areas or limiting the number of cattle.

The approach embodied in the SDCP marks a substantial improvement over the trajectory Pima County was on. During the three decades leading up to the adoption of the plan, the Board of Supervisors routinely approved subdivision proposals with only modest open-space requirements and paid little attention to the cumulative impact of individual projects. The consequences were evident: Tucson sprawled in every direction. The county did acquire some open space during this period. Between 1974 and 2004, it spent $125 million, including nearly $28 million from a 1997 voter-approved open-space bond issue, to buy 26,778 acres of open space (T. Davis 2002). But its approach was haphazard; there was no systematic effort to preserve the most biologically valuable land, restore critical ecological processes, or maintain a low-intensity development matrix between preserved land and Pima County's urbanized areas.

The SDCP is also precautionary; unlike many habitat conservation plans, the SDCP's Priority Biological Resources Map is *not* the result of a compromise between scientists' assessments of what is minimally necessary to conserve the region's remaining biodiversity and what development interests are willing to concede. Instead, it reflects experts' judgment about how much land, in what configuration, would be sufficient to conserve the region's biodiversity. Moreover, it values land not just where endangered species are now but also where they could be in the future if conditions improve. It also recognizes that water supplies are finite and that some riparian processes will need to be restored.

Implementing the SDCP

Because the SDCP's biological elements rest heavily on voluntary stewardship by the county, federal land managers, and large private landowners—especially ranchers—pro-environmental implementation will be crucial to its ultimate success in conserving biological diversity. Some signs are auspicious; since it began the SDCP planning process in 1998, Pima County's approach to development has shifted markedly. The county has initiated an aggressive land acquisition program and dramatically changed its approach to development permitting. Furthermore, the county is designing a sophisticated, multi-phased monitoring program and, in early 2008, completed its multiple-species conservation plan, which it hopes will add federal backbone to the implementation of the CLS. Several obstacles threaten Pima County's ability to conserve biologically valuable land, however. In the short run, the FWS's 2006 decision to delist the pygmy owl may jeopardize efforts to protect land in the fast-growing Northwest Side. Potential longer-term impediments include a change in county leadership—because the CLS guidelines are implemented in the course of rezoning, new leaders can disregard them if the political context changes; unwillingness by the state to make state trust lands available for conservation; insufficient funding for management and monitoring; and, as of late 2006, a state law that severely limits government's ability to regulate private property.

Assembling a Preserve through Development Approvals
The county's development approval process clearly became more pro-environment with the inception of Sonoran Desert conservation planning in 1998. That shift was formalized with the adoption of the revised Comprehensive Land Use Plan in December 2001. The most prominent example of the county's new emphasis was the rezoning of Canoa Ranch, which occurred just prior to the formal adoption of the CLS; after two years of wrangling, in March 2001 the board approved a compromise that allowed 2,199 homes on about 1,300 acres, as well as 153 acres of commercial development including three large shopping areas and a golf course—down from 6,100 houses and two golf courses in the original proposal. The board also agreed to buy and preserve 85 percent of the ranch for $6.6 million (T. Davis 2001b).

The CLS also has proven durable. Since the guidelines were formally approved in late 2001, they seem to have worked in two ways: by tempering developers' expectations about what the board would approve and the profits they could make, and by constraining the board's willingness to approve rezonings (T. Davis 2005). Between 2002 and 2004 developers requested eight rezonings inside the designated preserve, compared with 15 sought between 1999 and 2001. In addition, the number of rezonings approved dropped from 12 between 1999 and 2001 to three between 2002 and 2004. When it did approve rezonings, the board generally adhered to the CLS guidelines: from early January 2002 to the summer of 2006, 27 of the county's 33 zoning decisions were consistent with the CLS, and the remaining six decisions departed only slightly from CLS standards. By contrast with the past, beleaguered developers declined to challenge board decisions, which they saw as "done deals," and many vetted their proposals with environmentalists Carolyn Campbell and Christina McVie prior to submitting them to the Board of Supervisors.

In June 2005 the board grappled with a developer-backed request to modify the CLS guidelines and, in particular, scale back the requirement to conserve 75 percent of parcels within the CLS's multiple-use areas when the land is rezoned. Huckelberry himself advocated reducing the open-space set-aside to 65 percent, which he considered a more "reasonable" figure. The Sonoran Desert Coalition reluctantly agreed to support that standard, on the condition that the land would be permanently managed, protected, and monitored (T. Davis 2005). Although the Planning and Zoning Commission endorsed a 75 percent set aside by a 7–1 vote, the Board of Supervisors settled on 66.6 percent. At the same time, at the urging of the Coalition for Sonoran Desert Protection, the board unanimously *increased* set-asides from 75 percent to 80 percent on some landscapes important to the pygmy owl, including the Northwest Side's ironwood forest and the Altar Valley's mesquite flats. For land within the biological core, the guidelines did not change: developers must continue to set aside 80 percent of biological core areas and 95 percent of riparian areas.

The predictions of some opponents notwithstanding, it is notable that since the board's adoption of a more environmentally protective approach, development in Pima County has not ground to a halt. In its spring 2002 meeting, the county granted construction permits for developments on 30,000 acres, and in 2003 building activity skyrocketed in Tucson's southern suburbs. Builders received more than 8,000 permits to build houses during the fiscal year that ended on June 30, 2004, of which 3,443 were in unincorporated Pima County—a 43 percent increase over the previous year (T. Davis 2004b). Like other plans described in this book, the SDCP assumes substantial population growth—17,000 new residents each year, which would raise the population from 900,000 in 2000 to 1.2 million by 2025, according to Huckelberry (T. Davis 2001a). Rather than trying to limit the development necessary to accommodate growth, the plan simply redirects it toward less biologically sensitive areas.

Assembling a Preserve through Public Acquisition

In addition to requiring open-space set-asides as a condition of development, the county has used other means to conserve land within the CLS, such as the fee-simple acquisition of private land or development rights. In May 2004 more than 65 percent of county voters approved a bond issue of $174.3 million for land acquisition, most of it for properties within the CLS. Although the Chamber of Commerce came out against the bond measure, SAHBA declined to launch a campaign to defeat it after Huckelberry relinquished the county's prerogative to condemn land for acquisition—a tool he had used occasionally and had threatened often.[16] Huckelberry had also promised developers that the county would impose no new regulations on land use and that open space would include privately owned, publicly subsidized "working landscapes" (ranches) that did not grant public access. SAHBA spokesman Roger Yohem said that Huckelberry's concessions marked "a new era of communication between SAHBA and the county," adding that "trust levels have gone up tremendously" (T. Davis 2004a). Both the Tucson Association of Realtors and the business-oriented Metropolitan Pima Alliance supported the bond measure, and the Coalition for Sonoran Desert Protection, as well as numerous other local and national environmental organizations, engaged in a massive education campaign to drum up voter interest.

In preparation for the election, the Nature Conservancy and the Arizona Open Land Trust established a scheme for setting priorities among parcels to be acquired with the bond money. The two organizations produced a map of the 536,000 acres most likely to contain either some of the 55 vulnerable species covered by the plan or corridors between mountain ranges or highly valued habitat, such as cottonwood–willow or saguaro–ironwood. The Bond Advisory Committee appointed by the county then adopted that map, generating from it a list of properties it would try to acquire with the bond money. Despite these steps, the issue of which land to acquire with the bond money remained contentious after the initiative passed. Environmentalists insisted that the county acquire pygmy owl habitat in the Northwest Side, where the threat of development was imminent, despite the skyrocketing cost of land there. By contrast, Huckelberry—who wanted to get as much acreage as possible for the taxpayer dollar—advocated concentrating on large, outlying ranches. Further limiting the county's ability to acquire the land most threatened by development was a requirement, passed by the Board of Supervisors in April 2004, that bond purchase prices not exceed the county-appraised value of the parcel.

In hopes of defusing conflict over acquisitions, the county established a Conservation Acquisition Commission, a citizen advisory group that makes recommendations to the Board of Supervisors on individual parcels as they come up for consideration. Huckelberry also began repairing the rift with ranchers that had opened during the planning process by negotiating deals that would allow individuals to continue raising stock while transferring property ownership to the county (Huckelberry 2006; Poole 2007). By the fall of 2006 the county had spent nearly $69 million in bond money to acquire nearly 25,500 acres, as well as more than 86,000 acres of grazing leases. The county was considering another bond election in 2008, having estimated it would cost about $2.6 billion to save the 536,000 acres of biologically valuable land identified by the Nature Conservancy and Arizona Open Land Trust (T. Davis 2006).

Obstacles to Implementation

Although Pima County has established a pro-environmental trajectory, it faces several major threats to successful implementation of the SDCP's biological conservation element. In August 2003 the 9th U.S. Circuit Court of Appeals ruled unanimously that the FWS had not met its legal burden of proving the pygmy owl's Arizona population was significantly distinct from Mexico's, and sent the case back to the district court. In the spring of 2006 the FWS gave notice that it planned to delist the owl. Although county officials dismissed the importance of the delisting, saying the SDCP had always aimed to conserve biodiversity rather than any single species, environmentalists were concerned about losing their leverage to protect land in the Northwest Side, where skyrocketing property values discouraged public acquisition.[17]

Two other factors may impede acquisition of preserve land in the longer term. First, if the county declines to use condemnation, it will become more difficult for it to acquire land because landowners will demand higher prices as the remaining number of large parcels shrinks. Second, the state can thwart Pima County by refusing to allow the county to acquire state lands for conservation purposes. In November

2006 Arizona voters rebuffed Proposition 106, which would have allowed conservation of 700,000 acres of state land, including 300,000 acres in Pima County. Ranchers and Pima County home builders led a successful campaign against the measure, and several key environmental groups declined to campaign for it as well because they regarded it as too weak.

There are also potential obstacles to implementing land-use controls that restrain new development. Because the CLS guidelines are voluntary, a new Board of Supervisors or a turn in public sentiment could undermine efforts to limit development. (The county's multiple-species conservation plan—which in theory can provide a legal backstop—contains few concrete conservation measures and lacks the comprehensiveness of the SDCP.) Proposition 207, the state's regulatory takings initiative, may pose a more serious and immediate threat to implementation of CLS guidelines. The measure, approved in November 2006 by nearly two-thirds of Arizona voters, requires that government compensate landowners for any regulation that reduces the value of their property; if it declines to do so, it must forgo enforcing the regulation.

Finally, the county is only beginning to grapple seriously with the issue of management and monitoring, the weaknesses of which have seriously undermined other plans. For the county to achieve its biological goals, it will have to treat the models developed during the planning process as hypotheses to be tested and revised over time. Fortunately, in May 2007 the county received a $274,505 grant from the Department of Interior to design an ecological monitoring plan for the SDCP and its associated MSCP (Meltzer 2007). To facilitate genuinely adaptive management, the county aims to develop a more efficient and effective monitoring process than the traditional approach of simply counting species—one that will give managers early warnings if the plan is failing to conserve its species of concern. To that end, it is identifying a set of indicators that captures not only the status of threatened and endangered species but, based on conceptual models, the habitat components those species need for their survival.

Conclusions

The SDCP boasts many attributes that suggest it is not only a cut above its predecessors in Austin and San Diego but can substantially conserve and restore Pima County's biodiversity. Experts who have reviewed the CLS map say it comprises virtually all of the most biologically valuable land in Pima County. In fact, its output was remarkably similar to the results of the Nature Conservancy's parallel effort to designate a Sonoran Desert ecoregional assessment (R. Marshall 2006). The plan's particular emphasis on biological cores and riparian areas is consistent with those areas' extremely disproportionate ecological value. Its attention to multiple-use/recovery areas as buffer zones and its purposeful restoration of riparian processes reflect a precautionary approach. Furthermore, the county's precautionary mapping and its efforts to devise a sophisticated monitoring program reflect a genuine commitment to environmentally beneficial adaptive management.

Of course, the extent to which the SDCP actually preserves Pima County's biological diversity rests heavily on how the plan is implemented over time. To date, the county has largely adhered to its conservation guidelines; moreover, it has undertaken an aggressive effort to acquire large tracts of open space. In doing so, it has improved on the historic practice of evaluating rezoning proposals case by case without considering previous rezonings or cumulative impacts. Overall, the SDCP has shifted the status quo in Pima County from unfettered development accompanied by protection of isolated parcels to managed growth and landscape-scale conservation.

Ironically, the primary reason for Pima County's transformation is that planners did *not* adhere to a central tenet of ecosystem-based management: reliance on stakeholder collaboration. In fact, county leaders stated from the outset that their primary goal was to conserve biological diversity through a scientifically defensible process, not to come up with a plan that everyone could agree on. County leaders were abetted by a cohesive environmental community led by a politically savvy spokesperson who forged alliances with a variety of interests in the community, thereby ensuring public support for pro-environmental leadership. Relatively early in the process, a positive feedback emerged, in which the public's changing attitude toward development and the actions taken by political leaders reinforced one another. The county produced and disseminated pro-environmental arguments in easily digestible form, anticipating and rebutting home builders' claims in ways that made them appear cynical and self-interested. The public responded by reelecting pro-environmental supervisors and supporting a major bond issue to facilitate land acquisition.[18]

Also central to the county's ability to craft an ambitious plan was local policymakers' willingness to employ the regulatory leverage available to them. Early on, the Board of Supervisors made a critical decision to restrict development in the unincorporated county in order to ensure the plan would not be undermined by decisions made during the planning process. It also instituted a set of ordinances that minimized the environmental damage associated with the development that did occur.

A third—and counterintuitive—reason the SDCP is so protective is that, although it adopted a landscape-scale focus, the county declined to coordinate with several governmental entities that controlled land use in the planning area. County officials worked with federal agencies whose goals were compatible, but made only perfunctory efforts to engage the State Lands Department and municipalities within the region that might have pressed for a more permissive plan. Such a go-it-alone approach produced some hostility in the short run, though over time relations among jurisdictions thawed. Meanwhile, the county created a foundation of compelling scientific reasoning—captured in hundreds of technical reports—on which Tucson and other municipalities are now basing their own plans.

8

Re-creating Central Florida's Meandering Kissimmee River

Just as Pima County's Sonoran Desert Conservation Plan contrasts with the terrestrial habitat conservation plans in Austin and San Diego, the Kissimmee River restoration provides a useful counterpoint to the Comprehensive Everglades Restoration Plan described in chapter 5. In 1990 the state of Florida announced an ambitious plan to restore to its rambling course the midsection of the Kissimmee River, which the U.S. Army Corps of Engineers had finished straightening only two decades earlier. The final plan for the Kissimmee involves removing man-made impediments to historic patterns of water flow and instituting a regime of water releases more consistent with historical patterns. The plan's ultimate aim is to re-create the river's physical form and natural hydrologic processes, in hopes that doing so will prompt the reestablishment of river-floodplain habitats, enabling the fish, waterfowl, and wading birds that once inhabited the area to return. Although implementation of the restoration project is more modest than originally envisioned, it has already yielded tangible environmental benefits. By the summer of 2006, five years after the first phase of construction was complete, water was flowing through more than 15 miles of curving river; 11,000 acres of pasture had become floodplain wetlands; waterfowl and wading bird populations had reached the highest densities recorded since monitoring began; and game fish populations were rebounding.

The Kissimmee River Restoration Project demonstrates that landscape-scale efforts to restore ecological function *can* result in measurable environmental improvement, even in complex and highly degraded aquatic systems. Two aspects of the Kissimmee River restoration have contributed to its environmental effectiveness. First, planners have treated the river-floodplain ecosystem as a single entity, regardless of political boundaries. Landscape-scale thinking, in turn, caused scientists and engineers to focus on repairing ecosystem form and function, rather than trying to ensure the persistence of optimal numbers of particular species. It also brought about the development of more cooperative relationships among several agencies, particularly the Army Corps of Engineers (Corps) and the South Florida Water Management District (SFWMD), but also the U.S. Fish and Wildlife Service (FWS), the Florida Fish and Wildlife Conservation Commission (formerly the Florida Game and Fresh Water Fish Commission), and the Florida Department of Environmental Protection. Second, the restoration team has employed a genuinely adaptive approach to both planning and implementation. Engineers have worked with biologists to establish baseline conditions, devise and implement a variety of experimental restoration techniques, monitor the results of those interventions, and revise their approach based on the results of monitoring. Their findings have resulted in both learning and project adjustments.

The main reason for the Kissimmee River restoration's environmental effectiveness, however, is that it does not aspire to simultaneously restore the natural system *and* meet water users' current and future demands. Rather, it aims to achieve a single, preeminent goal—restoring the ecological integrity of the river-floodplain ecosystem—while accommodating most flood control and navigational demands. The program's unitary goal reflects its political origins: unlike the Comprehensive Everglades Restoration Plan and the California Bay–Delta (CALFED) Program, the Kissimmee River restoration did not emerge out of a collaborative effort to gain consensus among stakeholders. Instead, proponents built political will by conducting a campaign to raise public and high-level political awareness of the river's plight and the potential benefits of a large-scale restoration. Florida's political officials responded by establishing a set of strongly pro-environmental tenets that guided the project's designers and by using both regulatory leverage and political capital to increase the likelihood of success.

The Impetus for Ecosystem Restoration

Prior to the early 1950s the Kissimmee River was a "dark stream shadowed by old oaks and tall palms, flowing through the wetlands of central Florida for more than 100 miles, sometimes at depths of more than six feet, sometimes with hardly any water at all" (Greene 1983b). The river slowly wended its way south from its headwaters in Lake Kissimmee to Lake Okeechobee across a one- to two-mile-wide floodplain. A unique set of hydrological characteristics shaped the river-floodplain ecosystem's structure and functioning: Lake Kissimmee discharged water to the river continuously, but the flow varied widely within and across years, with peak flows typically occurring in late October, at the end of the summer–fall rainy season. Unlike most other river-floodplain systems, in which less than 20 percent of the floodplain remains permanently inundated, in some years as much as 80 percent of the Kissimmee River's floodplain was soaked for long periods; in fact, during peak flood conditions the river resembled a long, narrow lake (Toth 1995).

The unusual hydrology of the Kissimmee's historic floodplain created nearly 40,000 acres of wetlands in a mosaic of at least seven plant communities. Seasonal fluctuations in the water level produced a variety of habitats that supported 19 species of migratory waterfowl and 21 species of wading birds: in late fall, ducks and waders fed in the sloughs, potholes, and wet prairies on the upland edge of the floodplain; many of the same populations used the potholes, oxbows, and backwaters of the floodplain in winter, and the river and deepest marshes and cypress swamps near the river in spring (USEPA 1992). Prolonged flooding facilitated a close physical, chemical, and biological relationship between the river channel and its surrounding marshes. When water levels were high, fish moved freely between the river and the floodplain, where they found breeding, feeding, and nursery grounds; as the water gradually receded, the marshes yielded abundant supplies of prey for wading birds, including smaller fish and invertebrates. Many species that are now rare or endangered, such as the wood stork (*Mycteria americana*), frequented the river for feeding, nesting, or breeding.

The river's regular flooding regime was inconvenient for the region's human inhabitants, however, and in 1954—over the objections of the Florida Game and Fresh Water Fish Commission and the U.S. Fish and Wildlife Service—the U.S. Congress authorized the Corps to modify both the headwater lakes and the river itself as part of the Central and Southern Florida (C&SF) Flood Control Project (see chapter 5).[1] By the time the Kissimmee River portion of the C&SF project began, engineers had already cut an eight-mile section of canal, known as the Government Cut, immediately upstream of Lake Okeechobee, thereby creating an isolated river remnant called Paradise Run. Between 1962 and 1971 the Corps completed channelization of the river, transforming the sinuous stream into a drainage canal named C-38. When combined with the Government Cut, C-38 was a 56-mile-long, 30-foot-deep conduit, ranging from 300 feet wide upstream to 800 feet wide at the downstream end, that conveyed water from Lake Kissimmee to Lake Okeechobee. To facilitate control over water levels in the canal, the Corps divided it into five long pools, separated by dams and navigation locks, that stepped down water levels in six-foot intervals. The Corps also transformed the headwater lakes into a series of water storage reservoirs by widening the canals that connect them and building control structures to regulate their levels according to flood-control schedules and other operational rules. The primary purpose of this massive endeavor was to "relieve flooding and minimize flood damages, largely in the upper Kissimmee basin" (USACE 1985).

As a flood-control enterprise, the $32 million effort to tame the Kissimmee River and its headwater lakes was enormously successful: by controlling the flow of water, it protected thousands of square miles of land from both inundation and drought (Blake 1980). But its adverse environmental impacts began to manifest themselves even before the construction was complete. Excavating C-38 and depositing the spoil along its banks obliterated 35 miles of river channel and 7,000 acres of floodplain. In turn, the canal—which detractors called the "Kissimmee Ditch"—rapidly drained about 27,000 acres of surrounding marshland, much of which was then converted into pasture for cattle (Toth, Arrington, and Begue 1997). Pockets of wetlands remained at the lower end of each pool, but they were impounded and therefore didn't function like floodplain wetlands. The banks of the canal were largely barren, and the river's remnant oxbows stagnated and filled with organic debris.

Alterations in the river's physical form, combined with the stabilization of water levels, prompted a cascade of ecological responses that resulted in a 74 percent decline in the number of active bald eagle territories and a precipitous 92 percent drop in wintering waterfowl (Toth 1995). Both the number and the diversity of game fish also plummeted as limited marsh habitat and low dissolved oxygen levels rendered the modified system inhospitable to many of its former residents.[2] Channelization had more subtle consequences as well; for example, the composition of benthic invertebrates shifted, and what remained were species more commonly found in lakes, not rivers.[3] The impacts of the flood-control project were also apparent in the headwater lakes—particularly Kissimmee, Hatchineha, and Cypress: lower water levels eliminated the outer fringe of shoreline wetlands and drained adjacent upland marshes; in addition, the stabilized water level allowed berms of organic sediment to develop around the lakes and block fish access to shoreline habitat. Beyond the direct damages it caused, the flood-control project had secondary impacts. By encouraging development of the watershed, channelization arguably facilitated the drainage of more than 220,000 acres of wetlands within the lower basin, and the gradual filtering that acreage provided was replaced with the rapid runoff of nutrients from agricultural, commercial, and residential development (Toth 1990). Similarly, lowering lake levels made it economically feasible to drain surrounding marshes, which were converted first to agricultural, and later to residential and commercial, uses—a shift that seriously impaired water quality (V. Williams 1990).

This rapid and profound ecological transformation mobilized environmentalists, who argued that the flood-control project had been a disastrous mistake that should be remedied by returning the river to its original form. Environmentalists' initial rallying cry was that massive infusions of phosphorus and nitrogen from the newly channelized river were poisoning Lake Okeechobee, the vast and symbolic headwaters for the South Florida Everglades. Spurred by widespread outrage over eutrophication of the lake, in September 1971 Governor Reuben Askew convened the Conference on Water Management in South Florida, which brought together 150 experts from around the state. At the conclusion of that meeting, attendees issued a strongly worded statement to the governor that "The Kissimmee Lakes and marshes should be restored to their historic conditions and levels to the greatest extent possible in order to improve the quality of the water entering Lake Okeechobee" (A. R. Marshall 1971/1972). The following year the Central and South Florida Flood Control District (soon to become the South Florida Water Management District) held a public hearing to air concerns about environmental damage caused by Kissimmee River channelization. At the conclusion of that hearing the SFWMD's governing board recommended devising a plan to control land and water activities in the Kissimmee River Basin and establishing an interdisciplinary team to investigate the prospects for restoration (Blake 1980).

At the urging of Art Marshall, Florida's eminent and outspoken ecologist (see chapter 5), in April 1973 Governor Askew asked the legislature to address the problems of the Kissimmee River Basin because they were endangering Lake Okeechobee. In response, the legislature appropriated

$1 million for the Special Project to Prevent the Eutrophication of Lake Okeechobee. Research undertaken as part of that special project revealed that although the Kissimmee River was discharging nutrients into Lake Okeechobee, most of those nutrients came from cattle ranches and dairy farms just north of the lake.[4] Therefore, the project's final report rejected environmentalists' insistence on restoring the river as too expensive and ecologically uncertain. It argued instead for re-creating some marshes along the lower Kissimmee and instituting improved pasture management practices (Blake 1980).

Undaunted, a unified and determined environmental community—led by Johnny Jones, president of the Florida Wildlife Federation, as well as Art Marshall, environmental writer Marjory Stoneman Douglas, and the Sierra Club's Richard Coleman and Theresa Woody—demanded that the legislature approve full restoration of the Kissimmee. They continued to insist that channelization had been a mistake, and they touted the benefits of a river moving slowly through nutrient-filtering wetlands, while also highlighting the connection between the destruction of the Kissimmee's floodplain and fish and wildlife declines. The media continued to echo claims linking Kissimmee River restoration and water quality in Lake Okeechobee, despite a dearth of scientific evidence supporting such a relationship.

In 1976, after several years of unrelenting activism by some of the state's most influential environmentalists, the Florida legislature enacted the Kissimmee River Restoration Act (chapter 76–112, F.S.). The new law created the Coordinating Council on the Restoration of the Kissimmee River and Taylor Creek–Nubbin Slough Basin, known as the Kissimmee River Coordinating Council (KRCC), made up of five state department heads. The law gave the council an ambitious, pro-environmental mandate with an emphasis on reducing the intensity of management and relying more heavily on natural system dynamics. Specifically, it asked the KRCC to oversee the development of measures that would (1) restore the natural seasonal water-level fluctuations in the lakes of the Kissimmee River and in its natural floodplains and marshes; (2) re-create conditions favorable to increases in the production of wetland vegetation, native aquatic life, and wetland wildlife; and (3) use the natural and free energies of the river system to the greatest extent possible. In April 1977 the KRCC submitted its first annual report to the Florida legislature. It identified two restoration options: a modest plan to create a series of wetlands along the canal or a more expensive but environmentally superior alternative of partially backfilling C-38. Unable to make a decision in the face of determined opposition from landowners in the Kissimmee Valley, the legislature sent the issue back to the KRCC for additional study (Blake 1980).

Recognizing the importance of federal participation in any effort that would modify a Corps project, the following year Florida's legislative delegation persuaded Congress to authorize a Corps feasibility study to determine whether any modifications of the system of flood-control works were advisable to address problems with water quality, flood control, navigation, recreation, loss of fish and wildlife resources, or the loss of environmental amenities (USACE 1985). Among the modifications Congress invited the Corps to consider was restoring all or parts of the Kissimmee River below Lake Kissimmee. The Corps, which had only recently completed the flood-control project on the river, accepted its charge reluctantly. Nevertheless, for the next several years it worked with the KRCC to evaluate restoration needs, concepts, and options (Toth 1995).

Top-Down, Landscape-Scale Planning

Although the Kissimmee River restoration planning process that ensued featured landscape-scale thinking and adaptive management, planners did not rely on stakeholder collaboration to formulate the project's goals. Rather, both the planning for and the implementation of Kissimmee River restoration have been largely top-down, expert-driven processes backed by high-level political support. In 1984 Governor Bob Graham made the restoration a top priority of his administration. Invigorated by an environmentally ambitious mandate, an interagency team of ecologists, hydrologists, and engineers began work on a plan that would take a comprehensive approach to restoring the lower Kissimmee Basin. In 1990—after the team engaged in extensive modeling and experimentation, and made a series of modifications to quell opposition—state officials formally endorsed a holistic plan with a unitary goal: to restore the river's ecological integrity. Two years later Congress authorized federal participation in the restoration.

Building Political Momentum
In early August 1983, at the persistent urging of Johnny Jones and Art Marshall, Governor Bob Graham jump-started the restoration concept by launching his Save Our Everglades campaign, which called for restoring

the Kissimmee River, Lake Okeechobee, and Everglades ecosystems (see chapter 5). In his announcement, the governor placed the emphasis firmly on rejuvenating the healthy, functioning ecosystems that had existed prior to the C&SF Flood Control Project. Shortly thereafter, state officials announced they would hold three public hearings at which the Corps would present the KRCC's two options: a "partial backfilling" plan that would involve pushing spoil from the original excavation into 26 miles of the canal's middle stretch, forcing the river back into its historic course and thereby revitalizing 30,000 acres of wetlands; and a "non-dechannelization" approach that would leave the canal in place but re-create some of the adjacent marshes.[5]

By this point, it was clear that the Corps was very unlikely to participate in the restoration because of skepticism about its net economic benefits. Furthermore, although the idea of restoring the river was broadly popular in the state, it was encountering strong resistance at the local level, and as the public hearings got under way, journalists Juanita Greene and Randy Loftis (1983) warned that the Kissimmee River was about to become "cannon fodder in one of the region's wildest political fights." As predicted, ranchers, dairy farmers, recreational boaters, developers, local chambers of commerce, and many county and municipal officials turned out and vociferously opposed both of the proffered alternatives. Some characterized restoration as an unfair infringement on their ability to use the river or its floodplain. Others said they preferred the channelized river or portrayed the restoration as a boondoggle that threatened landowners' property rights and the local economy (Greene and Loftis 1983; Loftis 1984). Many simply distrusted water management officials, who only two decades earlier had touted the virtues of channelization, to get the project right (C. Lee 2007; Toth 2006). In the upper basin, the newly formed Upper Chain of Lakes Property Owners Association, primarily concerned about their own property values, led the opposition to the state's efforts to establish the headwater lakes' ordinary high-water line.[6] The association complained that the state was relying on "scientific theory" rather than "historical documented fact" to establish the line and thereby determine how much property it needed to buy in order to raise the level of the lakes (Greene 1983a).

Local objections and Corps resistance notwithstanding, on August 19 the KRCC endorsed the more ambitious partial backfill plan based on a synthesis of more than 70 studies on various aspects of restoration (Gruson 1983b). Citing overwhelming evidence of environmental benefits and no data to suggest the project would exacerbate flooding in the upper basin, the KRCC urged the state to move forward with restoration even if the federal government declined to participate. Three months later Governor Graham issued Executive Order 83–178, creating the Kissimmee River–Lake Okeechobee–Everglades Coordinating Council, which was composed of six state agency heads and succeeded the KRCC. Graham charged the new council with overseeing Kissimmee River restoration by coordinating interagency activity. As the legislature had with its predecessor, Graham gave the council a mandate that stressed moving toward a more natural, less intensively managed system. He asked for a plan that would "avoid further destruction or degradation of these natural systems; reestablish the ecological functions of these natural systems in areas where these functions have been damaged; improve the overall management of water, fish and wildlife, and recreation; and successfully restore and preserve these unique areas" (USACE 1992, 99).

Devising an Adaptive Approach Based on Integrative Science
Under the direction of project manager Kent Loftin, the SFWMD took up the challenge of crafting an approach to restoration that would be consistent with the ambitious objectives laid out in Governor Graham's executive order while reflecting the best available knowledge of ecosystem dynamics. To this end, Loftin established a process he hoped would generate both creative thinking and useful information about how different restoration approaches might provide ecological benefits without undermining flood control. He set up an interagency study team comprising biologists, chemists, hydrologists, and ecologists, and divided tasks into three areas: ecosystem restoration, flood control and other hydraulic engineering concerns, and other issues. Team members then advocated for their point of view and challenged each other to defend often conflicting interests (Loftin, Toth, and Obeysekera 1990). This process ensured that competing views were considered, thereby anticipating possible objections; it also created an unusual degree of interdisciplinary interaction and integration. The Sierra Club's Theresa Woody (1993, 204) explains that over time, "Engineers began to think like biologists, biologists began to understand aspects of engineering, and truly extraordinary dynamics were established among this interdisciplinary team."

Over the next six years the planning team devised a series of experiments and tests that would reduce the uncertainty associated with restoration. In late July 1984 the SFWMD initiated a $1.4 million field demonstration project to ascertain whether biological resources would rebound if engineers diverted water from C-38 through remnant river channels and inundated sections of drained floodplain. Prior to the inception of the project, the SFWMD, Florida Game and Fresh Water Fish Commission, and Department of Environmental Regulation signed a multiagency memorandum of agreement assigning joint responsibility for monitoring and evaluating the effects of the demonstration project, which primarily entailed building three steel weirs across C-38 in Pool B in conjunction with manipulating water levels. Staff scientists were stunned at the salutary effects of this small-scale experiment: wetland plant communities reestablished themselves, prompting large increases in wading birds and waterfowl, as well as forage fish, game fish, and invertebrates; the bottom of the old channel, which had filled with muck, was scoured clean; and sandbars reappeared (Booth 1992; Toth 1995).

Planners learned a great deal from the demonstration project. Most important, they discovered that wetland vegetation and wildlife would readily recolonize flooded pastures, and that the river itself would respond positively to the resumption of flow (Berger 1992). The experiment also showed that restoration could be compatible with flood control. At the same time, it revealed problems with using metal weirs to dam the channel, as the river dug a new course outside one of the weirs (Mulliken 1987). More important, the improvements were only temporary; although scientists initially found a twofold increase in the springtime catch rate of game fish species in the modified area, an unexpected subsequent decline in game fish runs revealed the importance of maintaining both minimum water levels and flow in late summer (Wullschleger, Miller, and Davis 1990).

In 1986, while evaluating the results of the pilot project, the SFWMD commissioned a group at the University of California at Berkeley, led by civil engineer Hsieh Wen Shen, to model—both physically and mathematically—a variety of restoration approaches. Essential to choosing among those options was a clear and agreed-on set of ecological criteria against which to judge them. For years, however, different agencies and interest groups had proposed competing objectives for the restoration. Ecologists Lou Toth and Nick Aumen (1994) explain that the initial emphasis on water quality led to plans for reestablishing the nutrient-filtration function once provided by the river's floodplains. Over time, however, reestablishment of the floodplain wetlands gained popularity as both a generic objective and a way to revive particular kinds of wildlife, such as waterfowl. Meanwhile, for proponents of river channel restoration the main goal was to bring back game fish species.

In 1988 the SFWMD convened a symposium of 150 local and national experts in hopes of reconciling those perspectives and establishing a scientific consensus on the purpose of the restoration. At that meeting it became apparent to participants that any restoration plan needed to recognize that

Natural ecosystems, like the prechannelization Kissimmee River, have a level of organization that transcends the optimal requirements of [their] individual components, and no criteria specifying individual species requirements, whether alone or in combination, would reestablish the complex food webs, river and floodplain habitat heterogeneity, and physical, chemical, and biological processes and interactions that determined the biological attributes of the former system. (Toth 1995, 57)

Symposium participants therefore converged on the holistic notion of restoring the ecological integrity of the Kissimmee River ecosystem. Doing so, by definition, required reestablishing "an ecosystem that is capable of supporting and maintaining a balanced, integrated, adaptive community of organisms having a species composition, diversity, and functional organization comparable to that of the natural habitat of the region" (Toth 1995, 58). Participants also made a convincing case that piecemeal restoration—in which some criteria are met in some segments while other criteria are met in others—would not accomplish restoration goals and might be of little or no value. The overarching aim, the experts agreed, should be to restore both the physical form of the river and its prechannelization hydrologic characteristics.

With these objectives in mind, Shen's Berkeley team thoroughly evaluated three restoration options: building a series of earthen plugs in C-38, installing ten new underwater steel weirs, or filling in 35 miles of the canal. After running a series of experiments and simulations, the team concluded: "In terms of satisfying the ecological and erosion criteria, the best plan of the three alternatives is to completely backfill a long, continuous reach of Canal C-38" (Crook 1989). In the process of testing this approach, Shen's team was able to allay many concerns raised by skeptics. For instance, the Corps had suggested that putting dirt back in the canal would cause tons of silt to be carried downstream to Lake Okeechobee, but Shen's studies showed that sedimentation could be controlled and that soil backfill could be stabilized to resist erosion by major flood flows (Shen, Tabios, and Harder 1994). Shen's team also provided evidence to support a more complete restoration by demonstrating that leaving remnant canal sections intact could severely impair restoration by causing high flow velocities, rapid recession of floodplain water levels, and inadequate floodplain inundation (USACE 1992).

Overcoming Political Obstacles

Even as the SFWMD and its consultants were overcoming the technical barriers to restoration, political impediments repeatedly threatened to derail the project. In particular, restoration planners continued to face determined resistance from local stakeholders and the Corps, as well as the prospect that additional development in the floodplain would preclude restoration. Rather than convening stakeholders and asking them to agree on a plan that would satisfy everyone, however, the experts tried to placate the political challengers while still achieving the goal of ecological restoration.

Local Opposition   One barrier to restoration was the insistence by landowners, cattle ranchers, dairy farmers, recreational boaters, local chambers of commerce, and county officials on maintaining flood control and navigation in the Kissimmee Basin. (These interests enjoyed the backing of the Corps, which observed that significant changes to flood control or navigation, both of which were authorized purposes of the Corps's Kissimmee River Project, would require congressional approval.) Recreational boaters objected that if dechannelization caused the river to drop below three feet at any point, it would be impassable to large boats. The SFWMD responded that data collected at the boat locks showed small fishing boats were by far the heaviest users of the river, and restoration planners had already agreed to treat their navigational requirements as a constraint. Moreover, the district pointed out, under its plan the river would be navigable 24 hours a day, except during droughts; by contrast, under the existing regime, boats could not navigate the river at night because the Corps did not open its locks at night (Anderson and Dewar 1990).

Maintaining flood control presented a more serious design challenge, however, and resulted in several reductions in the project's scale. Concerned that dechannelization at the river's northern end would impair flood-control capacity for the densely populated upper Kissimmee Basin (where it would have been prohibitively expensive to acquire land), planners limited their focus to the bottom two-thirds of the river. Similarly, dechannelization of Pool E would have inhibited floodwater collection capacity at the downstream end of C-38, so it was not seriously considered. In short, from the outset flood-control concerns restricted restoration to the 758-square-mile lower basin, which is primarily agricultural, and excluded the upper basin and most of the 622-square-mile Lake Istokpoga, both of which are heavily settled (see figure 8.1). Nevertheless, when the

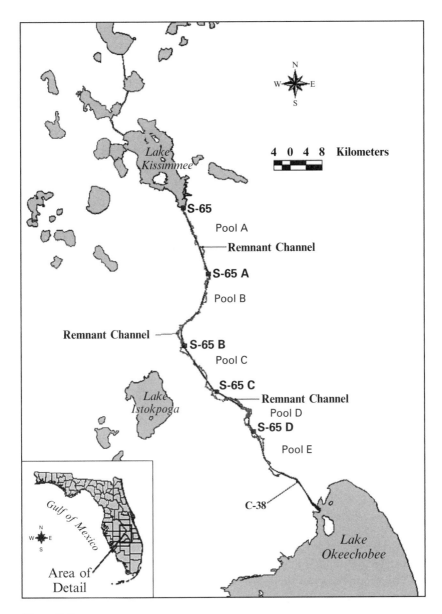

Figure 8.1
Florida's Kissimmee River

SFWMD announced its intent to conduct a demonstration project, some local residents strenuously objected. To assuage them, planners agreed to both modify the experiment and refrain from beginning the next phase of construction until it had monitored the results of the demonstration project for five years. In addition, they conducted a carefully designed series of high-discharge tests in the late 1980s to show that restoration in the lower basin would not worsen flood protection in the upper basin.

Critics raised flood-control and navigation concerns about proposals to adjust inflows to the river from headwater lakes as well. (Recreational boating had increased substantially after the flood-control project was finished, as people had built boat ramps, docks and marinas, and access channels to the lakes based on the constricted range of water level fluctuations.) Planners had ascertained they needed about 100,000 acre-feet of additional seasonal storage in the upper basin to meet flow requirements of the lower basin, if it were to be restored. Therefore they needed to establish that they could actually get this additional volume, without impeding navigation or reducing flood protection, by regulating the headwater lakes differently. Treating navigational and flood-control concerns as planning constraints, modelers simulated 21 regulation schedules and operation schemes for lakes Kissimmee, Hatchineha, and Cypress. They found that one alternative outperformed all the others in terms of minimizing the frequency of no-flow periods, re-creating the seasonal pattern of average and high flows, and increasing the frequency of lake levels higher than 50.8 feet—while also maintaining adequate levels of flood control. This alternative allowed lake stages to rise 1.5 feet above the existing schedule and had four discharge zones to facilitate continuous outflows that would vary with lake stages. Simulations suggested that if this new schedule were implemented, nearly 6,000 acres of shoreline wetlands would reestablish themselves around the three lakes, and more variable water-level fluctuations would reduce the formation of sediment berms, thereby increasing the accessibility of the shoreline habitats to fish (Toth, Arrington, and Begue 1997).

While modelers ran their simulations, a coalition of environmentalists worked assiduously to raise the salience of Kissimmee River restoration statewide and nationally, in hopes of ensuring that the determined opposition of politically powerful ranchers, boaters, and developers did not jeopardize the restoration. Led by chemist and tireless river advocate Richard Coleman, in February 1986 the Sierra Club unveiled a slide show that it planned to show to every candidate for state office. The club's goal was to train people's eyes to see that a channelized river does not "look right," the way a winding one does. To this end, the slide show contrasted the straight, civilized canal with the old, wild river and stressed that the natural system of periodic floods and wide expanses of marshland supported larger numbers and greater variety of wildlife and sent cleaner water into Lake Okeechobee. "What they [politicians] have been fed is that the Kissimmee Ditch is a beautiful river system. Is there anyone who has seen this slide show who believes that?" Coleman asked (Weiss 1986). Proponents of restoration also strove to mobilize South Florida's urban residents by linking the fate of the Kissimmee River to water quality in Lake Okeechobee and the health of the remaining Everglades. The Sierra Club's Theresa Woody told journalist John Mulliken (1988) that people in South Florida "need to realize that Kissimmee is the headwaters of the whole Everglades system, which recharges the Biscayne Aquifer—the source of the region's drinking water."

Consistently favorable media attention reinforced advocates' efforts to raise the visibility of Kissimmee River restoration. Throughout the 1980s the project garnered local, national, and even international coverage. For example, feature articles in the *Toronto Globe and Mail* (Immen 1984), *Economist* (Anon. 1989); *New York Times* (Gruson 1983a; Haitch 1987), *Wall Street Journal* (Slocum 1987), and *Christian Science Monitor* (Rheem 1987) described the restoration in detail and portrayed it as a pathbreaking effort that would yield tremendous water quality and habitat benefits by repairing one of the nation's worst environmental mistakes. Editorials supporting the restoration appeared regularly in the *St. Petersburg Times*, *Miami Herald*, *Fort Lauderdale Sun–Sentinel*, *Tampa Tribune*, and other major Florida newspapers.

Florida's political leaders both responded to and bolstered advocates' efforts to raise the salience of Kissimmee River restoration. They also moved aggressively to ensure that concerted opposition would not thwart the project. In hopes of defusing local ire and preventing development in the floodplain that could preclude full restoration, in August 1984 Governor Graham appointed a 33-member panel comprising representatives of local government; regional, state, and federal agencies; and agriculture, ranching, business, and environmental interests. He charged the newly formed Kissimmee River Resource Planning and Management Committee with reaching consensus on improved plans for land use, water quality, land acquisition, and economic development in the Kissimmee Valley. The governor indicated that if the committee failed, however, he would declare the entire valley, which included parts of five counties, an "area of critical state concern"—a move that would give the state the power to override local land-use decisions (Berger 1992).[7]

With the help of a professional mediator, the committee reached consensus (with one dissenting vote) on five broad goals: to maintain and improve water quality in the river and its tributaries; encourage economic development that does not interfere with protecting the river; assure reasonable flood control; maintain or enhance the river as a fish and wildlife habitat and aesthetically pleasing recreational waterway; and protect archaeological, historical, and distinctive cultural features of the river system (Mulliken 1985b).[8] The committee was deeply divided on the importance of restoring the Kissimmee River, however; in fact, many of its most influential members staunchly opposed the restoration. Moreover, the committee was unable to agree on specific land-use measures. In June 1986, after Graham extended its tenure, the committee did recommend a model ordinance that restricted development in the floodplain to "low-intensity agriculture," but only one of the five counties in the planning area adopted the ordinance, which did not specify what counted as low-intensity agriculture. In fact, the committee devoted most of its attention, and the bulk of its final report, to economic development and private property-rights concerns (Berger 1992; Loftin 2007; Whitfield 2007).

Despite its inability to produce substantive recommendations, the committee served two important political functions. First, it gave landowners and environmentalists a peaceful forum in which to talk, and helped defuse the explosive issue of property rights and land acquisition. Second, its inability to agree on land-use restrictions affirmed planners' suspicion that they could not rely on land-use ordinances to limit encroachment in the floodplain (Loftin 2007). The SFWMD governing board had already decided in March 1984 to begin buying land in the river's old floodplain, using money from the state's recently established Save Our Rivers Fund (Loftis 1984). But in 1985 the SFWMD formulated, and Governor Graham adopted, the Kissimmee River Strategy, known as the Seven Point Plan, which called for—among other things—expedited floodplain acquisition. Three years later the SFWMD board decided to continue acquiring land in the floodplain despite ongoing disagreements over the ordinary high-water line, figuring it would be cheaper to pay for the contested acreage than to settle the issue in court.[9]

Institutional Resistance   A second potential stumbling block was institutional resistance by the Corps, whose organizational culture was antithetical to restoration. In August 1984 the Corps's Jacksonville office released a draft of its environmental impact statement (EIS) that, as expected, did not support large-scale restoration of the Kissimmee River and encouraged a less ambitious approach (Mulliken 1985a). The majority of comments on the EIS criticized the Corps's conclusions. Nevertheless, the following year the Corps issued a final feasibility study and EIS concluding that although it would be possible to improve environmental conditions in the Kissimmee River, restoration did not pass the Corps's cost–benefit test and, in any case, was inconsistent with the agency's mission. (The Corps's analysis adhered to the principles laid out in the 1983 Economic and Environmental Principles and Guidelines for Water and Related Land Resources Implementation, which specified that the federal objective in water resource planning is to contribute to national economic development consistent with protection of the nation's environment.) Governor Graham responded by exhorting the Corps to reconsider its conclusion, describing the restoration as "a public goal of highest priority" (Mulliken 1985a). He continued to call for a change in the Corps's position, publicly backing a report by the U.S. Fish and Wildlife Service, issued in the spring of 1986, documenting severe habitat losses in the floodplain and recommending that the federal government assume a share of liability for environmental damages caused by channelization (Mulliken 1986).

In hopes of overcoming the Corps's resistance, the Sierra Club and Florida Wildlife Federation lobbied Congress to revamp the agency's mission to accommodate environmental restoration (Woody 1993). In response, Congress approved a provision in the 1986 Water Resources Development Act (section 1135) allowing the Corps to "fund design plans and construction modifications to existing…projects specifically for the purpose of improving the quality of the environment in the public interest." The act also instructed the Secretary of the Army to calculate the benefits of environmental enhancement to be at least equal to the cost of such measures. Having obtained these mandates, environmental activists then used them to lobby for authority to restore the Kissimmee River and to obtain a share of the $25 million Congress had set aside for repairs of Corps projects (Loftis 1987; Woody 1993). Although its budget subsequently included funding for the project, the Corps continued to balk, declining to spend the money until Bob Graham, who by then was a Senator, persuaded Congress to pass a bill requiring it to do so.

Settling on a Plan

Finally, in late 1989—after more than six years of experimentation, modeling, and political negotiation—the SFWMD announced that it would recommend the most ambitious of four restoration options it had evaluated. In its final report, issued in June 1990, the SFWMD elucidated its rationale: "Because stability and resilience are emergent properties of ecosystems, and not characteristics of component species populations, these features cannot be restored by simply summing or optimizing for requirements of individual species" (Loftin, Toth, and Obeysekera 1990, 13). Instead, the report said, the restoration should reestablish the prechannelization characteristics in as much of the river and floodplain as possible. To that end, the planning team proposed five criteria, as well as minimum thresholds for each: (1) continuous flow with duration and variability comparable to prechannelization records; (2) average flow velocities between 10.6 and 21.2 cubic feet per second when flows are contained within channel banks; (3) a stage–discharge relationship that results in overflow along most of the floodplain when discharges exceed 1,413 to 2,013 cubic feet per second; (4) stage hydrographs that result in floodplain inundation characteristics similar to prechannelization hydroperiods, including seasonal and long-term variability; and (5) extremely slow recession from the floodplain, at a rate that does not exceed about one foot per month. The team concluded that "complete" backfilling was the only alternative that would meet all five criteria simultaneously.[10]

The team went on to estimate the minimum area needed to reproduce the habitat diversity of the historic ecosystem and, after detailed examination of prechannelization floodplain maps, arrived at an area of about 25 square miles of river and floodplain—a result that could be achieved with about 15 miles of backfilling. Rather than adopting that minimum standard, however, the SFWMD proposed to fill as much of the canal as it could without jeopardizing flood control: between 25 and 30 miles. Its preferred alternative would—in conjunction with revitalization of the headwaters lakes—reestablish prechannelization hydrological characteristics along 52 contiguous miles of river channel and 24,000 acres of floodplain, and would restore the ecological integrity of about 35 square miles of river-floodplain ecosystem (Loftin, Toth, and Obeysekera 1990). A panel of peer reviewers, including eminent ecologist James Karr, affirmed the scientific soundness of the team's conclusions as well as the process by which they were reached.

In January 1990, Governor Bob Martinez surprised observers by endorsing the SFWMD's preferred approach—apparently convinced of the project's virtues but also hoping to get much of the credit while the federal government, which had already appropriated (but not spent) $6.3 million for Kissimmee River restoration, footed most of the bill (Silva and Dewar 1990). Advocates and state officials proceeded to engage in a full-court press to get the still-reluctant Corps on board. Although some Corps staff were interested in the restoration, senior officials resisted overturning a conclusion reached in a previous study, were uncomfortable with bowing to pressure from the state, and worried about setting a precedent that might lead to a wholesale reassessment of the Corps's flood-control mission—and hence might jeopardize its budget. Thanks to another round of intense lobbying by environmentalists and state officials, however, Congress authorized a second feasibility study by the Corps to determine whether modifications to the C&SF Project were necessary to "provide a comprehensive plan for the environmental restoration of the Kissimmee River" (WRDA 1990; PL 101–640). The mandate specified that the feasibility study should be based on implementing the level II backfilling plan specified in the SFWMD's June 1990 *Alternative Plan Evaluation and Preliminary Design Report.*

In early 1992 the Corps released its report on Kissimmee River restoration. To environmentalists' delight, the Corps had done an about-face. Its new study recommended an ambitious version of the state-supported plan: backfilling 29 miles of the canal, excavating nearly 12 miles of new river channel, and restoring flow to 56 miles of remnant river channel, thereby revitalizing nearly 29,000 acres of wetlands and, in turn, restoring 50 square miles of river-floodplain ecosystem (USACE 1992). The Corps noted it was recommending more extensive backfilling because doing so was likely to produce a more-than-proportional increase in environmental benefits. (Its analyses suggested the outputs-to-backfilling relationship would tend to increase exponentially rather than linearly.) The Corps forecast that its modified level II backfilling plan would take between 10 and 15 years to complete and cost $423 million. The Corps's position, a sharp reversal from its original stance, reflected a concerted effort by the agency to create a more environmentally friendly image (Cummins 1990). Colonel Terrence "Rock" Salt, head of the Corps's Jacksonville office from 1991 to 1994, embodied the new ethos and said his intent was to "have the Corps lead the way in protecting the environment" (Bair 1994).[11]

Since SFWMD planners had already dispensed with the most serious technical and political objections, in late 1992 Congress approved the

Kissimmee River Restoration Program, affirming its goal of "reestablish-[ing]...a river-floodplain ecosystem that is capable of supporting and maintaining a balanced, integrated, adaptive community of organisms having species composition, diversity, and functional organization comparable to the natural habitat of the region" (WRDA 1992; PL 102–580). Two years later the SFWMD and the Corps formally entered into an innovative project cooperation agreement that authorized a 50–50 cost-sharing partnership for the restoration program. The agreement assigned primary responsibility for land acquisition, restoration evaluation, and small-scale construction to the SFWMD, and required the Corps to take the lead on engineering design and major construction.

In the interim between the legislative authorization and the formal agreement with the Corps, state officials made several critical decisions about using their discretion and regulatory leverage to expedite land acquisition. In early 1993—after reexamining its policy of buying title to previously submerged land that it might, in fact, already own—the state decided it would give up ownership claims to former bottomlands, even if doing so meant buying land it already owned (AP 1993). The state also agreed to let the SFWMD seek condemnation of land by the Corps in federal court rather than requiring it to go through the more cumbersome and expensive state process. The mere threat of the federal eminent domain process, which is far less favorable to landowners, enabled the SFWMD to thwart efforts by recalcitrant landowners to extract inordinate purchase prices.

Despite its considerable leverage, the SFWMD board decided in 1992 to appease a group of disgruntled landowners by reducing the scale of the project. The Corps's feasibility study's suggestion that project-induced flooding could displace 356 homes had prompted the formation of a group calling itself Residents Opposed to Alleged Restoration (ROAR), whose goal was to block the project. Carolyn Thullberry, a retiree and the president of ROAR explained, "My function in life is to stop this backfilling. It's a waste of taxpayer money to spend $30,000 an acre...to build a swamp in Florida" (Sloan 1994). Faced with the arduous and expensive process of acquiring these properties, the SFWMD board reduced the length of backfilling from 29 to 22 miles—a compromise that reduced the area of floodplain to be restored by about 15 square miles. By protecting 85 percent of the homes threatened by flooding, and agreeing to buy out the remaining property owners, however, the board addressed most of ROAR members' concerns and defused the group's momentum.[12]

The Kissimmee River Restoration Plan

Notwithstanding the last-minute concession to landowners, as well as limitations on the project's scope from the outset, the final Kissimmee River Restoration Program is a relatively comprehensive and holistic effort to rejuvenate the lower Kissimmee Basin. The program, whose target completion date is August 2012, consists of 31 projects throughout the watershed (SFWMD 2004). Taken together, these projects aim to benefit 320 species of fish and wildlife by restoring ecological integrity to 40 square miles of river-floodplain ecosystem, including 43 contiguous miles of river channel and 27,500 acres of floodplain wetlands.[13] The program's fundamental premise is that "Reestablishment of the primary forces that once created and maintained the historic ecosystem will lead to the restoration of the complex attributes of the former ecosystem and ensure return and preservation of the river's resources and values" (Toth 1995, 58). Restoring those forces entails both re-creating the physical form of the river and floodplain and reinstating more natural inflow regimes from the headwater lakes.

To re-create the river's physical form, engineers must restore river and floodplain habitat obliterated by excavation and deposition of spoil, and remove berms and levees that obstruct the flow of water and other biological elements. In addition, to improve dissolved oxygen regimes, engineers must renew the connectivity between the river and its floodplain, and allow the river to flow for long, continuous stretches. To accomplish these ends, the plan describes acquiring nearly 67,680 acres of historic floodplain, backfilling 22 miles of C-38, recarving nine miles of river channel, and removing two dams and navigation lock structures (Toth 1995).

Reestablishing more natural inflow regimes, the intended result of the Headwaters Revitalization Project, involves both structural and nonstructural modifications to the existing flood-control project. The structural changes include maintaining and dredging a canal (C-35), widening two other canals (C-36 and C-37), and increasing discharge capacity at structure S-65 by installing two additional flood-control gates. The nonstructural work involves modifying the regulation schedule of S-65 and increasing the storage capacities of Lakes Kissimmee, Hatchineha, Cypress, and Tiger. To increase the lakes' storage capacities, the SFWMD anticipated purchasing about 20,800 acres of land around their shores and increasing maximum lake stages by about 1.5 feet (Toth 1995).

Planners made evaluation a central component of the Kissimmee River Restoration Program from the outset to ensure they could manage the restoration adaptively. In July 1991 the SFWMD appointed a seven-member scientific advisory panel to provide recommendations for a comprehensive ecological evaluation program (Koebel 1995). Based on a set of conceptual models and in consultation with the independent scientists, the SFWMD restoration evaluation team, led by senior ecologist Lou Toth, initially chose 42 indicators of project success. In response to the comments of peer reviewers, they reduced the number of indicators by combining some and dropping others for which suitable reference conditions did not exist. A final set of 25 performance measures included indicators that were expected to show reliable short- and long-term responses, were efficient to monitor, promised to provide useful information for managing the recovering and restored system, and reflected favored components of the ecosystem (Toth 2005). The team placed the highest priority on tracking the population density and reproductive success of species valued for their recreational, economic, or natural heritage attributes, such as game fish, wading birds, waterfowl, and threatened and endangered species. But they also included indicators of reestablished flow and reinundation of the floodplain, such as acres of wetlands and miles of river with improved dissolved oxygen levels. And they intended to track the productivity of the system's base components, such as aquatic invertebrates (insects, crayfish, snails, and clams), as well as nutrient concentrations and transport. In the upper basin, monitors devised a strategy to map the expansion of shoreline wetland plant communities around lakes Kissimmee, Hatchineha and Cypress to determine the impact of water-level fluctuation.

In terms of its potential for yielding environmental benefits, the Kissimmee River restoration marks a substantial departure from and improvement over the trajectory the Kissimmee Basin was on. Without restoration the flood-control system would have continued to degrade the basin's fish and wildlife. Water-level stabilization, combined with the continued absence of flow, would have perpetuated the deposition of organic matter in remnant river channels. The resulting low dissolved oxygen levels and disappearance of open-water wetland habitat would have threatened the viability of game fish species. Stable water levels would also have facilitated the buildup of plant litter, thereby accelerating the shift from a wetland to a terrestrial environment, eventually eliminating the wetlands that remained after channelization. As the wetlands disappeared, so would the remaining wading birds and waterfowl. Further development and land-use changes in the basin would have jeopardized the area's remaining natural resources (USACE 1992).

The plan is not only an improvement over the status quo, however; it also promises to deliver genuine environmental benefits. It is precautionary in that, although less ambitious than originally hoped, it aims to restore considerably more than the minimum 25 square miles necessary to restore the ecological integrity of the river-floodplain ecosystem. Moreover, although the program does not purport to eliminate human management of the river-floodplain ecosystem, it does reduce the intensity of management on the river itself and redirects management of the headwater lakes toward greater consideration of ecological integrity. And, although buffering was not built into the plan, some buffering did result from the land acquisition process because the SFWMD purchased land beyond the floodplain to simplify boundaries for landowners. Moreover, Florida's stormwater management rules limit the development potential of the land adjacent to the floodplain because much of the area has meager drainage and contains isolated wetlands (Loftin 2007).

Implementing the Kissimmee River Restoration Program

Although the plan was not the product of collaboration with stakeholders, implementation of the Kissimmee River Restoration Program has proven durable and has adhered to the adaptive approach established by experts during the planning process. In April 1994 the Corps initiated a second demonstration project—a "test fill" of C-38, in which they filled about 1,000 feet of the canal with nearly 4.5 million cubic feet of spoil—designed to resolve some technical uncertainties that remained from the 1990 pilot project. Consistent with earlier studies, the intervention produced minimal erosion. The test fill also alleviated concerns that backfilling might generate high turbidity and suspended solids, and that reestablishment of flow would flush accumulated organic deposits from remnant river channels. Furthermore, by 1997 SFWMD biologists were reporting an increasing number of wading birds and the return of nesting colonies to the restored oxbow.

With the results of the test fill in hand, in 1999 the Corps began the first phase of the restoration. Completed in February 2001, this phase included moving more than 325 million cubic feet of earth to backfill 7.5 miles of C-38. Journalist Steve Newborn (2000) described the construction scene: "Huge earth movers...crept across the bank like yellow ants, each dumping tons of soil carved from the high artificial banks [spoil deposits from when the canal was dredged] back into the channelized bed." Engineers also demolished one water control structure and boat lock, and they recarved 1.25 miles of new river channel using old aerial and satellite photographs to plot the course of the oxbows they wanted to rebuild (SFWMD 2007). These activities reestablished about 15 miles of river and restored about 11,000 acres of floodplain (P. J. Whalen et al. 2002).

Even before phase I was complete, scientists were astounded by the biological recovery they observed. According to Lou Toth, several signs indicated the river was coming back to life: wetland-loving blue-winged teal and white ibis had returned. The smartweed and Cuban bulrush that once choked the river remnants had retreated. White sandbars were forming on the river's inside bends, "a signal of a healthy river." Scientists were especially encouraged by the sight of dying wax myrtle, which had thrived in the drained flooplain but could not tolerate the newly inundated soil. Toth remarked, "It's almost: Add water, instant marsh" (Santaniello 2000). "From a bird standpoint, it's been amazing," said Stefani Melvin, an environmental scientist with the SFWMD. "I've seen five new species that I haven't seen before.... This past winter you could go out there and see 500 blue-winged teal come off the marsh. It was really amazing after being out there for years and seeing no ducks" (Newborn 2000).

By December 2002 the restored section of the river was showing even more signs of improving health. Journalist Curtis Morgan (2002) described it thus:

Two years ago, the riverscape largely amounted to a drainage canal, straight as a highway and nearly as lifeless. Now, a thin ribbon of water dark as molasses curls around glistening ponds and soggy marsh. Gators soak up sun on sandbars. Flocks of white wading birds rise at the hum of [SFMWD ecologist] Toth's helicopter overhead, then flutter down again like snowflakes.

Biologists cautioned that it would take five to seven years before they could fully assess the rebound of wildlife, but already fish were jumping, and eight types of shorebirds not seen previously were showing up in surveys, along with larger flocks of egrets and herons. Marsh plants were appearing as well, as dormant seeds responded to the infusion of water (Koebel 2007; Morgan 2002).

Although work slowed after the completion of phase I, restoration proceeded, and by August 2004, 14 of the 31 projects in the Kissimmee River Restoration Program were complete, seven were in the planning phase; six were in the design phase, and four were in the construction phase. Less than two years later, the SFWMD celebrated an important milestone: the governing board had unanimously approved the purchase of the last parcel needed to complete the final phases of the restoration. Acquisition of the last 12,000 acres brought the total to 102,061 acres.[14] The second phase of construction, which began in June 2006, involves backfilling another 1.9 miles of canal, beginning at the northern end of the phase I project area. It also entails removing three weirs and excavating a small section of new river channel to reconnect about half a mile of continuous river channel. Once this phase is complete, in October 2009 engineers will begin backfilling another 12.5 miles of canal, reconnecting more river channels, and removing one more water control structure, thereby restoring more than 14,000 additional acres of floodplain.

As construction proceeds, monitoring studies indicate that phase I work has wrought beneficial environmental changes. An evaluation by the SFWMD detected increases in dissolved oxygen in restored river channels, substantial decreases in accumulated organic sediment in the channels, more bass and other sunfishes, and increased use of the river and floodplain by shorebirds, wading birds, and waterfowl (SFWMD 2006). In 2007 the SFWMD reported the highest densities yet recorded of both ducks and long-legged wading birds on the restored floodplain, as well as dramatic increases in species richness, including a jump from two to ten native shorebird species, many of which depended on the sandbars and shallows that were destroyed by the river's channelization (Koebel 2007).[15] In terms of water quality, total phosphorus loads and concentrations have increased, but scientists expected this effect—the spoil placed back in the canal contained some naturally occurring phosphorus—and believe it is temporary (SFWMD 2006). The most important indicator of improving water quality is the reemergence of aquatic invertebrates that are the building blocks of the river-floodplain ecosystem. Those passive filter feeders, which are common in free-flowing rivers but accounted for just 1 percent of all aquatic invertebrates in the channelized system in the early 1990s, represented 30 percent by 2006, and scientists expected them to become even more abundant (Cusick 2006b).

In addition to construction, the SFWMD—with the assistance of its agency partners and stakeholders—continues to engage in evaluation, modeling, and planning activities related to the restoration. Ongoing studies aim to better identify the mechanisms that drive individual components of ecosystem response to restoration. As part of a $4.5

million analysis, scientists are planning pilot studies to evaluate vegetation response to the headwaters revitalization project. The SFWMD is also conducting the Kissimmee Basin Hydrologic Assessment, Monitoring, and Operations Study (KBMOS) to determine how much water is necessary for the river restoration and how much is available for other uses. To develop measures for evaluating the relative performance of different plans, the district is drawing on the expertise of engineers, scientists, hydrologists, hydrogeologists, and operations personnel from the Corps, FWS, EPA, Florida DEP, and others, as well as soliciting input from local government officials and stakeholders. Finally, the district is devising a long-term management plan for the Kissimmee Chain of Lakes (KCOL). During 2006 the district worked with its agency partners to develop a conceptual model of a generalized KCOL lake (G. Williams et al. 2007).

Notwithstanding progress in construction, evaluation, and planning, the full benefits of restoration are contingent on the Headwaters Revitalization Project.[16] For example, although the number of wading birds has increased, reproduction has not followed suit. Scientists suspect it may take years for prey populations to reach levels that would support breeding colonies; they also suggest that the timing of floodplain inundation and recession may not be appropriate for rookery formation until the headwaters project is complete (G. Williams et al. 2007). The headwaters project was delayed by negotiations over two fish camps on the rim of Lake Kissimmee—a total of 300 homes that could be swamped by rising water levels—as well as by modifications to the plan made in response to directives from the SFWMD governing board. Some observers worry that the SFWMD has not made sufficient provisions—in terms of acquiring land and making structural changes—to accommodate the additional water storage needed to institute new lake-level regulation. As land acquisition discussions dragged on, in June 2001 engineers instituted an interim operation schedule for S-65 in an effort to improve water deliveries to the Kissimmee until the headwaters project is complete. The interim schedule allocates water for discretionary releases in order to meet the ecosystem's need for continuous flow. It does not, however, raise the high pool stage, and so does not permit the natural flows ultimately anticipated; nor does it provide the benefits to the littoral zone that the headwaters project is supposed to create.

Beyond obstacles to completing the headwaters revitalization, several other factors threaten the restoration program's ability to deliver long-term ecological benefits. To date, the district and the Corps have used the Kissimmee River as a model of success, and that has helped ensure its continuation. It is possible (if unlikely) that funding for the project could dry up, as advocates stop lobbying for it and shift their attention elsewhere. More ominous, however, is Florida's growing demand for water. Demographers predict that in the coming decade Florida's population will grow another 21 percent; regulators predict that, between those residents and the millions of tourists who flock to the state each year, total demand for water will rise by a billion gallons, to 9.3 billion gallons a day (C. Barnett 2007). Water utilities are already eyeing the Kissimmee's headwater lakes in anticipation of meeting that demand. Related to the water management threat is the burgeoning development in the Kissimmee Basin. There are 37 major subdivisions in the planning or construction stage around the headwater lakes. But the SFWMD is confined to encouraging developers to try to minimize their impact on the water supply since, like CERP, the Kissimmee River restoration plan does not explicitly incorporate land-use decision making, and buying land outside the floodplain is beyond the project's scope.

Conclusions

Although constrained by the need to maintain flood control for landowners who have settled the original floodplain, the Kissimmee River Restoration Program nevertheless appears likely to restore the ecological vitality of a substantial portion of the lower Kissimmee Basin. In part this is because planners employed an adaptive, landscape-scale approach. Because they ignored political boundaries and focused on ecosystem form and function rather than individual elements, planners devised a holistic scheme that aims to restore the ecological integrity of a large swath of the lower Kissimmee Basin. The project's holism, in turn, compelled engineers and scientists within and across state and federal agencies to coordinate their objectives and activities. Because they have used adaptive management, in the form of multiple pilot projects, to test hypotheses about how the system would respond to different kinds of intervention, planners have been able to home in on an approach that appears capable of producing the desired results.

But the primary explanation for the environmental effectiveness of the Kissimmee River restoration is that, by contrast with the initiatives described in the first part of this book, ecological restoration is its paramount goal.

Ernie Barnett (2002), formerly of Florida's Department of Environmental Protection, explains: "We've had the best success [moving forward with restoration] where we have had really clear, defined marching orders and were able to move out, like the Kissimmee River restoration." The main reason for the program's single-mindedness is that planners did not ask stakeholders to reach consensus on its goals. Instead, a pro-environmental goal emerged as a result of conventional politics: proponents of restoration built sufficient political will that state officials felt compelled to support a large-scale restoration. When necessary, they exerted regulatory leverage to ensure that land acquisition would proceed expeditiously. Because of the program's clear purpose, experts were able to focus on achieving that aim, and they exhibited enormous ingenuity in doing so. In turn, media coverage of the positive impacts of well-designed pilot projects helped proponents continue to build support for the restoration, energized those who were responsible for implementing it, and converted many former skeptics.

Interestingly, despite its origins in adversarial politics, the Kissimmee River restoration exhibits the three features that are typically associated with stakeholder collaboration. Its experimental, adaptive approach is so innovative that it has become the model for river restoration. The project has also gotten accolades for its use of the best available science: it has been designed, implemented, and overseen by highly qualified experts. And finally, the restoration has proven durable. Nearly ten years into implementation, work continues on reestablishing the sinuous, slow-moving river that once delighted fishermen and boaters and provided habitat for central Florida's waterfowl and wading birds.

9

Making History in the Mono Basin

Just as the Kissimmee River Restoration Program provides a useful contrast with the Comprehensive Everglades Restoration, a comparison between the Mono Basin restoration and the California Bay-Delta Program is also enlightening. A critical turning point for the Mono Basin came in 1994, when California's State Water Resources Control Board (SWRCB) issued a decision that required the city of Los Angeles to restore the health of the remote Mono Basin. The city had tapped the basin's streams as a source of freshwater for 50 years—with devastating ecological consequences. But Water Rights Decision 1631 required the L.A. Department of Water & Power (DWP) to reduce its water withdrawals drastically, until Mono Lake reaches 6,392 feet above sea level—25 feet below its prediversion level of 6,417 feet but well above its historic low of 6,372 feet, hit in 1982. The board also made the DWP's licenses to divert Mono Basin water conditional on restoration of four of the five streams that feed the lake. The results of the SWRCB's orders have been salutary: despite a series of dry years in the early 2000s, by August 2007 Mono Lake had risen nearly ten feet—to 6,384 feet—and its native habitats and wildlife were returning; moreover, thanks to increased flows and streamside restoration work conducted by the DWP, the basin's creeks were recovering from decades of abuse.

The environmental benefits of the Mono Basin restoration are partly attributable to its landscape-scale focus. That focus emerged over time as demands for restoration of the lake and its feeder streams spawned integrative scientific assessments, which in turn laid the foundation for a comprehensive approach to ecological restoration. The effort to address the basin's problems at a landscape scale also brought together the state and federal agencies with interests in the basin, all of which ultimately backed an environmentally protective approach. The restoration's adaptive elements have enhanced its effectiveness as well. Although there are no explicit provisions for adjusting the lake-level target set by the state, the SWRCB did reserve the right to reduce the DWP's diversions in light of new evidence. More important, the stream restoration plan requires the DWP to modify its tactics based on the results of monitoring. In response to this mandate, the agency has undertaken several experimental projects aimed at increasing stream flow, with the caveat that it will change course if its preferred methods fail to attain agreed-on restoration targets.

The primary explanation for the environmental effectiveness of the Mono Basin restoration, however, lies in its unitary goal of restoring ecological integrity—and in the politics that yielded that result. Over more than a decade, advocates engaged in a highly effective campaign to raise public awareness of and concern about Mono Lake's deterioration. They also filed a series of lawsuits, the impact of which was to sharply curtail the city's ability to divert water from the Mono Basin. The combination of legal decisions and growing public support shifted the balance of power dramatically in favor of restoration proponents and propelled their definition of the problem—that Los Angeles's diversions had seriously harmed the Mono Basin's ecology—to prominence. As a result, the policy question became not whether but how much to reduce the DWP's diversions. When the SWRCB finally resolved that question in an environmentally protective way, the DWP acquiesced—in part because it had already been working with environmentalists on finding ways to reduce the city's dependence on Mono Basin water.

The Impetus for Restoration

For at least three-quarters of a million years, Mono Lake has been a feature of the arid, mountainous landscape of northeastern California. Set in a closed basin east of the Sierra Nevada crest, the lake's shore is stark and windswept; its most famous features are the limestone tufa towers that rise from the lakebed.[1] Historically, water entered Mono Lake by way of three major creeks—Mill, Lee Vining, and Rush—and, because the lake has no outlet, left only by evaporation. Over time, the lake became naturally saline—about one-half again as salty as the sea—as well as alkaline and sulfurous, thanks to its volcanic surroundings. It supported abundant life, especially in the spring and summer, when algae and other microscopic organisms proliferated. The specially adapted brine shrimp and alkali fly that feed on the algae in turn provided sustenance for millions of migrating birds, including one of the largest known nesting colonies of California gulls. The lake's larger tributary creeks supported lush bottomlands that featured cottonwood forests, stands of ponderosa pine, multiple meandering stream channels, backwater ponds, and wet meadows. Waterfowl and migratory songbirds thrived in these bottomlands, as well as in lake-fringing springs, wetlands, and protected lagoons. (Canaday 2007; Cutting 2006; Hart 1996)

Although white settlers began colonizing and creating extensive sheep-grazing pasture in the area around Mono Lake during the late 1800s, the most serious threat to the Mono Basin ecosystem arose in 1912. At that time, the city of Los Angeles began acquiring property and water rights in the Mono Basin in hopes of extending the Los Angeles Aqueduct north from the Owens Valley. The project languished for years, but in 1930 the DWP finally convinced L.A. voters to pass a $38.8 million bond issue that would enable it to finish its land purchases in the Mono Basin and begin construction.[2] In 1934 the city obtained permits to divert up to 200 cubic feet per second (cfs) from four of the basin's five streams and to store up to 94,000 acre-feet in city-owned reservoirs (Bass 1979). Completed in 1941, the waterworks consist of a diversion dam on Lee Vining Creek; a buried conduit that transports water from the two Rush Creek tributaries, Parker and Walker creeks, to the Grant Lake reservoir on Rush Creek, eight miles above Mono Lake; and a stretch of pipe that carries water south from Grant Lake, burrowing under the Mono Craters for another 11.5 miles to the East Portal on the upper Owens River (see figure 9.1) (Hart 1996).

For a decade after L.A.'s diversions began, the lake dropped about a foot each year but remained within its historic range of fluctuation. In the mid-1950s, however, the lake's surface fell below 6,405 feet, triggering a cascade of ecological consequences. As the lake level continued to decline, wetlands around the shoreline dried up and became inhospitable to waterfowl; as the lake became more saline, brine shrimp became smaller, and alkali flies became smaller and fewer in number. Starved of water, tributary streams shrank or disappeared altogether, and their surrounding vegetation withered.

In 1965, toxic dust storms arose, another side effect of the lake's dwindling size: the receding shoreline left alkali flats that dried up and blew away in strong winds. Then, in the winter of 1966–1967, a torrential

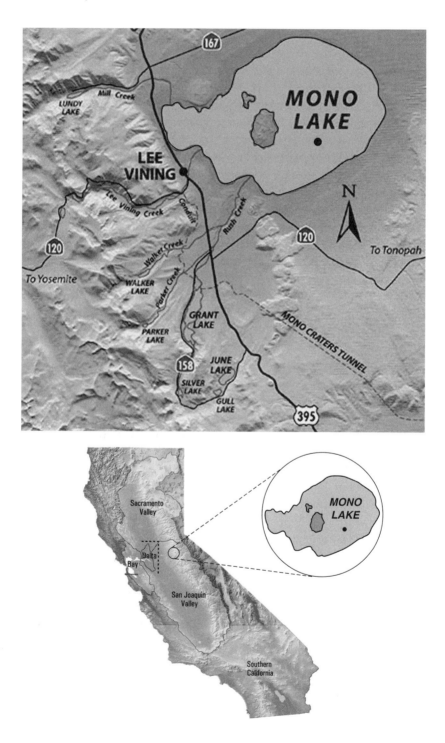

snowmelt barreled down Rush Creek, drastically altering its topography and leaving former side channels stranded above the newly incised main channel. A 1969 snowmelt had similarly devastating effects on Lee Vining's creekbed. The combination of dropping lake levels and stream incision drained the delta lagoons that had provided important open-water habitat for waterfowl. Low lake levels also led to the formation of a land bridge that enabled coyotes to cross from the mainland to Negit Island and decimate the gulls' breeding grounds. The trout fishery on Rush Creek vanished, along with 90 percent of the ducks and geese that used to frequent the basin.[3] (Cutting 2006; SWRCB 1994a)

Magnifying the injuries to the basin, in 1970 L.A. completed a second aqueduct parallel to the original one and doubled its average diversions to nearly 100,000 acre-feet per year.[4] As a consequence, the lake began shrinking even more rapidly, sinking by as much as two feet per year and becoming increasingly saline. In 1982 Mono Lake stood at a historic low of 6,372 feet, 45 feet below its prediversion level. Its volume had decreased by 50 percent, its surface area had shrunk by more than 25 percent, and it had become three times as salty as the ocean. Because the stream banks had lost their stabilizing vegetation, floods during the wet years from 1982 to 1984 caused extensive erosion. Subsequently, diminished flows and widened channels prevented overbank flooding, which further reduced the extent and vigor of riparian vegetation and wetlands.

Although locals protested L.A.'s increased diversions, few people believed anything could be done to stop the DWP juggernaut (Steinhart 1980). Strengthening the city's claim to Mono Basin water, at the beginning of 1974 the SWRCB had quietly converted what had been interim permits issued to the DWP into permanent licenses. Around the same time, however, the seeds of a profound challenge to L.A.'s water rights were sown. After doing some cursory research on Mono Lake and its environs, ecologist David Gaines became enamored of the area. He encouraged some colleagues to get a small research grant that would enable a group of undergraduates to initiate the first detailed survey of the lake's brine shrimp, flies, and, most important, birds. In the process of learning about the ecology of the basin, the students—who called themselves the Mono Basin Research Group—became passionately committed to ensuring its survival. In 1978, having failed to get an existing environmental

Figure 9.1
Mono Basin Watershed and Waterworks

group to take up the cause, Gaines, his future wife Sally Judy, ornithologist David Winkler, and several friends founded a new advocacy organization: the Mono Lake Committee. Independently, but during the same period, Tim Such, a student at the University of California, Berkeley, had also become interested in Mono Lake as a result of research he had done in college. While Gaines, Winkler, and others undertook a public relations campaign, Such began pursuing a legal theory on which to base a challenge to L.A.'s Mono Basin diversions (Hart 1996).

Adversarial Politics Yields Protective Goals

Inchoate challenges to the DWP's dominance crystallized in the late 1970s, when newly mobilized environmentalists, soon to be joined by fishing advocates, launched a two-pronged campaign of lawsuits and public-awareness events. Their efforts spawned several integrative scientific assessments, which in turn generated a more holistic understanding of the Mono Basin ecosystem and a more coherent and environmentally protective approach to it among the agencies with interests in the region. Advocacy and litigation also brought about a steady shift in the balance of power in favor of environmentalists, accompanied by a rise to preeminence of the view that L.A.'s diversions had wrought serious damage in the Mono Basin ecosystem. Within the context of this shifting balance of power, the DWP and Mono Lake Committee engaged in a long-running negotiation aimed at identifying ways to replace the water L.A. stood to lose if environmentalists prevailed.

Building the Political Will for Restoration
Appealing for public support in hopes of building the political will to prevent the lake's demise was one of the main tactics employed by the Mono Lake Committee and its allies. Their objective was to retain at least 70 percent of the approximately 100,000 acre-feet taken from the basin each year (Boyarsky 1986a). To this end, Mono Lake Committee spokespeople warned of the dire impacts of continued diversions. "The dispute is not whether diversions are going to kill the lake, but at what point they're going to kill the lake," said David Gaines (Lawrence 1985). Activists devised a multifaceted campaign to create attachment to the lake and dramatize its plight: they organized the annual "bucket walk," in which they manually transferred water from the diverted streams to Mono Lake; sponsored bicycle trips carrying water from DWP headquar-

ters in L.A. to the lake; and presented slide shows throughout the state. The Mono Lake Committee took care to insist it was a friend, not a foe, of Los Angeles. To reinforce this point, it set up an office in the city and began hosting programs aimed at urban residents, including field trips to the basin. As important, committee leaders emphasized their willingness to help L.A. find ways to replace lost Mono Basin water. They insisted from the outset, however, that they would agree only to a solution that did not involve tapping another environmentally sensitive source, such as northern California's Sacramento–San Joaquin Delta. They argued that Los Angeles could make up for the lost water by stepping up its conservation and recycling.

Almost immediately after it was launched, the campaign to save Mono Lake attracted high-level political attention. In December 1978 Huey Johnson, Secretary of Resources under Governor Jerry Brown, convened an Interagency Task Force on Mono Lake and charged it with developing a plan "to preserve and protect the natural resources in Mono Basin, considering economic and social factors" (Interagency Task Force 1979, 1). Task force members came from the DWP; the state Water Resources and Fish and Game departments; the U.S. Forest Service, Bureau of Land Management (BLM), and Fish and Wildlife Service; and Mono County. (Johnson did not invite private stakeholders to participate in the discussion, but the Mono Lake Committee nevertheless submitted information and encouraged the public to write letters.)

From the outset, most task force members accepted the Mono Lake Committee's view that L.A.'s diversions were harming the lake, and they focused on finding replacement water for the city. In May 1979 the task force held public hearings to vet the alternatives it was considering, after which it chose an option that, according to its calculations, would have raised the lake to 6,388 feet by cutting L.A.'s exports to 15,000 acre-feet a year. The plan echoed the Mono Lake Committee's assertion that the city could make up for its lost water through conservation and recycling. Although a dozen of the state's newspapers urged it to abide by the Task Force's recommendations, the DWP rejected the majority's approach as "simplistic" and managed to prevent passage of a bill that would have codified it (Hart 1996; Steinhart 1980).

The DWP's repudiation of the task force plan was consistent with its position that there was no definitive scientific evidence of environmental problems at Mono Lake—a stance it adhered to throughout the 1980s. For instance, in a letter to the *Los Angeles Times* on April 17, 1986, Paul Lane, general manager and chief engineer of the DWP, asserted: "Mono Lake's ecosystem continues to be healthy and productive, which was the case even in 1981, when Mono Lake's water level reached its lowest level in recent history." LeVal Lund, chief of aqueducts for the DWP, acknowledged that the lake would continue to drop, but said it would stabilize in 80 to 100 years at about 40 square miles (in the mid-1980s it covered about 60 square miles, down from 85 square miles in 1940) and disputed claims that the environment would be spoiled as a result. "Our philosophy," he said, "is that there is still going to be a unique and scenic location" at Mono Lake. "It certainly will be much smaller, but you have to look at it as balancing. There are people on Earth and people need water" (Lawrence 1985).

The department also tried to shift attention to the trade-offs facing the people of Los Angeles, arguing that reducing its Mono Basin diversions would cause water shortages and that replacing lost water and power would be exorbitantly expensive. Duane Georgeson, assistant general manager of the DWP and the principal spokesman for the department during the 1980s, estimated that Mono Basin water accounted for a hefty 17 percent of L.A.'s water supply and claimed that giving up 70,000 acre-feet would increase ratepayers' bills by about $50 per year, costing the city $32 million annually. "Our customers by and large think they are paying too much for their water and electricity right now," he pointed out (Boyarsky 1986a). Furthermore, he observed, each acre-foot of Mono Basin water generated electricity equivalent to five barrels of oil, and it would cost five barrels of oil worth of electricity to pump replacement water from the north. "That is a pretty poor trade-off these days," said Georgeson. "I believe the security of an affordable, high-quality water supply to the people of Los Angeles should be as important as the fact that a picturesque saline lake in the high desert is declining" (G. Young 1981, 514).

Representatives of the DWP were unabashed in their rhetoric in part because the agency's power seemed immutable. According to historian Norris Hundley, Jr. (2001, 340), it rested on "a powerful array of interlocking interest groups that went far beyond the city." Among the DWP's allies were the immense Metropolitan Water District of Southern California (MWD), which did not want to see the city increase its share of that agency's water, and San Joaquin Valley farmers, who feared that any increases to L.A. would come at their expense. In the early 1980s, however, a series of court decisions catalyzed a discernible shift in political alliances: over time the DWP became more isolated, and state and federal officials began lining up behind the pro–Mono Lake position.

Hastening that shift was the abundant, favorable attention to Mono Lake and its defenders in both the local and national media. In the early 1980s the lake was on the cover of *Life* and featured prominently in *National Geographic, Harper's, Smithsonian, Audubon,* and *Sports Illustrated.* Echoing the Mono Lake Committee's storyline, the media—both in California and nationally—depicted the lake as teeming with life and emphasized what would be lost if it were not protected. Many journalists took a more extreme position, however, enabling the Mono Lake Committee to maintain its more moderate stance. For example, Page Stegner (1981, 69) wrote in *Harper's*: "To allow the life-sustaining element of an environment as biologically, ecologically, geologically, and aesthetically rich as this to flow into the Pacific Ocean through a sewer pipe is an act of criminal negligence on a national scale." Some in the media portrayed Los Angeles as a profligate villain and depicted the struggle between environmentalists and the DWP as a David and Goliath battle. Even the *Los Angeles Times* deplored the DWP's intransigence and observed in a July 26, 1989, editorial:

The [DWP's] entire institutional history and singular purpose is to get the water. It will yield to environmental constraints only when forced to do so by the law and not just because it is the right thing to do. By then, of course, the damage may be irreversible. The department will not give proper consideration to the moral and ethical issues involved unless forced to by a higher authority: the Board of Water and Power Commissioners, the City Council or the mayor.

Litigation Favors Restoration Proponents

In addition to the salience campaign, proponents of Mono Basin restoration relied on litigation as a means of gaining leverage over the DWP, and their successes in court steadily bolstered their position—both legally and among the public—while eroding that of the intransigent agency. In May 1979 the National Audubon Society, the Mono Lake Committee, and Friends of the Earth sued the DWP, claiming its diversions had increased the salinity of Mono Lake to the point that the survival of the brine shrimp, as well as the gulls that depend on them, were in jeopardy. Building on the legal investigations of Tim Such, environmentalists' lawyers devised an innovative argument that the state should intervene under the common-law doctrine of the "public trust," according to which the state is obligated to protect the public's interest in submerged lands (*National Audubon Society v. Los Angeles*). To the astonishment of most observers, on February 17, 1983, the California Supreme Court issued a landmark ruling in support of environmentalists' claim: in a 6–1 decision, the court declared that the public trust existed in Mono Lake's scenery, ecology, and human uses; that it had not been properly considered in the past; that it *should* be taken into account; and that Los Angeles's water rights were subject to revision (*National Audubon Society v. Superior Court*, 33 Cal.3d 419).[5] The justices did not prescribe a particular remedy; instead, they turned the matter back to state agencies and lower courts for settlement. Nevertheless, the decision established the principle that the public trust, although not paramount, had to be preserved "so far as feasible" in all water allocations.[6] In doing so, it gave legitimacy to environmentalists' definition of the problem and set the stage for a reallocation of water.

In addition to empowering environmentalists, the 1983 public trust decision changed the dynamics of a congressional debate over efforts to protect the land surrounding Mono Lake. In 1984, after several years of inaction on the issue, Congress established the 118,000-acre Mono Basin National Forest Scenic Area within the Inyo National Forest—a move that raised the national profile of Mono Lake and enhanced the federal interest in preserving the basin's environment.[7] In December 1986, the U.S. Ninth Circuit Court of Appeals upheld a finding by Superior Court Judge Lawrence Karlton that the federal government, not the state, owned 12,000 acres of land around Mono Lake and adjacent to federal property that had been uncovered as the lake level had fallen—a ruling that put the federal government in an even more advantageous position vis-à-vis Los Angeles (Anon. 1986).

Although crucial to enhancing environmentalists' clout, the *Audubon* suit was just the first of several legal victories for restoration proponents. In mid-November 1984, two sportfishing advocates—Dick Dahlgren and Barrett McInerney—filed a second lawsuit (*Dahlgren v. Los Angeles*) claiming that the SWRCB's predecessor, the Division of Water Resources, had issued diversion licenses to L.A. in violation of a 1937 provision of the Fish and Game Code (section 5437) that prohibits the dewatering of creeks below dams.[8] (Fly fisherman Dick Dahlgren was moved to file the suit after observing brown trout that apparently had been flushed over a spillway from Grant Lake reservoir and had reestablished themselves in Rush Creek after a series of wet winters in the early 1980s.) On November 20 Judge David Otis of Siskiyou County issued a temporary restraining order, under which the DWP was required to release 19 cfs into Rush Creek until the DWP and California Department of Fish and Game completed a joint study to determine how much water had to flow through the creek to sustain the fishery (Taylor 1986).

Then, in August 1986, after brown trout reappeared in Lee Vining Creek, the Mono Lake Committee filed suit to prevent the DWP from shutting off those flows as well, citing both the Fish and Game Code and the public trust doctrine. On August 12, 1986, Mono County Superior Court Judge Edward Denton granted a temporary restraining order requiring the department to release 10 cfs down Lee Vining Creek until the case was tried. For the DWP, the implications of these two lawsuits were alarming; if it eventually lost both the Rush Creek and Lee Vining cases, it would have to relinquish 21,000 acre-feet of water, about one-fifth of its Mono Basin supply (Taylor 1986). Also worrisome for the DWP was an even more substantial legal challenge: Audubon, the Mono Lake Committee, and California Trout had come up with yet another claim, this one based on Fish and Game Code section 5946, which mandates that state regulators include in any diversion license a requirement to release water in accordance with section 5937. Their lawsuit, *California Trout v. State Water Resources Control Board*, asked the Third District Court of Appeals to rule the permanent water diversion licenses granted to L.A. in 1974 illegal and void. In July 1986 Judge Lloyd Phillips of the Sacramento County Superior Court upheld the DWP's permits, but the plaintiffs appealed the ruling.

In any case, by this time the combination of largely favorable rulings and a highly effective publicity campaign had begun to pay off. The Forest Service estimated that the magazine coverage drew more than 250,000 people to the area in 1988 (Roderick 1989). By the fall of 1989 the Mono Lake Committee had an estimated 18,000 members and a $700,000 budget. "Save Mono Lake" bumper stickers were a common sight in both northern and southern California—and had been spotted as far away as Sweden. Even the DWP's Duane Georgeson acknowledged, "They've done a pretty good job of mobilizing public opinion" (Roderick 1989). The elected leadership of Los Angeles was articulating a position consistent with that of the media and, apparently, public opinion. For example, in his 1986 gubernatorial campaign, Los Angeles Mayor Bradley declared: "I believe Los Angeles is ready to do its part to preserve Mono Lake. It appears this goal may best be addressed by increasing and regulating flows into Mono Lake so that the lake can be stabilized in a healthy environmental state" (Boyarsky 1986b). Other public officials were expressing newfound interest in saving the lake as well; for example, by 1986 City Councilman Zev Yaroslavsky—who had once supported the DWP's assertion that the lake's gulls and shrimp would adapt to lower lake levels—was advocating preservation (Hart 1996).

Scientific Assessments Bolster Environmentalists' Claims

A series of integrative scientific assessments completed in the late 1980s further bolstered environmentalists' definition of the problem and narrowed the range of plausible solutions. In August 1987 the National Research Council (NRC) of the National Academy of Sciences released *The Mono Basin Ecosystem: Effects of a Changing Lake Level*, a study of the Mono Lake ecosystem that had been ordered by the U.S. Congress three years earlier. According to this authoritative synthesis, if Los Angeles continued taking its maximum diversions, it would destroy the Mono Basin ecosystem. The NRC report forecast that alkali flies and the birds that feed on them would begin to suffer as the lake surface sank below 6,370 feet, and at 6,360 feet the effects would become acute. For brine shrimp and their predators, the impacts would set in at about 6,350 feet, at which point there would also be no protected nesting area for gulls. Although the report did not recommend a particular lake level, it implied the surface should be kept above 6,370 feet and emphasized that, whatever level was chosen, a ten-foot buffer should be added against drought—suggesting a minimum level of 6,380 feet (NRC 1987).

Both sides seized on the NRC study as supporting their positions: Martha Davis, executive director of the Mono Lake Committee, said, "The report confirms what we have been saying—there is a problem with the diversions, and if [they] continue we will lose the ecosystem"; at the same time, the DWP's Duane Georgeson pointed out that the report found "no threat to the ecosystem today" (Boyarsky 1987). A second report made the DWP's position less tenable, however. In April 1988 the Community and Organization Research Institute (CORI) at the University of California, Santa Barbara, released a report, *The Future of Mono Lake*, that had been requested by the California legislature in 1984. Unlike the NRC panel, CORI had underwritten some new research on the basin, as a result of which it arrived at somewhat less optimistic conclusions: it forecast decreases in algae, alkali flies, and brine shrimp below 6,375 feet and death by 6,352 feet, which could be reached as soon as 2012. Relying on the work of geomorphologist Scott Stine, the CORI report proffered a new rationale for stemming the lake's decline. Stine had discovered a submerged terrace around the lake that had been cut at the lake's record low of 6,368 feet, reached at about 150 A.D. He argued that if the lake fell below this "nick point," its feeder streams would "begin a new cycle of downcutting, further damaging stream habitats and making a sort of badlands of the shore" (Hart 1996, 124). If the lake's level subsequently rose, the rebound would devastate many tufa groves. To provide a buffer sufficient to ensure the lake never reached the nick point, CORI recommended keeping the surface at 6,382 feet or higher in normal runoff years, which would entail reducing L.A.'s diversions by about 42 percent (Botkin et al. 1988).

Shortly thereafter, a third technical analysis made a case for even more protective measures, rendering the DWP's position even less defensible. In September 1988 the Forest Service released its draft scenic area management plan, which advocated maintaining the lake surface at between 6,377 feet and 6,390 feet. To attain this objective, the agency said, the city needed to reduce its diversions from the Mono Basin by 75 percent. At a series of public hearings the DWP vociferously objected to this conclusion, again citing "a very high cost in terms of replacement water…to the city of Los Angeles" (Stewart 1988). Nevertheless, supporting a lake-level goal of 6,390 feet became the official Forest Service position.

More Legal Setbacks for Los Angeles

Adding to the weight of the scientific reports in the late 1980s was a series of judicial rulings that affirmed environmentalists' definition of the problem and even more tightly circumscribed the range of possible solutions. On October 20, 1987, the Third District Court of Appeals heard arguments in *California Trout v. State Water Resources Control Board*, and on May 23, 1988, it reversed the 1986 Superior Court decision and ruled the licenses held by the DWP for diversion of Mono Basin water were illegal. The DWP asked for a rehearing, and in late January 1989 the appeals court issued a slightly revised ruling: the justices observed that, given the 30 years between construction of the first and second aqueducts, the second one should be considered an entirely new project and therefore subject to the fish-flow requirements in Fish and Game Code section 5946. Rather than simply extinguishing L.A.'s water rights, however, the court ordered the SWRCB to modify the city's licenses to accommodate the necessary stream flows. This decision, which came to be known as *CalTrout 1*, weakened the DWP's position substantially because it made clear that the city was not legally entitled to at least some of the water it had been diverting.

Shortly after the *CalTrout* case concluded, El Dorado County Superior Court Judge Terrence Finney began reviewing three closely related cases concerning the public trust in Mono Basin—*Audubon v. Los Angeles*, *Dahlgren v. Los Angeles*, and *Mono Lake Committee v. Los Angeles*— that had been consolidated into one: the *Mono Lake Water Rights Case* (or Coordinated Proceedings). One of the environmentalists' lawyers, Bruce Dodge, immediately asked Judge Finney to halt exports of water out of the Mono Basin until the lake had risen above the minimally protective level of 6,377 feet, and on June 15, 1989, the judge obliged by issuing a temporary restraining order. Environmentalists then requested a preliminary injunction, in response to which the DWP proposed instead continuing diversions and observing the impact on the lake. Judge Finney rejected the DWP's proposal and prohibited the agency from causing the lake to fall below 6,377 feet for the remainder of the runoff year ending in March 1990 (Ellis 1989). The judge did agree, however, to turn the final water rights allocation over to the SWRCB, where the DWP anticipated exerting more influence than it had in court.

Meanwhile, although litigation had enabled environmentalists to secure more water for Mono Lake and its streams, natural changes threatened the basin's health, lending urgency to demands for a more comprehensive restoration. In a series of wet winters during the early 1980s the DWP— with storage at capacity—had been forced to release water into Mono Lake. As a result, by 1986 the lake surface had risen to 6,381 feet. That year, however, the state entered what turned out to be a six-year drought (see chapter 6), and by the spring of 1989 the lake stood at 6,375 feet— almost exactly where it had been when the restoration battle began. Even after the DWP stopped exporting water from the Mono Basin on June 15, 1989, in response to the court order, the lake surface continued to fall.

With lake levels dropping and the SWRCB dragging its feet on modifying L.A.'s licenses to require higher stream flows, as required by the *CalTrout 1* ruling, Audubon and CalTrout went back to the Third District Court of Appeals in hopes of expediting a remedy. On February 23, 1990, the appellate court asked Judge Finney to oversee the SWRCB's implementation of *CalTrout 1*. In addition, in a ruling known as *CalTrout 2*, the court directed the SWRCB to add to L.A.'s revised diversion licenses a clause requiring the DWP to "release sufficient water into the streams from its dams to reestablish and maintain the fisheries which existed in them prior to its diversion of water." Given the beating the streams had taken, this provision constituted a mandate for ecological restoration, not just additional water—another coup for environmentalists. In the spring of 1990 Judge Finney held a series of hearings aimed at determining how that restoration should proceed. Throughout those hearings the DWP tried to minimize its responsibility by arguing that no one really knew the historic conditions of the Mono Basin streams. But environmentalists' lawyers—drawing on records kept by retired Fish and Game biologist Elden Vestal, who for more than a decade (1939–1950) had kept voluminous notes on the state of the streams and their trout fisheries—succeeded in creating a convincing portrait of the prediversion streams.

In June 1990 Judge Finney took the first step toward restoration by issuing a ruling that specified a minimum flow schedule that would double the existing flow in Rush Creek and increase Lee Vining Creek's flow sevenfold. Finney's schedule, which would stand until the Department of Fish and Game completed the flow studies ordered by Judge Otis in 1986, guaranteed about 60,000 acre-feet for the lake—enough to maintain its surface at between 6,368 feet and 6,375 feet. Finney also ordered the parties to the lawsuit to sit down together and begin planning for stream restoration.

While that process got under way, Judge Finney tackled the separate but related issues of stream diversions, lake levels, and the public trust. In advance of his summer 1990 hearings on those issues, both the Forest Service and the California State Lands Commission filed legal briefs on the side of the Mono Lake Committee. At the same time the DWP—unconvinced by the scientific assessments to date—devised a strategy to counter environmentalists' efforts to extend the existing court order by once again presenting testimony from its own biologists, who would argue that the existing lake level did not pose any threat to the ecosystem. According to DWP engineer Dennis Williams, "Our biologists have assured us that the lake ecosystem is very healthy at current levels, that migratory birds have an abundant food supply available, the brine shrimp are in the lake in large numbers and there are numerous islets for the gulls to nest on" (Ellis 1990).

The hearings that ensued featured weeks of testimony by experts who engaged in highly technical debates over every conceivable aspect of Mono Basin ecology. Apparently undaunted by the welter of contradictory claims, on April 17, 1991, Judge Finney issued another strongly pro-environmental decision: for the third time in three years, the judge affirmed the injunction on the DWP's diversions and reiterated that lake levels had to be allowed to rise to 6,377 feet to ensure protection for gulls before L.A. could be allowed to take any water. In explaining his ruling, Finney emphasized the need for precaution in the face of uncertainty: "After hearing all the experts' testimony concerning the California gulls at Mono Lake," Finney wrote, "there is simply too much we don't understand about the California gull and its nesting habits to allow us to roll the dice by land-bridging the islets and hoping for the best" (Ellis 1991a).

Having resolved the public trust-related cases, Judge Finney turned his attention back to the stream restoration. In compliance with his order, the DWP, the Mono Lake Committee, Audubon, CalTrout, and the Department of Fish and Game had set up a Restoration Technical Committee (RTC), comprising one member from each entity, to design the stream restoration. The RTC had agreed it would take only those actions on which there was unanimous agreement, but that requirement soon became problematic. Restoration of the minimally impaired Parker and Walker creeks had been relatively straightforward. But in 1992, when the RTC turned its attention to the badly damaged Lee Vining Creek, the DWP balked at the cost and proposed that recovery ought to occur naturally, not as a result of human intervention (Hart 1996).[9] Judge Finney was not persuaded, however, and he ordered work on the Lee Vining Creek to proceed. To avoid similar delays on the devastated Lower Rush Creek, Finney expanded the RTC to include three unaffiliated scientists and terminated the unanimity requirement.

Efforts to Collaborate

Even as they engaged in a highly charged series of court battles, the Mono Lake Committee and the DWP were meeting outside the courtroom and away from the media in a collaborative process that, although it did not settle the controversy, allowed the parties to brainstorm potential solutions. Spurred by the momentous 1983 public trust decision, the collaborative process began in March 1984 with a daylong conference in which participants laid out their positions. From there the Mono Lake Group—consisting of representatives of the Mono Lake Committee, the DWP, the L.A. Water and Power Commission, and the mayor's office—began a conversation aimed at ending the impasse on Mono Lake.[10] The discussion quickly foundered on the question of a desirable lake level, however, so the parties turned to the more tractable question of how Los Angeles might replace lost Mono Basin water if it became necessary to do so.

For several years the Mono Lake Group met regularly but made little headway. In December 1987, however, as legal rulings increased the pressure on the DWP, the group agreed to sponsor a study of alternatives to diverting water from the Mono Basin. Within six months, Tom Graff of the Environmental Defense Fund had located two irrigation districts in the San Joaquin Valley that were interested in an exchange. If the city helped them finance best management practices, the districts would sell the city their excess water. Although the deal was attractive, it had several political drawbacks.[11] The most immediate difficulty was financial, however, and the group turned its attention to finding state and federal money to subsidize replacement water for L.A.

In 1989, after extensive discussions with the Mono Lake Group, Assemblyman Phillip Isenberg of Sacramento agreed to sponsor a financial measure that would help the DWP replace some of its Mono Basin supply. The Environmental Water Act of 1989 (AB 444) allocated $60 million for this purpose but made the money conditional on a joint application by the DWP and the Mono Lake Committee, thereby establishing the two as equal partners in finding a solution (Roderick 1989). Yet despite the promise of state funding, the Mono Lake Group struggled to agree on a spending strategy. A breakthrough appeared likely in 1990, after Mayor Bradley appointed three new, environmentally oriented members to the five-member Board of Water and Power Commissioners, and one of them, Mike Gage, became the principal negotiator on behalf of the DWP. Although Gage was willing to endorse a minimum lake level of 6,377 feet—a position considerably more generous than what many DWP staff supported—and Martha Davis got the Mono Lake Committee to modify its target level from 6,388 feet to 6,386 feet, neither side was willing to adjust its position sufficiently to close a deal (Ellis 1991b; Jones 1991).

In 1992, with the AB 444 money rapidly disappearing, discussions resumed; within the year, however, talks again had become acrimonious and eventually broke down. Even as negotiations faltered, the Mono Lake Committee persisted in exploring funding options for the city, and in 1992 it secured a provision in the federal Reclamation Projects Authorization and Adjustment Act authorizing money to pay one-quarter of the cost of some Southern California water recycling projects, including 120,000 acre-feet to offset reduced water diversions from the Mono Basin (C. A. Arnold 2004).

Adversarial Rulemaking

As the Mono Lake Group's negotiations ground to a halt, the SWRCB's decision drew nearer. By this time the PR campaigns and lawsuits unleashed during the 1980s and early 1990s had facilitated a dramatic shift in the balance of power between the DWP and proponents of restoration. The courts had made it clear that the SWRCB had considerable flexibility to accommodate environmental considerations in setting the agency's water rights. An unrelated ruling, the June 1986 *Racanelli* decision (see chapter 6), had affirmed that the state had the authority to reduce existing water rights in order to ensure that the public interest in water was protected. Political officials in L.A. and at the state level had weighed in publicly on the side of restoration. Therefore, although the DWP continued to resist, it was clear that the question facing the SWRCB was not whether to cut L.A.'s diversions but by how much. At the urging of senior environmental scientist Jim Canaday, the board went one step further and decided to address the health of the entire Mono Basin ecosystem. With that end in mind, its approach was first to determine the flows needed to protect fish, and second to ascertain the amount of water and other resources necessary to protect the public trust at Mono Lake and the surrounding area (SWRCB 1993).

In advance of the hearings, the SWRCB had asked its consultants, Jones & Stokes Associates, to prepare a draft Environmental Impact Review of the proposal to modify Los Angeles's water rights. That process was complicated by uncertainty and disagreement about what project the report would evaluate and what the baseline environment (and hence the restoration target) ought to be—the date of the Superior Court's temporary injunction (1989) or just before the DWP's diversions began (1941). Despite these difficulties, in late May 1993 the SWRCB released its 1,800-page *Draft EIR for the Review of the Mono Basin Water Rights of the City of Los Angeles* (DEIR)—a massive compilation of the most up-to-date scientific information on the Mono Basin ecosystem. After examining seven alternatives ranging from no diversion to unlimited diversion, the DEIR concluded that the best option, when compared to a 1989 baseline, was one that maintained the lake at a mean height of 6,385 feet. Under this scenario, Los Angeles would have to cut its Mono Basin exports to 44,000 acre-feet per year, about half the long-term average (D. E. Murphy 1993). The DEIR noted, however, that an even higher level of 6,390 feet, with water exports reduced to 30,000 acre-feet per year, would be superior when compared to a 1941 baseline: that level would submerge enough of the lake's alkali rim to end the dust storms and would provide more protection for brine shrimp. (The only drawback would be that some newly revealed tufa towers would probably fall when the lake level rose above 6,390 feet.) Reiterating the Mono Lake Committee's long-standing contention, the DEIR also said that whatever lake level was chosen, reclamation and conservation could easily compensate for Los Angeles's lost water, and the energy loss would be trivial.

The DEIR's assertion was supported by the fact that cutting off Mono Basin diversions had, to date, had a negligible impact on Los Angeles, despite the fact that it occurred in the middle of a severe drought. A combination of conservation and some additional water from the Metropolitan Water District made up for the loss of Mono Basin supplies.[12] Just as the Mono Lake Committee had predicted, Los Angeles had reduced its water use by more than 20 percent between March 1991 and April 1992, after the city instituted a vigorous conservation program. Moreover, there had been no political backlash; in fact, a January 1991 poll showed that the vast majority of Californians had cut their water use at home and were willing to accept mandatory water-use rules (Roderick 1991). Even after the drought ended and the city lifted its restrictions, residents continued to use about 15 percent less water than they had previously (SWRCB 1994a).

According to environmental writer John Hart (1996), the DEIR prompted a decisive shift in thinking among lake proponents, who had become progressively more willing to imagine a genuine restoration rather than incremental improvements that would merely avert ecological collapse. Hart describes a July 1993 meeting during which several prominent scientists engaged in a lively discussion—not about whether the target lake level should be raised to 6,390 feet but about whether 6,390 feet was high enough. Ultimately, the Mono Lake Committee adopted a position of 6,390 feet or higher, but many of its allies went further. Based on the results of its computer modeling, the Great Basin Unified Air Pollution Control District supported a level of 6,392 feet; because of its interest in re-creating waterfowl habitat, the Department of Fish and Game endorsed a level of 6,405 feet.[13]

Despite its impressive scientific foundation, the DEIR did not prompt the DWP to reassess its position; instead, the agency adhered to its stance that an average lake level in the 6,370s would suffice, and that stream flows should be set well below the recommendations being prepared by the Department of Fish and Game. The DWP continued to reject assertions that the low lake levels had harmed the Mono Basin ecosystem and to argue that reduced diversions would be extremely costly for city ratepayers. As a consequence, the 43 days of quasi-judicial hearings that stretched from October 1993 to February 1994, like the preceding 16 years of legal battles, featured an array of experts presenting evidence and debating every aspect of the Mono Basin ecosystem. By this point, however, almost all the institutions with an interest in Mono Basin had formally declared their support for the goal of raising the lake to 6,390 feet or higher, and Governor Wilson's administration had endorsed that objective as well.

As the SWRCB hearings were getting under way, newly elected mayor Richard Riordan and his appointee to head the DWP, as well as four new Water and Power commissioners, called on the Mono Lake Group to resume negotiations. In December 1993 Los Angeles Councilwoman Ruth Galanter brokered a deal among the parties in which most of the $36 million that remained in the Environmental Water Account authorized by AB 444 would go to the East Valley Reclamation Project, and the rest would go to smaller reclamation and conservation projects; in exchange the DWP would relinquish its claim to at least 41,000 acre-feet per year of Mono Basin water (Cone 1993). The deal provided far less water than scientists believed would be sufficient to restore Mono Lake. It did, however, mark a new turn in California water politics: as legal scholar Tony Arnold (2004, 22) observes, it was "the first time Los Angeles voluntarily relinquished any of its water rights in favor of an alternative source." According to some (Canaday 2007; M. Davis 2007), this deal—and the negotiations that preceded it—played a crucial role in persuading the DWP to accept the unprecedented cut in its supplies that the SWRCB eventually required. Others are not so sure. The DWP had been soundly defeated in every forum, its political overseers were aligned against it, and the organization's leaders may simply have concluded that continuing to fight was futile (Hanna 2007).

The Mono Basin Restoration: Planning and Implementation

The SWRCB's September 1994 decision set environmentally protective levels for both Mono Basin stream flows and Mono Lake; in addition, it required the DWP to prepare and implement ambitious stream and waterfowl habitat restoration plans. In consultation with the Mono Lake Committee and independent scientists, the DWP came up with an adaptive approach that aims to restore a more naturally functioning, dynamic, and self-sustaining hydrologic ecosystem in the basin. Although it does not purport to require a comprehensive restoration, the SWRCB's mandate and the plans devised to comply with it do address the ecosystem's two most serious problems: they curb L.A.'s diversions, and they try to repair the historic damage those diversions have wrought. Furthermore, other initiatives beyond the DWP–financed restoration, such as the Mill Creek restoration plan and efforts to prevent inappropriate development in the basin, render the overall approach more comprehensive. Also, despite the fact that it is largely a product of adversarial politics, the restoration has proven to be durable. The DWP has completed a variety of infrastructure changes aimed at reorienting the system to provide water for the natural system, and it has begun coordinating water releases with Southern California Edison, which manages the hydropower facilities in the area, to more closely mimic the prediversion hydrograph. In addition, despite periodic setbacks and disagreements, the DWP has worked cooperatively for more than ten years with the Mono Lake Committee and independent scientists to monitor results and revise its restoration plans. Their efforts have yielded measurably beneficial outcomes: notwithstanding a series of dry years in the early 2000s, ecological conditions in the Mono Basin have improved markedly.

Mono Lake Restoration: The SWRCB Sets the Goals

On September 16, 1994, the SWRCB published its final EIR, in which it endorsed a plan that was even more protective than the alternative favored in the draft report. With Decision 1631 the water board established a target average lake level of 6,392.6 feet, primarily to reduce the occurrence of alkali dust storms in the basin.[14] It decided against the even higher 6,405-foot target that scientists said would have maximized waterfowl habitat. It justified its choice as necessary to preserve the south tufa groves, which it deemed an important aesthetic and recreational resource, but it also wanted to avert the most draconian cuts to L.A.'s water supply. Nevertheless, the target lake level provided a substantial buffer against prolonged drought: 6,390 feet is the lower limit of the range of levels for which aquatic productivity at Mono Lake is relatively high, and allowing the lake to fall below 6,388 feet would have negative but hard-to-quantify effects; however, the target lake level is more than 20 feet above the topographic "nick point" of 6,368 feet (SWRCB 1994a).

In addition, Decision 1631 established a schedule for reaching the target lake level that imposes the risk of failure on water users, rather than the Mono Basin ecosystem. It prohibits any diversions by the DWP until the lake reaches 6,377 feet, at which point the agency can divert only 4,500 acre-feet per year until the lake level reaches 6,380 feet; after that, it can divert 16,000 acre-feet per year until the lake reaches 6,391 feet. At 6,392 feet and above, the DWP can take all water in excess of the required fish flows, which should yield an average of 30,800 acre-feet per year. After reaching 6,391 feet, even if the lake level starts falling again, L.A. will be allowed to take as much as 10,000 acre-feet, as long as flow requirements are met. If the lake falls below 6,388 feet, however, DWP diversions must cease (SWRCB 1994b). As for meeting the city's water needs, the SWRCB noted that L.A.'s successful conservation efforts to date suggested its citizens had made permanent changes in their consumption patterns. It cited other means of offsetting water losses, such as greater use of local groundwater, reclamation, and obtaining additional supplies from the Metropolitan Water District of Southern California.

Beyond devising a plan to achieve the desired lake level, the SWRCB set channel maintenance and flushing stream flows that were consistent with the protective recommendations of the Department of Fish and Game. It also prescribed in-stream flows that, although slightly below Fish and Game's recommended levels, were well above those suggested by the DWP. The board also included provisions for oversight, adjustment, and enforcement of its orders, saying: "The SWRCB shall have continuing authority to require modification of restoration activities as appropriate and to modify stream flow requirements as necessary to implement restoration activities. Modification of stream flow requirements may reduce the amount of water available for export" (SWRCB 1994b).

Experts anticipated it would take 20 years or more for the lake to reach the legally mandated stabilization level, and as long as 50 years for the streams to recover. In the meantime, however, they predicted that a rising lake would effect a variety of changes in the Mono Basin ecosystem. The lake itself would eventually reoccupy some of the area it had lost, bringing it closer to U.S. Route 395. Some of the tufa towers would fall, and others would become islands. Negit Island, which periodically had been linked to the mainland by a land bridge that emerged as the lake level fell, would regain its isolation. The alkali band on the lake's eastern rim would shrink, and wetlands would become more common along the shore, creating new habitat for ducks (Hart 1996).

By contrast, had the status quo continued—that is, had L.A. been allowed to continue the diversion levels it had established prior to the 1989 injunction—the DWP would have taken, on average, about 85,000 acre-feet per year (73 percent of the total surface runoff), leaving about 32,000 acre-feet for the lake. As a result, the lake surface gradually would have fallen to, and then fluctuated around, 6,355 feet. The diminished stream flow and lower lake level would have had a host of devastating ecological consequences beyond what had already occurred in the basin. Riparian and lake-fringing vegetation would have been damaged and shoreline wetlands would have disappeared, further reducing the available habitat for ducks and migrating birds. Gull nesting sites would have been compromised when the land bridge connecting the lake's islands to the shore reappeared. Brine shrimp populations would have plummeted, restricting the food supply for birds. Air quality would have been impaired by severe alkali dust storms. (SWRCB 1993)

In response to the final EIR, DWP spokespeople reiterated the department's long-held position that it would depend heavily on the Metropolitan Water District—and consequently the fragile Sacramento–San Joaquin Delta—to make up lost water supplies (Bancroft 1994a). Jim Wickser, assistant general manager of the DWP, pointed out that the city was already spending about $38 million each year to replace lost Mono Basin water with increased purchases from the Met. He noted that L.A. residents had already cut their water consumption through conservation and reclamation, but needed the unused supply to accommodate population growth. Unconvinced by the formal scientific assessments, he added: "We have almost 20 years of history where the lake has been at or below that elevation, and the brine shrimp and birds are at higher numbers than ever recorded in the past" (Cone 1994). Wickser's belligerent rhetoric notwithstanding, after a hurried meeting with Martha Davis, who pledged to continue helping the DWP find outside funding, the department decided not to appeal the SWRCB's order (Hart 1996). On September 28, the SWRCB held a triumphant news conference to announce its unanimous (5–0) support for Decision 1631 (Bancroft 1994b).

Negotiating the Stream and Waterfowl Habitat Restoration Plans

In addition to requiring the DWP to cut back its diversions in order to raise the lake and enhance stream conditions, the SWRCB instructed the agency to prepare a plan for further stream restoration, as well as one for duck habitat, by November 1995. Following the advice of geomorphologist Scott Stine and others, the board insisted that planners include some active restoration measures. In February 1996 the DWP submitted its draft plans. After receiving extensive critical comments on those drafts, the SWRCB scheduled a hearing, but—in response to a request by environmentalists and agency representatives for an opportunity to resolve their differences with the DWP—the board postponed the hearing twice, finally rescheduling it for January 18, 1997. After six days of evidentiary testimony the parties agreed once again to sit down together, and in May 1997, after another round of negotiations and two more days of testimony, they announced a settlement. Signing on to the settlement were some entities that had been involved in Mono Basin policymaking for years: the DWP, the Mono Lake Committee, the National Audubon Society, CalTrout, the U.S. Forest Service, the State Lands Commission, the California Department of Fish and Game, and the California Department of Parks and Recreation. More recently involved signatories were the BLM, which manages wilderness in the region; the Trust for Public Land; and Arcularius Ranch. People for Mono Basin Preservation, a new local group formed to oppose some aspects of the proposed restoration, declined to participate in the settlement negotiations.

The overarching goal of the settlement, which included a Stream Restoration Plan and a Waterfowl Habitat Restoration Plan, was to re-create the critical ecological processes that shaped the evolution of the Mono Basin ecosystem. To this end, the proposed Stream Restoration Plan aimed to provide stream flow regimes that would allow naturally functioning, dynamic, and self-sustaining hydrologic systems to reemerge. A crucial aspect of this approach is providing high peak flows, which shape stream channels, transport and deposit sediment, spread seeds, scour pools, and provide the energy needed for streams to reestablish their natural processes. Although everyone concurred with these basic principles, the parties disagreed on the appropriate level of peak flows, as well as the best way to attain them, given that all the Mono Basin creeks have dams or diversion facilities that moderate their natural flows. The difficulties promised to be particularly acute on Rush Creek because the DWP's Grant Lake does not have an outlet capable of releasing the level of peak flows that scientists had recommended. Rather than modifying the dam, which would be costly, the DWP proposed diverting water from the Lee Vining Creek. Environmentalists were skeptical about this scheme, especially after an initial attempt failed. But they agreed to let the DWP try the so-called Rush Creek augmentation method, with the proviso that if monitoring showed it was not working the agency would build an outlet in the Grant Lake Dam (Hopkins 1997c).

The Stream Restoration Plan also included provisions for opening side channels on Rush Creek. As part of its earlier restoration efforts, the DWP had already opened seven channels that had been plugged with gravel or had been abandoned because of stream degradation. But the settlement proposal targeted five more channels for reopening in order to spread water, provide fish habitat, and raise the water table along its course. The hope was that vegetation would recolonize the area rewatered by newly opened channels and, over time, become self-sustaining.

Finally, the settlement asked the board to allow the DWP to use an adaptive approach, in which restoration activities would be adjusted in response to conclusions reached on the basis of monitoring. The DWP had voluntarily initiated monitoring in 1997, in advance of the Water Board's final orders, but the settlement laid out agreed-on monitoring activities for each of the four streams, their scope and duration, the protocols to be used for gathering data, and the methods for analyzing that data. The settlement proposed that monitoring continue until 2014 (the year hydrologic models predicted the lake would reach its target), when the SWRCB is scheduled to review the Mono Basin's recovery to judge whether it meets the settlement's "termination criteria." Those criteria include acreage of riparian vegetation, including mature trees of sufficient diameter, height, and location to provide woody debris in streams; length of main channel; channel gradient; channel sinuosity; channel confinement; variation of longitudinal thalweg elevation; and the size and structure of fish populations.[15]

Reaching agreement on the Waterfowl Habitat Restoration Plan was more difficult. In particular, conflict arose in response to a proposal by the DWP—based on the advice of its scientific experts and supported by environmentalists—to take water from Wilson Creek, an irrigation ditch, and return it to Mill Creek, which empties into the northwest corner of Mono Lake. Although local operators had been diverting Mill Creek for hydropower and irrigation, Los Angeles had never tapped it. As a result, the underlying physical structure of its bottomland had not been incised the way Rush Creek's and Lee Vining's had. Moreover, Mill Creek was once the lushest of the basin's streams; its valley bottom supported an assemblage of forests, wetlands, and meadows that was unusual in the high Sierra desert landscape. Given all this, proponents thought Mill Creek offered a prime restoration opportunity. But some local residents feared that rewatering Mill Creek would ruin fishing in Wilson Creek and dry up historic ranches, particularly the 1,031-acre Conway Ranch, and they formed a group called People for Mono Basin Preservation to fight the proposal. In hopes of averting a bitter dispute between the Mono Lake Committee and locals opposed to the Mill Creek proposal, in early 1997 the SWRCB's Jim Canaday suggested that a group of stakeholders establish the Conway Ranch Evaluation Working Group (CREW) to discuss the fate of the ranch, which had senior rights to Mill Creek's water, and the allocation of Mill Creek water more generally. (Hopkins 1997a, 1997b, 1997c; Little 1997)

In October 1998, while the CREW negotiations were still under way, the SWRCB issued its final orders regarding the DWP's Mono Basin Restoration Plan (Orders 98-05 and 98-07). The water board's orders take a generally protective approach. The Stream Restoration Plan affirms the basic principles of the settlement—a focus on "reestablishing natural processes and historic conditions, rather than former landscapes" (Cutting 2006). They also adopt most of the settlement's specific provisions: they delineate peak (or spring restoration or channel maintenance) flows based on Fish and Game biologists' recommendations, to be in place until scientists can agree on more accurate flows; require the DWP to open five side channels in the floodplain; mandate the rehabilitation of the Rush Creek return ditch; and call for evaluating and implementing methods for passing sediment down Walker, Parker, and Lee Vining creeks below the diversion structures.[16] They require other active restoration measures as well, such as planting riparian vegetation, removing invasive tamarisk along lower Rush Creek, and placing large woody debris in creeks. They also include some buffering provisions, such as prohibiting livestock grazing within the riparian corridor and limiting vehicle access in sensitive areas near streams. And they prescribe an adaptive approach, specifying that "Stream monitoring shall evaluate and make recommendations, based on the results of the monitoring program regarding the magnitude, duration, and frequency of the stream restoration flows necessary for the restoration of Rush Creek; and the need for a Grant Lake bypass to reliably achieve the flows needed for the restoration of Rush Creek" (SWRCB 1998).

Reluctant to wade into the controversy surrounding the Waterfowl Habitat Restoration Plan, the SWRCB declined to mandate Mill Creek restoration. According to the water board (SWRCB 1998):

It is apparent from the testimony and other evidence presented by [People for Mono Basin Preservation] that many Mono Basin residents view Wilson Creek and the resources dependent upon it from a distinctly different perspective than is reflected in the waterfowl scientists' report. Rather than seeing Wilson Creek as an unnatural, historic artifact to be disregarded in pursuit of restoring "natural conditions," the record shows that many Mono Basin residents view Wilson Creek, and the resources dependent upon its flow, as being an invaluable part of their heritage with benefits to fish, wildlife, recreational users, and the scenery.

The board also rejected a proposal in the settlement supported by environmentalists to establish a new foundation with responsibility for overseeing waterfowl habitat restoration, on the grounds that doing so would abrogate its own responsibility for oversight. Instead, the orders posit that the single most important feature for waterfowl is a higher lake level, which will re-create shoreline habitat. In addition, they require the DWP to contribute as much as $275,000 to restoring habitat at County Ponds, natural depressions on the Forest Service–owned DeChambeau Ranch— the preferred option of many Mono County residents.[17] Finally, the orders require that the DWP participate in an interagency controlled burn program to reestablish open water at springs around the lake's shore, should such a program be established, and conduct a comprehensive program of waterfowl use surveys and habitat monitoring.

In December 1998, the DWP released its draft implementation plan, which pulled the SWRCB's directives into a single administrative document that specifies timelines and procedures for activities such as channel rewatering, revegetation, and aerial photography. In addition, the DWP voluntarily added a provision for semiannual meetings to discuss end-of-season reports and the upcoming season's activities. Notwithstanding the DWP's more cooperative stance, implementation initially encountered setbacks. For example, as of the spring of 2000, neither Rush nor Lee Vining Creek was regularly receiving peak flows consistent with the SWRCB's 1998 restoration orders because a dispute between the DWP and the Department of Fish and Game over habitat requirements and regulatory authority had stalled work on the Rush Creek return ditch for a year and a half (Reis 2000). Moreover, the Rush Creek augmentation method preferred by the DWP was not working and was, in fact, having negative side effects on Lee Vining Creek (Reis 2000).

Over time, however, most differences were ironed out, and restoration projects moved forward. In 2001 the DWP and Fish and Game reached agreement on how to increase the capacity of the Rush Creek return ditch; in 2003 the DWP completed work to eliminate seepage and strengthen its walls; and in 2004 engineers successfully tested the ditch, which now operates at its full capacity of 380 cfs. In 2005 and 2006 the DWP continued to experiment with augmenting Rush Creek's wet-year peak flows. In 2007 the DWP opened two Rush Creek side channels, making them perennial at the insistence of the Mono Lake Committee— in the process going beyond the recommendations of the SWRCB scientists. (On the scientists' advice the DWP indefinitely deferred opening the remaining three Rush Creek side channels.) It also began operating the newly upgraded Lee Vining Creek diversion facility, enabling bypass of sediment, and began developing a sediment bypass method for Parker and Walker creeks (in the meantime doing a manual "dredge and place" operation). Throughout the 2000s the DWP continued to monitor fisheries, streams, stream flows, and waterfowl habitat (LADWP 2006; McQuilkin 2007).

The major outstanding restoration issue in the Mono Basin concerns Mill Creek, which the DWP never diverted and hence is not required to restore. In 1998 the Trust for Public Land acquired the Conway Ranch, thereby preventing a 440-unit commercial and residential development that had been approved for the site, but differences over north basin water rights remained unresolved because CREW had failed to reach agreement. In 2001 nine interested parties again began negotiating north basin water rights and Mill Creek restoration—this time in the context of the Federal Energy Regulatory Commission (FERC) relicensing of the Lundy Hydroelectric Project. In 2002 the parties agreed to create a technical team that would help them conduct a joint fact-finding exercise, because they felt disputes about information were continuing to hinder the discussion. Two years later the majority of the parties reached a settlement, which they submitted to FERC in hopes of shaping the relicensing decision. For environmentalists the key element of the settlement was a commitment by Southern California Edison to upgrade the Mill Creek return ditch. Because that ditch had not been maintained at capacity, 70 percent of Mill Creek's flow routinely had been shunted to Wilson Creek, far in excess of legal water rights (McQuilkin 2005). The DWP, although it had not participated in the settlement negotiations, weighed in favorably, whereas Mono County and the People for Mono Basin Preservation declined to sign the agreement. In November 2007 FERC issued its decision: although it did not require the Mill Creek return ditch upgrade, Southern California Edison intends to do it anyway.[18]

Ecosystem Reawakenings

Even before the DWP finished crafting its board-ordered plans, the stream restoration projects begun in response to the court orders of the late 1980s and early 1990s were already bearing fruit. In November 1995 the *San Francisco Chronicle* reported jubilantly that Walker Creek was flowing again, and "A magnificent slice of California, between the volcano-scarred desert ranges of the Great Basin, and the tall, snow-filled granite of the Sierra, [was] awakening from a sort of time capsule and returning to exuberant life" (Petit 1995). In August 1996 the *Christian Science Monitor* reported that signs of renewal were also evident in Lee Vining Creek, which for the previous 50 years had been a dry landscape of tumbleweed and dying trees (Sneider 1996). By 1998 the cottonwoods planted in 1993 along Lee Vining Creek were up to 12 feet tall, and stream channels opened in the mid-1990s had raised water tables in some sections of the bottomlands. Two years after workers opened a side channel in Rush Creek's bottomlands, ducks and fish had returned, and the aquatic plant *Elodea*—a sign of good water velocity and substrate—had reestablished itself (Anon. 1998a). In the summer of 1998, hot July weather melted a large winter snowpack, raising Sierra reservoirs and spilling high flows down Mono Basin streams. In a demonstration of their stewardship, the DWP and Southern California Edison reservoir operators enhanced the environmental benefits of the extra water by coordinating their releases (Reis 1998).

Mono Lake was exhibiting signs of health as well. By 1999, after four consecutive wet winters, the lake had reached 6,384 feet, having risen nearly ten feet since the 1994 decision. There were visible changes in the shoreline as a result: dust storms had abated, lake-fringing lagoons had increased, and alkali fly habitat had grown (Reis 1999). By 2001 the "meromixis" that set in during 1995 and temporarily reduced the lake's productivity had begun to abate, and by the spring of 2004 the effects were showing up in the improving reproductive success of gull populations (Hite 2005).[19] Ironically, because meromixis results from the rapid infusion of freshwater into a hypersaline lake, it ended as conditions became drier (in 2004, when the level fell to 6,381 feet) but reappeared again in the summer of 2006, when the lake reached 6,385 feet following a wet spring. (Scientists expect meromixis will become less frequent as the lake rises and its salinity decreases.)

Because of natural variability, as well as experiments with different restoration techniques, the Mono Basin's ecological recovery has not been linear. Nevertheless, although it rises and falls depending on the season and the amount of rain- and snowfall each year, the lake's surface is holding steady or rising. As of late 2007, roughly 20 miles of Mono Basin streams have been rewatered, and riparian vegetation structure is changing, though only gradually (Reis 2007). Scientists have discerned a significant positive trend in waterfowl numbers since 1996—although they advise caution when interpreting short-term population trends, given the natural variability of waterfowl populations (LADWP 2006). Although it is occurring slowly, the signs of an ecological revival—such as the appearance of the rare willow flycatcher and yellow warbler—are unmistakable. Thus, journalist Jane Kay reported in July 2006:

The lake is teeming with brine shrimp and alkali flies that feed the birds. Bright green native grasses grow right down to the lake, now large enough to cover the once-exposed lake bottom. The surging waters cover the old land bridges that had allowed coyotes to eat gull eggs and baby birds.... The tributaries of Lee Vining and Rush Creeks are gushing mountain streams filled with brown trout, and willows flourish on the edge along with the resurgence of Jeffrey pines. Sprouting up are buffalo berry bushes and woods' roses, prized by the willow flycatcher. The songbird known as "the ivory billed woodpecker of Mono Lake" disappeared, then suddenly reappeared as water returned to dry creeks.

Ongoing Vigilance and Outreach by the Mono Lake Committee

The effects of environmental restoration of the Mono Basin are evident, but threats continue to arise, and a combination of vigilance and ongoing efforts to create an environmental stewardship ethic are crucial to ensuring the comprehensiveness and long-term environmental benefits of the effort. One of the Mono Lake Committee's functions is to scrutinize the DWP's compliance; although the department generally has worked amicably with environmentalists and the county, and has demonstrated a genuine commitment to restoration, its enthusiasm cannot be taken for granted. For example, in 1998 the DWP wrote to the California Air Resources Board (CARB), asking it to redesignate the basin as an air quality attainment area under the Clean Air Act—a move that would undermine the SWRCB's air quality rationale for the Mono Lake level requirements (Spivy-Weber 1998). The Great Basin Unified Air Pollution Control District responded that a redesignation would be premature, but the DWP countered by criticizing the district's monitoring. Meanwhile, CARB published a list of areas it was recommending that the EPA redesignate, and it included the Mono Basin.

As important as its watchdog role are the Mono Lake Committee's efforts to ensure restoration-compatible land use in the basin. In 2002 environmentalists fought off a proposal by CalTrans to widen and straighten nearly three miles of U.S. 395. The highway already runs as close as 250 feet to the lake's fragile shore, a distance that will close to 100 feet when the lake reaches its target level. After the Mono Lake Committee raised the alarm about the project, which would have destroyed shoreline wetlands being restored under the SWRCB order, supporters flooded Governor Gray Davis's office with more than 2,000 requests to scale it back (Anon. 2002; Romney 2002). In 2005 CalTrans abandoned the project in the face of entrenched opposition and declining funds.

Even as the CalTrans threat receded, however, a series of development proposals jeopardized conservation efforts. (Ironically, because environmentalists' PR campaigns have made the lake a more popular destination, the demand for resort accommodations has increased.) In November 2004 Bill Cunningham, owner of 120 acres within the National Forest Scenic Area on the lake's western shore, proposed building 30 resort homes on his property. The Forest Service ruled in 2003 that the development would be "incompatible and detrimental to the integrity of the scenic area" (D. Thompson 2004), and county planners assumed that the Forest Service's land regulations precluded development. But a legal opinion issued in early 2004 held that the county should process the application under its own, less restrictive zoning and let the Forest Service enforce its rules separately. After several years of intense negotiations, the Mammoth Mountain Ski Area purchased most of the property with the intent of trading it for a parcel owned by the Forest Service at the base of Mammoth Mountain (Boxall 2005; McQuilkin 2008). In the meantime, the Mono Lake Committee was struggling to fend off a development threat at Cedar Hill, a 3,748-acre parcel northwest of the lake that constituted 20 percent of the private property in the Mono Basin. The land was for sale, and a developer had proposed building a subdivision on it. In 2005, however, at the behest of the Mono Lake Committee, the Wilderness Land Trust acquired the tract, and in mid-2007 donated it to the BLM.

In addition to monitoring implementation of the SWRCB's orders, seeking creative solutions to problems not addressed by those orders, and trying to avert hazards to the basin's long-term ecological health, the Mono Lake Committee and its allies strive to foster environmental stewardship—both locally and farther afield—in order to ensure that protection of the Mono Basin endures. Locally, the Mono Lake Committee's headquarters hosts an informational display and provides tours of the basin. The Committee also supports a Web site that boasts an extraordinary breadth of materials on current research and the history of the restoration. And it continues its outreach to Los Angeles: in the summer of 1994 it created an Outdoor Experience program that brings young people from Los Angeles to Mono Lake to do trail restoration work and learn about the basin's ecology. In 1995 the DWP became a partner in the venture and agreed to lease a house on L.A.-owned property to the committee to serve as a base camp for the program. By 2003 the program had served more than 2,000 youth, most of them from L.A., and many from disadvantaged, inner-city backgrounds. They, along with hundreds of visiting schoolchildren, helped plant or water newly planted trees along the decimated lower reaches of Rush and Lee Vining creeks (Miller 2003).

Conclusions

The Mono Basin restoration has yielded a variety of environmentally beneficial policies and practices, many of which have translated into tangible environmental improvements. Most important, the SWRCB's 1994 decision required the DWP to revamp its Mono Basin infrastructure completely, in order to give priority to delivering water to the natural system. The order also created substantial incentives for the DWP to comply and put the burden of proof on the agency to demonstrate why it cannot deliver the promised results (Vorster 2006). As of 2007 the DWP has spent about $60 million (Hanna 2007), and its efforts have transformed a declining and increasingly brittle ecosystem into one that appears likely to recover some measure of resilience. Reduced diversions by the DWP have allowed the lake level to rise higher than it would have if the status quo had continued. The release of increased base and peak stream flows, combined with active restoration projects, has improved the vitality of four of the basin's five creeks. The area's bird populations, which are the most obvious indicators of ecological renewal, are thriving: ducks, gulls, and other migrating species are returning to the basin; rare songbirds, such as the willow flycatcher and yellow warbler, are turning up in surveys for the first time in many years.

The restoration's effectiveness is partly a result of its landscape-scale focus. Landscape-scale thinking emerged over the course of the controversy, largely as a result of a series of integrative scientific assessments prepared in an effort to resolve the controversy. Those reports did not convince entrenched opponents in the DWP but did enlighten and galvanize officials in other agencies with an interest in the basin, bringing them together in support of a comprehensive restoration. The use of adaptive management has also yielded substantial benefits. It has allowed stream restoration projects to move forward despite some ongoing disagreements among scientists over the best approach. In the Mono Basin, adaptive management has not been merely an excuse for accepting minimal protection measures, as it has in other cases, in part because the SWRCB has given the scientific team an extraordinary amount of autonomy. In addition, the program operates in the context of clear objectives and oversight: if the results of the DWP's experimental methods do not meet the agreed-on termination criteria, challengers can appeal to the SWRCB to enforce its orders.

But the effectiveness of the restoration is primarily a consequence of its clear and unitary goal of restoring the Mono Basin ecosystem to an ecologically healthy condition. That goal, in turn, was the product of an aggressive litigation and persuasion campaign combined with the tenacious insistence by the Mono Lake Committee on helping to solve L.A.'s water problems. Interestingly, although the process leading up to the SWRCB's decision was almost entirely adversarial, the Mono Basin restoration, like the Kissimmee River restoration, exhibits several features typically associated with collaboration and flexibility. It is grounded in the best available science, as well as the local knowledge of long-term residents and especially of retired biologist Eldon Vestal. It is innovative: it requires active measures, not just adjustments in diversions; prescribes increased conservation and reclamation rather than simply acquiring water from other ecosystems; and institutionalizes a cooperative planning process with an ongoing supervisory role for the SWRCB. It has proven to be durable: the DWP chose not to appeal the water board's decision, and has been implementing it conscientiously for more than a decade. Over time, as new DWP personnel with no history in the conflict joined the restoration effort, relations improved between the agency and its former adversaries. In fact, in several instances the DWP has demonstrated genuine stewardship by going beyond what is required by the SWRCB.

The restoration's durability may be at least partly attributable to the collaborative process that proceeded in parallel with the adversarial contest over restoration goals. As the negotiator on behalf of environmental interests, Martha Davis combined a principled stance on behalf of the lake with an inexhaustible willingness to talk. Her strategy was "to figure out how to win, but win in such a way that there was a solution set that was easier for the city to accept than continuing to fight" (Martha Davis 2007). Davis believes the decade-long effort to resolve the conflict in a collaborative forum made it easier for the DWP to accept the SWRCB's decision and laid the groundwork for ongoing and productive cooperation between the DWP and its former adversaries. That said, it is clear that the negotiations themselves occurred only because restoration advocates had so effectively used more conventional tactics to create substantial uncertainty for the DWP and to persuade political leaders—in city government, the legislature, and the courts—to institutionalize environmentalists' definition of the problem and prescribe environmentally protective solutions.

10

Ecosystem-Based Management and the Environment

All seven of the initiatives described in this book have yielded concrete policies and practices that are likely, over time, to produce some environmental benefits. Each has prompted the creation of a deeper and more holistic understanding of how local ecosystems work, which in turn has fostered a more widespread recognition among policymakers and stakeholders of the relationships among the landscape's ecological elements and functions. Without exception, the programs have furnished participants with a rationale for raising large sums of money that have been used to acquire ecologically valuable land or undertake activities aimed at restoring ecological functions. And each has empowered environmentally oriented personnel within agencies and jurisdictions, some of whom have tried to institutionalize more environmentally beneficial practices. Only some, however, have yielded policies and practices that are likely to conserve and restore biological diversity and, therefore, ecological resilience (see tables 10.1 and 10.2).

Based on a comparison of the seven cases, a landscape-scale focus appears to be an important catalyst for the adoption of more protective policies and practices. In every case, trying to address problems at a landscape scale prompted planners to adopt more comprehensive approaches to environmental problem-solving and led to new forms of coordination among disparate agencies and jurisdictions. The beneficial effects of collaborating with stakeholders and of flexible, adaptive implementation are less evident, however. In cases where policymakers deferred to stakeholders to set goals, the policies and practices that emerged appear unlikely to conserve or restore ecological health because, to gain consensus, planners skirted trade-offs and opted instead for solutions that promised something for everyone. The resulting plans typically feature

Table 10.1
Terrestrial EBM Results

	Austin BCCP	San Diego MSCP	Pima County SDCP
EBM Attributes			
Landscape-scale Focus	Yes	Yes	Yes
Stakeholder Collaboration	Yes	Yes	No
Flexible, Adaptive Implementation	Yes	Yes	Yes
Intermediate Outputs			
Integrative Science & Comprehensive Planning	+	+	+++
Inter-Agency/Jurisdiction Coordination & Consistency	+	+	++
Trust, Transformation & Innovation	+	+	++
Agreement on & Grounding in Best-Available Information	+	+	+++
Durable Implementation	+	+	++
Stewardship & Going Beyond Legal Minimum	+	0	++
Learning & Adjustment	0	0	NA
Overall Outputs/Outcomes			
Environmentally Protective Plan	+	+	+++
Environmental Improvements	0	0	0

Key: - decline from status quo; 0 no discernible change or a mixed bag; + minimal increase; ++ moderate increase; +++ substantial increase.

management-intensive approaches with little buffering. As a result, they impose the risk of failure on the natural system. A commitment to flexible, adaptive implementation has not compensated for the failings of these environmentally risky plans and, in fact, has sometimes exacerbated them. Adaptive management has not translated into a willingness to alter policies in the face of new information, partly because minimalist plans actually provide little room for adjustment, but also because management and monitoring are insufficiently funded, and learning by scientists does not translate automatically into management changes. Flexible implementation has allowed managers with missions that are incompatible with ecological restoration to resume resource-user-friendly practices when political conditions shift.

By contrast, when policymakers—elected officials, administrators, or judges—endorsed an environmentally protective goal and used

Table 10.2
Aquatic System EBM Results

	CERP	CALFED	Kissimmee River Restoration	Mono Basin Restoration
EBM Attributes				
Landscape-scale Focus	Yes	Yes	Yes	Yes
Stakeholder Collaboration	Yes	Yes	No	No
Flexible, Adaptive Implementation	Yes	Yes	Yes	Yes
Intermediate Outputs				
Integrative Science & Comprehensive Planning	++	+	++	++
Inter-Agency/Jurisdiction Coordination & Consistency	++	+	++	++
Trust, Transformation & Innovation	0	0	+++	++
Agreement on & Grounding in Best-Available Information	+	+	+++	+++
Durable Implementation	0	0	++	+++
Stewardship & Going Beyond Legal Minimum	+	0	++	++
Learning & Adjustment	+	0	+++	+++
Overall Outputs/Outcomes				
Environmentally Protective Plan	+	+	+++	+++
Environmental Benefits	0	0	++	++

Key: - decline from status quo; 0 no discernible change or a mixed bag; + minimal increase; ++ moderate increase; +++ substantial increase.

regulatory leverage to prevent development interests from undermining that objective, the resulting policies and practices are more likely than their counterparts to conserve or restore ecological integrity. A willingness by political leaders to make ecological health the preeminent aim can change the balance of power and alter perceptions of what is politically feasible. When restoring ecological health is the paramount goal, planners are more likely to approve, and managers to implement, approaches that rely less on energy-intensive manipulation and more on enhancing the ability of natural processes to sustain themselves—even if doing so imposes costs on some stakeholders.

A Landscape-Scale Focus

According to the optimistic model of ecosystem-based management (EBM), a landscape-scale focus prompts the creation of an integrative scientific assessment that, in turn, enhances awareness among stakeholders and policymakers of the relationships among ecosystem elements and processes. In theory, such an improved understanding spurs the development of a more comprehensive plan—one that explicitly takes those interrelationships into account. In addition, the optimistic model predicts that efforts to plan at a landscape scale will foster cooperation among agencies and jurisdictions, resulting in a more consistent and coherent management approach. Pessimists worry, however, that economic considerations will dominate landscape-scale planning efforts, reducing their comprehensiveness. They also fear that institutional barriers such as competing missions, concerns about maintaining turf, and the pursuit of economic development will undermine interagency and interjurisdictional coordination. The cases suggest that a landscape-scale focus does, in fact, yield the environmental benefits anticipated by optimists. But those advantages may be tempered by the factors identified by pessimists, particularly if planning is done in collaboration with stakeholders.

Integrative Science and Comprehensive Planning

Consistent with the optimistic model, in six of the seven initiatives described in this book, a landscape-scale focus prompted the creation of an integrative scientific assessment, and that assessment enhanced awareness among stakeholders and policymakers of the relationships among ecosystem elements and processes. In Austin, San Diego, and Pima County, reputable scientists prepared assessments that first described what was known about the habitat requirements of key species. They then delineated the characteristics of terrestrial reserves that would be sufficient to conserve the remaining biological diversity in the designated planning area. To ensure the long-term biological effectiveness of the proposed reserves, scientists prescribed retaining buffers between set-aside land and developed areas; they also recommended measures for conserving or restoring ecological processes likely to be critical to the survival of key species. Similarly, scientists prepared integrative assessments for the Everglades, Kissimmee River, and Mono Basin that identified the key elements of aquatic systems, the relationships among them, and the ecological drivers that sustain those interactions. Even the CALFED science program, which explicitly declined to construct a whole-system model, fostered research that aimed to yield insights into poorly understood relationships rather than perpetuating a focus on individual elements.

The causal connection between integrative scientific assessments and comprehensive planning is more complicated than the optimistic model suggests, however. In every case, the preparation of an integrative assessment did correspond to the creation of an approach to environmental problem-solving that was more comprehensive than the status quo. When consensus among stakeholders was a prerequisite, however, planners resisted protecting some important elements or processes in order to avoid imposing costs on powerful interests. In both Austin and San Diego, for instance, planners constrained preserve boundaries and characteristics and eliminated protection for some of the most valuable and vulnerable habitat in response to development pressure. Similarly, Comprehensive Everglades Restoration Plan (CERP) and CALFED planners conspicuously avoided measures that would curtail water users' allocations. They also declined to address land-use issues, despite widespread recognition that the health of aquatic systems is intimately related to decisions about the development of adjacent land; in fact, a general unwillingness to incorporate land-use decision making into planning for aquatic systems almost certainly perpetuates the very death-by-a-thousand-cuts that landscape-scale planning is supposed to avert.

By contrast, in the Pima County, Kissimmee River, and Mono Basin cases, where plans emerged from more conventional politics, policymakers made fewer concessions to development or user interests. Like their counterparts in stakeholder-driven processes, policymakers did accommodate political and economic considerations, thereby rendering their approaches less than fully comprehensive. For example, in the Kissimmee River case the state curtailed its restoration ambitions because it was unwilling to buy out landowners in the most densely populated segment of the floodplain. And in the Mono Basin case, the state chose a target lake level that was lower than the historic high in order to avoid cutting off Los Angeles's Mono Basin water supply altogether. Nevertheless, in all three of the comparison cases policymakers embraced measures that imposed substantial costs on some interests in order to increase the likelihood of genuine environmental improvements: the Pima County Board of Supervisors adopted the Conservation Lands System devised by its Science and Technical Advisory Team, without modification, over the vocal objections of the development community; despite the resistance of property owners and water users, the Kissimmee River and Mono Basin projects seek to restore those ecosystems' physical features and flow regimes in order to reinstate more self-sustaining biological processes.

Coordination among Agencies and Jurisdictions

As predicted by the optimistic model, a landscape-scale focus also led to an increased propensity among agencies and jurisdictions to coordinate their planning and adopt more consistent approaches to environmental management. Interestingly, however, in the four cases where planners sought stakeholder consensus, cooperation has attenuated over time, as agencies and jurisdictions have capitalized on changes in the political context to pursue their existing missions. For instance, the City of Austin and Travis County were forced to work together to construct the Balcones Canyonlands Conservation Plan (BCCP) and, because they obtained a single permit, remain jointly responsible for its success. Nevertheless, policymakers declined to create an overarching authority that considers the preserve as a whole for the purposes of management and monitoring. As a result, ongoing coordination among the preserve's landowners is minimal, and there have been heated disputes between the city and county over the inclusion of particular parcels within the preserve.

In San Diego, the initially high level of cooperation among agencies and jurisdictions within the Multiple Species Conservation Plan (MSCP) planning area declined markedly over time. The planning effort began as a joint endeavor among 11 local jurisdictions, as well as the Navy, the California Department of Fish and Game, and the U.S. Fish and Wildlife Service (FWS). As the planning process moved forward, however, two municipalities dropped out, as did the Navy after developers targeted its land for conservation. The shift to subarea planning further reduced coordination among the remaining participants—a shift to which the FWS gave its imprimatur by issuing separate permits to each jurisdiction. Policymakers' reluctance to create a regional entity to facilitate interjurisdictional consistency in monitoring and management further eroded the MSCP's coherence.

Pima County provides an instructive comparison: there, county officials initially eschewed coordinating with the state of Arizona and the municipalities of Tucson, Marana, and Oro Valley, opting instead to maintain an environmentally protective stance that might have been diluted by a more cooperative process. (The county did work with federal partners, Arizona Game and Fish, and the Tohono Nation because those entities shared its more environmentally protective philosophy.) The county's go-it-alone approach exacerbated existing interjurisdictional tensions in the short run. By asserting a strong position and funding the science to support it, however, the county laid the groundwork for long-term cooperation. The municipalities have since used the county's detailed technical analyses as they craft their own habitat conservation plans, and, in fact, interjurisdictional coordination has been increasing in Pima County.

Like their terrestrial-system counterparts, the CERP and CALFED planning processes began with a burst of interagency coordination, as federal and state agencies with divergent missions sought to replace conflict and disjointed management programs with conciliation and more coherent approaches. In both cases, mechanisms for joint operations created during the planning process have ensured some level of ongoing cooperation. On the other hand, because such interaction is voluntary, agencies have defected when their existing missions were inconsistent with the direction of the program. For example, in northern California the state and federal water projects declined to coordinate their decisions with the wildlife agencies when doing so facilitated exporting more water from the Delta. In South Florida, interagency tensions resurfaced after the federal government approved CERP—although a more serious complaint has been that efforts to coordinate have resulted in an endless series of meetings and few actions.

By contrast, in the Kissimmee River and the Mono Basin a landscape-scale focus guided by a single goal has led to interagency cooperation—but only after initial resistance. In central Florida the Army Corps of Engineers—reluctantly at first, but more enthusiastically after Congress issued a federal mandate—has been working alongside the South Florida Water Management District to dechannelize the Kissimmee River. In the Mono Basin, the Los Angeles Department of Water and Power (DWP) vigorously defended its prerogatives in court and the media for more than a decade. Once the State Water Resources Control Board (SWRCB) issued a clear ruling requiring the DWP to restore Mono Lake and its feeder streams, however, the agency began working closely with the SWRCB, Southern California Edison, the Forest Service, and even its former adversary, the Mono Lake Committee, to devise and implement a restoration plan.

Collaborating with Stakeholders

According to the optimistic model, collaborating with stakeholders is likely to result in more effective environmental protection than the conventional regulatory process for three reasons. First, deliberation that aims for consensus fosters trust among former adversaries, as well as transformation in their perceptions of their interests and, hence, a greater capacity to devise innovative approaches to environmental conservation and restoration. Second, in a collaborative forum participants are likely to incorporate local ecological knowledge and learn from, rather than bicker about, the scientific basis for planning, and therefore to devise a plan that is consistent with what scientists believe is necessary to conserve or restore an ecosystem. Collaborative planning promises a third benefit as well: once formulated, plans are actually implemented because those who devise them are committed to their realization. Pessimists worry, however, that collaborating with stakeholders prevents planners from addressing the root causes of ecosystem decline and instead yields vague, lowest-common-denominator solutions that unravel during implementation. A careful analysis of the cases described in this book suggests that the beneficial effects of stakeholder collaboration are mixed at best. Although collaboration can improve relationships among stakeholders, the newly created trust is fragile and does not necessarily translate into transformed interests or a greater willingness to adopt or implement environmentally protective policies and practices. Furthermore, collaboration does not necessarily lead to a plan grounded in the best available science or to durable implementation.

Trust, Transformation, and Innovation

Consistent with the optimistic model, in the four cases where policymakers convened stakeholders with the aim of achieving consensus on a landscape-scale plan—the BCCP, MSCP, CERP, and CALFED—trust among most participants increased, at least temporarily. Many participants also gained a broader view of their own interests and became more sympathetic to the concerns of former adversaries. There is little evidence that stakeholders' interests were genuinely transformed, however; in all four cases, intractable disputes over baseline levels of environmental protection and insistence by all parties on formal assurances make clear that some level of distrust and value differences persisted over time. Moreover, there is minimal support for the notion that collaboration led to the development of innovative mechanisms. In the two instances where novel mechanisms *were* adopted, they emerged as a means to break a bargaining impasse rather than as a result of brainstorming among trusting, transformed individuals. In Austin, Assistant City Manager Joe Lessard came up with tax increment financing as a way to facilitate Travis County's participation in the BCCP; similarly, high-level policymakers brokered the adoption of the Environmental Water Account in order to cement agreement on CALFED. (Moreover, although the latter mechanism clearly made the water supply more reliable for users, its environmental benefits have been minimal at best, providing a reminder that innovative is not necessarily equivalent to environmentally beneficial.)

In fact, with respect to trust and transformation, much of the evidence in the four cases where planning depended on stakeholder consensus is more consistent with the pessimistic than the optimistic model of EBM. To increase their chances of reaching agreement, consensus-oriented groups tended to include only "reasonable" participants and to marginalize those who espoused more "extreme" views. Because perceived reasonableness is a function of proximity to the status quo, environmentalists who espoused substantial changes in management emphasis were disadvantaged by collaborative approaches. Furthermore, although relationships among them improved, many participants nevertheless described stakeholder negotiations as more akin to bargaining than to deliberation, particularly as they began trying to make plans more specific. Consistent with the pessimistic model, stakeholder groups tended to avoid issues likely to provoke serious disagreement and to mask such differences by using vague language—a decision that ultimately haunted implementation.

Finally, as Thomas Stanley, Jr., (1995) warns, reliance on stakeholder collaboration was associated with an unwillingness to confront the fundamental causes of the ecological decline that prompted EBM in the first place. Stakeholder-driven processes clearly did not result in the kind of philosophical transformation envisioned by some EBM proponents, from a utilitarian view of nature to one in which human behavior is more appropriate to the living systems of which humans are a part and on which their survival depends. Instead, in all four cases where policymakers insisted on agreement among stakeholders, planners redefined the problem: in the terrestrial cases the problem was that Endangered Species Act (ESA) enforcement was impeding development; in the aquatic cases, the problem was inadequate storage and "wasted" water. The resulting plans promised to solve these problems by expanding the pie so that it was possible to meet the demands of humans and the needs of natural systems simultaneously—in other words, by furnishing win–win solutions that perpetuate the unsustainable consumption patterns that prompted the problem-solving effort in the first place.

Grounding in the Best Available Information
Stakeholder collaboration did not seem to increase the use of local ecological knowledge, as predicted by the optimistic model; in fact, in all
seven cases planners incorporated local knowledge when they perceived
it as relevant and ignored it when they did not, regardless of the decision-
making process used. In all three terrestrial cases planners drew on the
knowledge of some local experts in formulating their assessments, but
they marginalized others. For example, in San Diego two local academics
repeatedly asked for greater attention to seasonal wetlands in the MSCP,
but decision makers declined to accommodate their concerns. (The
plan's inadequate protection of wetland-dependent species ultimately
provided environmentalists with the ammunition for a successful lawsuit.) Ranchers in Austin, San Diego, and Pima County felt that planners ignored their perspective and relied on erroneous science; as a result,
the Farm Bureau actively opposed all three plans. Collaboration did not
markedly enhance the use of local knowledge in the aquatic-system cases
either. Both CALFED and CERP operate large, formal science programs
that are difficult for those who bear local knowledge to penetrate. On
the other hand, judicial and quasi-judicial decisions in the Mono Basin
case turned on local knowledge: the field notes of a local Fish and Game
biologist, as well as the recollections of people who had lived in the basin
for decades, formed the basis for environmentally protective rulings by
judges and the State Water Resources Control Board.

Nor did stakeholder collaboration, as the optimistic model predicts,
put an end to bickering among stakeholders over the science or lead to
plans grounded in the best available science. In San Diego developers
repeatedly challenged the assessments of the MSCP's scientific consultants and then, finding themselves unable to steer the official science in
the direction they preferred, commissioned their own experts to devise an
alternative preserve design. In formulating options for the Everglades restoration CERP scientists discounted long-standing scientific claims about
the importance of sheet flow, partly because their models were poorly
designed for testing flow-related hypotheses but also because accepting
them would have entailed choosing between environmental and user benefits. The CALFED science program skirted discussions of trade-offs—
and hence disputes over science—by framing scientific questions in ways
that were unlikely to threaten the status quo. (How can we continue
exporting water without jeopardizing the survival of endangered fish species?) Only when the ecological crisis deepened, rather than abated, did

the program support a more wide-ranging inquiry into the causes of the Delta's collapse. That inquiry, in turn, highlighted precisely the trade-offs planners had hoped to avoid (and prompted disputes over science) by implicating water diversions as the main cause of ecological problems in the Bay–Delta.

It is noteworthy that the Sonoran Desert Conservation Plan (SDCP), the Kissimmee River restoration, and the Mono Basin restoration are more recognizably grounded in precautionary interpretations of the available science than the plans that relied more heavily on stakeholder collaboration. In Pima County, for example, the Board of Supervisors declined to chisel away at the Conservation Lands System devised by the SDCP's Science and Technical Advisory Team to mollify development interests and ranchers. Similarly, in central Florida experts devised and implemented an ambitious plan to restore the Kissimmee River's ecological integrity, with little interference from political officials or stakeholders. And in the Mono Basin the courts and the water board repeatedly rebuffed the environmentally risk-tolerant claims of scientists hired by the Department of Water and Power.

Durable Implementation

The evidence also does not support the prediction of the optimistic model that collaboration ensures durable implementation. Instead, more consistent with the pessimistic model, implementation exposed many of the differences papered over during the collaborative planning processes, as stakeholders sought to prevent or modify projects that threatened their interests. In the MSCP and CALFED cases, lawsuits filed by disgruntled stakeholders have seriously undermined the basis of plans, thereby disrupting their implementation. According to a 2006 court ruling, San Diego's MSCP underprotects vernal pool species and must be revised. Similarly, in northern California, litigation by both agricultural and environmental interests has challenged CALFED's legitimacy; the most recent lawsuits by environmentalists and fishing interests have forced substantial reductions in water diversions. In South Florida, after a brief period of unity among stakeholders, legal and administrative disputes have virtually paralyzed implementation of CERP.

In several cases government officials, not stakeholders, have exhibited the strongest commitment to implementation, sometimes in the face of stakeholder resistance. For example, in Travis and San Diego counties managers and elected officials have lobbied tirelessly for state and federal financing and have worked to increase public awareness of and support for their preserves. Even a firm commitment among some government officials may be insufficient to ensure implementation, however, particularly in the face of chronic funding shortages. Travis County, for example, has instituted a financing mechanism that guarantees a steady source of money for management and monitoring but does not provide sufficient resources to acquire the rest of the preserve. Meanwhile, the city of Austin—which has completed its acquisition—is struggling to find the money to manage and monitor its land. The city and county of San Diego face a similar quandary.

Comparison among all seven cases suggests it is not necessarily stakeholder consensus on a plan that enhances implementation, but the willingness of policymakers to institute a predictable regulatory framework that both requires and rewards more protective management approaches. In Pima County, for example, the adoption of a protective approach to land-use regulation has reinforced civic pride in the community's environmental ethic, which in turn has reduced the wiggle room for county supervisors as they make zoning and permitting decisions. In the Kissimmee River restoration, the combination of a clear mission and the *results* of pilot projects—such as immediate and striking increases in bird and fish populations—have bred commitment to the restoration among policymakers, engineers, and the public. In the Mono Basin case, since the SWRCB established a clear set of standards, the DWP has worked assiduously to meet them. In that case as well, visible improvements have inspired loyalty to the restoration.

Flexible, Adaptive Implementation

Optimists believe that reliance on nonregulatory mechanisms fosters stewardship, and is therefore likely to result in efforts to undertake measures that are more environmentally protective than the law requires. In addition, adaptive management promotes continuous learning and a willingness by managers to adjust their approaches in response to new information. Pessimists argue that a variety of factors—such as turnover of personnel, entrenched agency missions, and resistance to change among managers—impede stewardship and learning from new information and, furthermore, that reliance on flexible implementation allows laggards to engage in business as usual without consequences. Consistent with the pessimistic model, reliance on nonregulatory mechanisms has not necessarily fostered efforts to go beyond the legal minimum; rather, when plans have multiple goals, development interests and their allies have sought ways to avoid compliance or comply minimally. Similarly, despite concerted efforts by scientists, none of the four full-fledged EBM initiatives have actually implemented adaptive management. On the other hand, the Kissimmee River and Mono Basin cases suggest that, when used in service of a single goal, adaptive management *can* yield environmentally beneficial results.

Going beyond the Legal Minimum

There is little evidence of increased stewardship by stakeholders in response to flexible implementation in the four cases of full-fledged EBM. Throughout the BCCP and MSCP planning processes, development interests staunchly resisted efforts to do more than required to get a "take" permit; more to the point, during implementation they have sought minimal compliance, not stewardship. For example, in Austin many developers have obtained individual "take" permits from the FWS, rather than going through the BCCP, in hopes of cutting a better deal; yet such project-by-project permitting is precisely what the BCCP sought to replace. In South Florida water utilities have resisted requiring users to pay more for water, and water managers institute conservation measures only in times of drought. Agricultural users have adopted best management practices but have done so largely in response to water quality litigation, not CERP. In northern California, prudent urban water utilities *have* taken steps to conserve and recycle water, as well as to diversify their water supply portfolios, but they have done so primarily in recognition that climate change and levee problems threaten the reliability of the Bay–Delta water supply, rather than in response to CALFED. Agricultural users have also adopted water-saving measures, largely in response to mandatory cutbacks, while continuing to fight for their water rights.

By contrast, in Pima County, which adopted—and to date has largely adhered to—a set of stringent but nonbinding development guidelines, many developers now routinely consult with environmentalists before taking a rezoning proposal before the county's Board of Supervisors. Although neither the Kissimmee River nor the Mono Basin restoration relies on nonregulatory mechanisms to achieve its goals, both have proceeded largely unimpeded by resistance from stakeholders. In fact, in the Mono Basin case the DWP has on occasion taken steps that exceeded what it would have been required to do by the state water board. For example, in 2006 the DWP opened two stream channels on Rush Creek and made them perennial at the behest of the Mono Lake Committee, even though the stream scientists (whose opinion weighs most heavily with the water board) supported a less ambitious approach.

Learning and Adjustment

The evidence on the extent to which a stated policy of adaptive management has resulted in learning and policy adjustment is also mixed. A variety of factors have impeded the use of adaptive management in the four cases of full-fledged EBM: inability to agree on a baseline level of environmental protection, reluctance to allot money for monitoring, unwillingness to create institutions that can coordinate collection and analysis of data across jurisdictions and agencies, and political constraints on adjusting policies and practices. Of even greater concern, in each case policymakers used a commitment to adaptive management to justify minimally protective plans, noting that they could adjust current practices if monitoring revealed them to be inadequate to restore ecological health. But the options for making such adjustments are so highly constrained that it is virtually inconceivable that managers can respond effectively to new information. In both the BCCP and MSCP cases, for example, there are few options for protecting additional land beyond what is already designated for acquisition, and the "no surprises" provision—which limits landowners' liability—only exacerbates the challenge. CALFED's Environmental Water Account guarantees water for users while often failing to meet the needs of fish. Similarly, in most of their projects CERP planners have little wiggle room for adjustment on behalf of the environment because water users and flood control-recipients enjoy guaranteed protection.

It is too early to assess Pima County's propensity to manage adaptively, although the county's Conservation Lands System provides considerable buffering and leeway for adding land. In the Kissimmee River and Mono Basin cases, on the other hand, managers have already demonstrated the utility of an adaptive approach. Test projects have helped refine restoration techniques, resolved disagreements over the impacts of particular methods, and revealed the importance of re-creating the flow of water to the recovery of the aquatic system. The productive use of adaptive management in these cases suggests that it may be more readily employed when the overarching goal is clear and experimental projects can be designed to resolve uncertainties about how best to achieve it. (CERP's recent adoption of incremental adaptive management for its environmentally beneficial Decomp Project further supports this notion.)

The Importance of Political Leadership and Regulatory Stringency

Comparisons among the seven cases described in this book suggest that setting goals through stakeholder consensus reduces the likelihood that EBM will conserve or restore ecological health because doing so perpetuates, rather than mitigates, the existing imbalance of power between development and environmental interests. That imbalance shapes many aspects of the political context of negotiations, from the availability of local regulatory tools for managing growth to the language that is typically used in discussions of growth management. In each of the four cases where plans were developed collaboratively, environmentalists began the process with considerable clout, thanks to the existence of a regulatory hammer provided by the ESA or the Clean Water Act. Over time, however, development interests regained their ascendancy. In part this is because their abundant economic resources translated into greater staying power. But development interests also had the status quo on their side. Preventing major change is considerably easier than trying to revamp the procedures, incentives, and management cultures of multiple agencies and jurisdictions. Being aligned with the status quo confers discursive advantages as well: in the United States, the language typically employed in local planning debates favors economic growth and development. Local officials' dependence on property-tax revenues and long-standing affinity for growth only enhance development interests' superiority.

Some proponents of EBM have acknowledged but downplayed such power differences or have hoped that by transforming interests and rendering agreement on science, collaboration would result in more effective and durable plans. But the evidence in the cases described above belies such expectations. Collaborative processes did not ameliorate power differences; in fact, those differences permeated stakeholder negotiations. Furthermore, when plans were devised collaboratively, a commitment to flexible, adaptive implementation has neither facilitated learning and adjustment nor fostered efforts among stakeholders to exceed legally required levels of protection. Instead, the absence of enforcement mechanisms has enabled pro-development stakeholders to avoid taking protective measures when doing so has conflicted with their immediate economic interests.

The comparison cases suggest, however, that landscape-scale ecological conservation and restoration initiatives *can* yield measurable environmental benefits if elected officials, judges, or administrators insist on the preeminence of environmentally protective goals and establish clear regulatory boundaries within which stakeholders can negotiate. In these cases, development interests have been far less able to co-opt the process and more inclined to accept reductions in their resource allotments. The most notable instance of pro-environmental leadership was Pima County's SDCP process, where from the beginning public officials expressed their commitment to an environmentally protective plan and used all the regulatory and rhetorical tools at their disposal to promote such an outcome. Development interests tried to circumvent the local process by alerting allies at the state level and using the courts, but they failed to derail the SDCP, thanks to the unwavering commitment of local officials. Similarly, in the Kissimmee River case Florida governors, as well as experts and high-level administrators at the South Florida Water Management District, adopted and maintained pro-environmental positions despite the resistance of development interests. And in the Mono Basin case, a series of pro-environmental judicial rulings, as well as the decision of the state water board, firmly established an environmentally protective regime.

What, then, causes such pro-environmental leadership to emerge? In the cases described above, environmentalists used litigation and public campaigns to empower public officials who wanted (or were willing) to endorse an environmentally protective definition of the problem. In all three cases, environmentalists' success depended not only on their tactical skills but also on the extent to which the available science provided a firm foundation for advocacy and whether existing laws and regulations furnished legal hooks. In each case, the environmental community was unified behind tenacious and skilled policy entrepreneurs. In the Pima County and Mono Basin cases, environmental advocates also took pains to build broad coalitions that included neighborhood and social justice groups. In the Kissimmee River case, restoration advocates were among the state's most prominent environmental spokespeople, and they focused on creating linkages between the river and other highly valued resources, particularly Lake Okeechobee and the Everglades. Of course, although environmentalists can increase the likelihood that environmental leadership will emerge, they cannot guarantee it. Their success also depends on the occurrence of favorable focusing events, the tactical choices of their opponents, and other factors.

Alternative Explanations and the Role of Complexity

These findings will not be popular among the many scholars and practitioners who believe that collaboration with stakeholders and flexible implementation are essential components of EBM. They will rightly point out that although the comparison cases I have chosen are matched in some important respects, they vary in other, unaccounted-for ways, leaving open the possibility that factors besides political leadership and regulatory stringency explain the observed differences in outputs and outcomes. For example, the evidence suggests that federal support can enhance efforts to formulate and implement environmentally protective plans. In five of the seven cases, the active involvement of Interior Secretary Bruce Babbitt gave additional political impetus to landscape-scale planning initiatives. In the terrestrial cases the federal government played a particularly important role: it supplemented local efforts by creating wildlife refuges, national monuments, and national conservation areas; in Austin and San Diego federal dollars also bolstered local funding for preserves. But in two of the three most ambitious efforts—the Pima County SDCP and Mono Basin restoration—state and local officials proceeded with only modest financial support from the federal government.

A second factor that may affect the protectiveness of EBM initiatives is the condition of the local or regional economy. Each of the cases spans at least a decade, during which governments experienced both flush and tight economic conditions. Although those conditions did influence both the formulation and the implementation of plans, their impacts were mixed and offsetting. During periods of economic growth state and local governments had more money to buy and manage land, but private demand increased simultaneously, making it more difficult to acquire property or limit water use. During economic downturns pressure on a region's natural resources abated, giving policymakers more leeway to buy or designate protected areas, but governments struggled to raise funds for land acquisition and had to contend with concerns about a dwindling property tax base.

The third, and most compelling, alternative explanation for the variation in protectiveness of outputs and outcomes across cases is complexity, either ecological or political. One kind of complexity arises out of the configuration of the target landscape. Of the three terrestrial cases, San Diego was the most advanced in terms of development and so had the fewest options for preserving large swaths of land. The Austin area was substantially less developed than San Diego but did not have large tracts of state- or federally owned land on which to base its preserve. Pima County, by contrast, contained a substantial amount of not-yet-developed and publicly owned land, most of which was potentially eligible for preservation. By the same token, CERP and CALFED are among the most complicated of the nation's EBM initiatives: they encompass large regions and aim to restore ecosystems that have experienced extensive human modification that has altered natural processes in fundamental ways. Both the Kissimmee River and Mono Basin ecosystems have also been extensively modified, and restoring them presents technical challenges, but both lie in relatively undeveloped locations, so planners have had more latitude in restoring them. In short, the configuration of the target landscape clearly contributes to the ease or difficulty of conserving or restoring its ecological health.

The relative *organizational* simplicity of the comparison cases—which involve fewer agencies and jurisdictions—also appears to have contributed to the ability of individual leaders to emerge and exert regulatory leverage over recalcitrant stakeholders. Like a single goal, clear lines of authority and assignment of responsibility facilitate decisive action. By contrast, trying to coordinate numerous entities with no single agency or jurisdiction at the helm diffuses authority in ways that can impede progress. That said, complexity is at least to some extent an artifact of the problem-solving approach. The effort to satisfy all stakeholders and "balance" numerous goals leads almost inevitably to elaborately constructed "win–win" solutions that are extremely difficult to execute. Moreover, as Deborah Stone (2003) points out, arguments about complexity are often linked to claims that it is impossible to assign blame for failure (or responsibility for remediation), and hence make it easier to avoid taking action. "In politics," Stone observes, "models of complex cause often function like accidental or natural cause. They postulate a kind of innocence, because no identifiable actor can exert control over the whole system or web of interactions" (196).

The point here is that willingness to exert leadership and regulatory leverage in service of an overarching, environmentally protective goal is not the only cause of a program's environmental effectiveness. It does, however, appear to be a necessary ingredient. And it is the causal factor that advocates and policymakers can most readily affect.

Generalizing

Although they don't provide a roadmap, the lessons drawn from this comparative analysis should help us think about the likely outcomes of other large-scale, multiple-species habitat conservation planning and aquatic-system conservation and restoration initiatives. Examples of the former include the Coachella Valley in California, Clark County, Nevada, and Washington State HCPs. Examples of the latter include efforts to restore the ecological resilience of California's Owens Valley, the Klamath Basin, the Puget Sound, and the Louisiana coastal wetlands. This analysis may also help advocates and policymakers decide whether and how to proceed as new opportunities for EBM arise. Given the heterogeneity of EBM and of collaborative, place-based environmental problem-solving more generally, I can draw only tentative conclusions based on the detailed exploration of seven cases. On the other hand, my findings are buttressed by the arguments of theorists who urge caution in embracing collaborative planning, as well as by some recent empirical studies.

Some prominent political theorists have warned that stakeholder collaboration will not ameliorate, and in fact may be distorted by, power imbalances in the political–economic context of negotiations. They highlight several mechanisms that appeared repeatedly in the cases described above. For example, Joshua Cohen and Joel Rogers (2003, 251) counter the suggestion that reasoned deliberation will mitigate power differences, explaining that "The problem of generalizing deliberation is not that subordinate groups are unable to hold their own in deliberations, but that those with power advantages will not willingly submit themselves to the discipline of reason if that discipline presents large threats to their advantage." Environmental philosopher Robyn Eckersley (2002) points out that powerful parties in a collaborative process may exert more than "communicative power." In practice, she adds, one party may dominate another because it has more effective means of force at its disposal, such as the power of the state, the private power to make threats or offer inducements, or the more subtle power that comes with being the dominant cultural or ethnic group in a society. She adds that proponents of collaborative problem-solving do not really consider the impact of the ascendancy of neoliberalism and the increasing salience of the rights of corporations relative to citizens, both of which condition and constrain the negotiating margins of local and regional environmental policymaking. In short, Eckersley says, "The greatest weakness of [collaborative problem-solving] is that it has a tendency to be conservative, to take too much as given, to avoid any critical inquiry into 'the big picture' and to work with rather than against the grain of existing structures and discourses (such as those that are prevalent in real-world liberal democracies) and facilitate 'interest accommodation' in the context of the prevailing alignment of social forces" (65). Planning scholar Susan Fainstein (2000, 458) raises a similar caution, concluding that "Ideas can give rise to social movements that in turn change consciousness, ultimately resulting in the adoption of a new public policy, but this is more than a matter of negotiation and consensus building among stakeholders.... The aroused consciousness that puts ideas into practice involves leadership and the mobilization of power, not simply people reasoning together."

Empirical studies of environmental planning and policymaking also provide support for claims about the propensity of collaborative planning to yield conservative solutions and the hazards of pursuing multiple goals simultaneously. For example, in her detailed analysis of the Quincy Library Group (QLG), Sarah Pralle (2006) finds that the decision by key activists to plan collaboratively in a local forum led planners to redefine the problem as forest fires, rather than excessive logging, causing ecological degradation—a move that defused conflict and allowed for a solution that gave something to everyone without necessarily addressing the region's core environmental problems. Pralle notes that the focus on process disarmed environmental challengers, who found it difficult to combat the "overwhelmingly positive characterization" of local, collaborative decision making. Supporters of the QLG derided as extremists activists who questioned the benefits of collaboration or continued to employ conventional political tactics. Pralle observes that "In a world of polarized interest groups and partisan gridlock, policymakers may be more than willing to settle for outward signs of consensus rather than true political compromises" (202). By contrast, she finds that environmentalists in Clayoquot Sound succeeded in scaling back British Columbia's pro-development forest policy by using conventional (salience-building) techniques to redefine the problem and expand the conflict into the international arena.

Richard Norton (2005) disparages the implementation of North Carolina's highly touted Coastal Area Management Act, which aims to promote both environmental protection and economic growth. He argues that

If a fundamental reason for promoting more and better local planning is to ensure effective regional growth management, then a policy of balancing economy and environment in a "fair" way will not do enough. To yield truly sustainable outcomes—sustainable in the sense of sustaining the ecological systems upon which we depend for our survival—fair outcomes need to be sought within the constraints of first ensuring the protection of those ecological systems (67).

Norton goes on to observe that it is fruitless to try to engineer away the ecological consequences of equitable growth and, moreover, that a focus on process, rather than substance, will not ensure environmentally adequate outcomes.

Other scholars have discerned the importance of political context in determining the adoption of environmentally protective measures, although they have not focused specifically on power relations. For example, in his study of the politics of dam removal, William Lowry (2003) finds that ambitious river restoration is more likely to be undertaken when the political context is receptive—that is, in places where the decision-making venue is tolerant of change, the costs of maintaining the status quo are high and apparent, and there is widespread acceptance of scientific information on the benefits of dam removal. Lowry hints at, but does not pursue, the fact that these characteristics are matters of perception that are at least partly subject to manipulation by strategic advocates. (He does suggest that "Participants and analysts alike increasingly recognize that the policy-making process can be significantly different when traditional adversaries are communicating and building trust rather than resorting to demonizing and litigating" (203). He does not investigate the empirical validity of this claim, however.) Lowry also finds restoration more likely when the number of jurisdictions involved is limited and a unidimensional change is required.

My findings also seem to resonate with the results of some investigations of multi-state EBM—though it is perilous to draw parallels with cases that have been analyzed using different criteria. Barry Rabe (1999) points out that progress in cleaning up the Great Lakes Basin has been halting because "The multi-institutional system that has evolved remains vulnerable to uneven—and ever-changing—levels of commitment from its respective institutions" (251). Moreover, he finds that "The links between numerous policy initiatives remain very tenuous, often dependent on policy entrepreneurs who cannot be relied on to serve as permanent champions" (251). And finally, "Measurement of environmental quality in the basin continues to prove elusive, with no standard metric in place to serve as a reliable and comprehensive evaluation tool" (251). Many of the same features characterize the oft-cited Chesapeake Bay Program. After more than 30 years of collaborative work aimed at reviving the Chesapeake Bay, the population of native oysters, which traditionally filtered vast amounts of the bay's water, has collapsed; aquatic grasses, an important indicator of overall bay water quality, cover roughly one-tenth of the acreage they occupied historically; and the population of the famous blue crab is well below half its potential. Most observers blame the program's voluntary, collaborative approach, which allows states, localities, and stakeholders to defect when their immediate economic interests are threatened (Ernst 2003; Horton 2003).

In his impressionistic assessment of his experience with Platte River collaborative watershed planning, John Echeverria (2001) argues that from the outset, planning was "heavily weighted in favor of parochial economic interests" (560). He notes that the political context in which the process began was inhospitable to environmentally protective measures and largely determined what was possible. Echeverria claims that water users and political leaders in the river-basin states embraced the Platte program because it gave them more say in the outcome than a purely federal process would have; it also enhanced the opportunities to argue for taxpayer help in paying project mitigation costs. He concludes that "If the process succeeds in generating any type of program to address Platte River management issues, the solution will almost certainly be a failure, both in absolute terms and relative to what could reasonably be achieved through traditional regulation or other, more innovative approaches" (560).

Notwithstanding these disappointing results, there are clearly instances in which collaborative efforts succeed in conserving or restoring natural resources. Commentators often cite the Malpai Borderlands Group and the Applegate Partnership as cases in point, but there are others as well. Unlike the cases investigated in this book, however, these initiatives seem to feature most of the characteristics posited by Elinor Ostrom (1990) and her colleagues as essential to effective local, collaborative management of common pool resources: appropriators believe they will be harmed if they do not adopt rules to govern use of the resource; are affected in similar ways by proposed rules; value the continued use of this common property resource (discount rates are low); face low information, transformation, and enforcement costs; share generalized norms of reciprocity and trust; and constitute a relatively small and stable group. In addition, the target resource is in sufficiently good shape that efforts to protect it will confer benefits, there are valid and reliable indicators of system health, the flow of resources is relatively predictable, and the system is sufficiently small to allow knowledge of external boundaries and internal microenvironments (Ostrom 2001). Such conditions are increasingly rare, however, particularly in the United States, and most do not hold at a regional scale.

As Ostrom (2001) notes, we will be better able to design institutions that reliably yield environmentally beneficial outcomes if we improve our understanding of how scale and homogeneity affect outcomes. For example, the scale at which participants most readily reach agreement may not be the most appropriate scale for addressing environmental problems. Furthermore, although it may be easier to mobilize a small group for collective action, only a larger group may have sufficient financial and political resources to actually succeed in meeting the objectives of collective action (Agrawal 2000). Consistent with these observations, policymakers have experimented with institutional designs, such as nested hierarchies, in the Chesapeake Bay Program, the Great Lakes, and other programs. The analysis above suggests that if such initiatives are going to yield genuine environmental benefits, they must be backed by rules that establish precautionary floors while allowing for—and in fact encouraging—efforts to exceed the legal minimum.

Conclusions

All the cases described in this book demonstrate movement toward more environmentally beneficial management. But only in those cases where public officials circumscribed the planning process by articulating a strong, pro-environmental goal and employing regulatory leverage are the resulting policies and practices likely to conserve biodiversity or restore damaged ecosystems. In response to this argument proponents of collaboration are likely to point out that marginal, incremental improvements may well trigger more substantial changes in the long run. For example, according to Mark Imperial (1999) research on common pool resource management suggests that taking small steps may allow participants to garner political support and develop their capacity for managing complex problems over time. There are hazards associated with an incremental approach, however; over time, as more claims are made on a resource, and new stakeholders assert their "rights" to use the resource, it becomes increasingly difficult to roll back excessive consumption. Moreover, incremental changes in policies and practices may dampen demands for better environmental management—either by producing insufficient improvement or by creating the sense that a problem has been addressed. Interestingly, in the cases described in this book, it was *major* change—sufficient to produce tangible benefits—that triggered positive political feedbacks (Baumgartner and Jones 1993).

The fact that the most environmentally beneficial results emerged as a result of conventional politics should not suggest the simplistic conclusions that the old way is always the best way. It is important, however, to separate valid critiques of the conventional regulatory approach from politically driven complaints designed to enhance the power of development interests. Interest in consensus-building and other dispute resolution mechanisms grew as confrontation over environmental policy escalated during the Reagan administration, after the business community and conservative activists mobilized to contest the first wave of environmental laws and regulations (Amy 1990). But anti-environmentalists are not the only ones who have embraced collaborative approaches. Many regard them as important means for democratizing decision making. Others, as Susan Fainstein points out, turned to collaboration out of dissatisfaction with the previous generation of planners, who used their expertise to impose an order that was inattentive to social justice and environmental concerns. Ironically, collaboration took hold just as a new generation of more consultative and environmentally oriented experts emerged. Of course, this is one reason those who oppose environmental regulation prefer bottom-up approaches: they can no longer count on experts to promote their beliefs. For many political officials, asking stakeholders to generate consensual plans is self-serving: it relieves them of the burden of making politically risky decisions.

In deciding how to proceed, environmental activists should be aware of the trade-offs associated with participating in time-consuming collaborative efforts that divert scarce resources from other activities that can stimulate and support pro-environmental leadership, such as campaigns to raise public concern about environmental problems and litigation to assure the legal status of environmental protection. Before engaging in collaborative efforts, which contain conflict and redefine problems, they should consider the ripeness of a controversy and, in particular, the extent to which their power is stable and embedded in political and economic structures. In many instances, environmentalists are likely to be more effective if they expend their limited resources building coalitions with low-income communities, workers, and social justice advocates, who are also disadvantaged by devolved, collaborative approaches (Foster 2002).

Policymakers inclined to support environmentally protective policies face similar strategic choices in deciding whether and how to establish stakeholder processes. Planning scholar John Forester (1989) cautioned in the 1980s that if planners present themselves as neutral mediators, they may encourage premature consensus-building when empowerment and organizing strategies (prenegotiation strategies) are more appropriate. He encouraged planners to reach out to those in the community with fewer resources and help them to participate fully. He suggested that planners "Anticipate political–economic pressures shaping design and project decisions and compensate for them, anticipating and counteracting private raids on the public purse by, for example, encouraging coalitions of affected citizens' groups and soliciting political pressure from them to counter other interests that might threaten the public" (155–156).

One important benefit of mobilization and litigation is their ability to create the uncertainty necessary to yield the shifts in the balance of power needed for productive deliberation. As Cohen and Rogers (2003, 252) explain:

The acknowledgement of pervasive, persistent, and profound uncertainty, and the associated recognition of mutual dependence, may throw into question our sense of our own interests. After all, even the powerful come to see their own fate as dependent on securing the willing cooperation of others, as the fate of the weak depends on the willing cooperation of the strong.

But, they add, pervasive uncertainty—sufficient to make substantial differences of power ineffective—is a rare and special case. Given that, "The benefits of deliberation may well require direct efforts to address inequalities of power" (Cohen and Rogers 2003, 253).

Institutionalizing creative national-level policies may be the single most effective way to empower environmentalists vis-à-vis development interests, unify localities in service of an environmentally protective goal, and therefore facilitate collaborative problem solving that actually delivers on its promises. There are numerous ways the federal government can provide a backstop for landscape-scale efforts. One of the most effective ways to combat sprawl and reduce resource waste is to protect wetlands and habitat from development. To this end, Congress should strengthen, not relax, the ESA and the Clean Water Act, and provide the FWS, in particular, more resources with which to reward stewardship while penalizing noncompliance. Legislators could fund habitat conservation with the proceeds from a mechanism that raises money nationally, such as a surcharge on electricity consumption or water use—both of which are directly linked to destruction of natural areas. Congress should also greatly expand the conservation measures in the Farm Bill to encourage stewardship rather than the production of commodities. And federal legislation should eliminate dual mandates in the federal resource agencies; they have been disastrous for the environment and only marginally helpful for those who make their living off natural resources. We should encourage these agencies to engage fishermen, ranchers, and loggers in policymaking, but only once the preeminence of an environmentally protective mandate has been established. Many readers, including those who consider themselves environmentalists, will object that the "political climate" is inhospitable to such measures. But the vast majority of Americans benefit handsomely from environmental protection. The challenge for proponents of stronger national policies is to increase public awareness of how existing policies improve their quality of life and argue persuasively the need for new ones.

In short, the findings described above should not be construed as disparaging efforts to involve stakeholders in planning processes, but rather as affirming the importance of undertaking such negotiations within a hospitable context. The first step is to promote a protective regulatory framework and strong pro-environmental leadership to ensure collaboration is deployed in ways likely to improve environmental outcomes. As Robyn Eckersley (2002, 50) points out, in view of the respective strengths and limitations of critical advocacy and pragmatic mediation, our "real-world democracy" would be even poorer if it were made up only of mediators, or only of advocates. The tension between the two is, in fact, healthy, because it steers democratic deliberation away from policy paralysis, on the one hand, and policy complacency on the other. "Democracy," she points out, "is about *arguing* as well as *making decisions* and advocates and mediators play different but invaluable roles in each of these phases" (Eckersley 2002, 66). The point, then, is that both advocacy and mediation are important functions; the trick is deciding which strategy to deploy when.

Notes

Chapter 1

1. By the late 1990s the phrase "ecosystem-based management" had largely replaced the original term "ecosystem management," as proponents sought to emphasize management of human activities within ecosystems, rather than management of ecosystems themselves.

2. Dewitt John was one of the first to describe a set of emerging approaches to environmental problem-solving that stand in contrast to the traditional regulatory process. John (1994, 7) coined the term "civic environmentalism" to characterize initiatives whose central animating idea is that "Communities and states will organize on their own to protect the environment without being forced to by the federal government. . . . Civic environmentalism is fundamentally a bottom-up approach to environmental protection." Others have since used a variety of terms to denote variants of the phenomenon John described, including community-based environmental protection (USEPA 1999), landscape-based planning (B. L. Johnson and Campbell 1999), sustainable community-building (Mazmanian and Kraft 1999), collaborative conservation (Cestero 1999; D. Snow 2001), grassroots ecosystem management (Weber 2000), empowered participatory governance (Fung and Wright 2003), collaborative environmental management (Koontz et al. 2004), and adaptive governance (Brunner et al. 2005).

3. More precisely, Leach and Sabatier (2005) find that neither trust nor social capital explains the implementation of restoration projects, except to the extent they promote agreements that lead to projects. Raymond (2006) finds that trust is not essential to overcoming collective action problems.

4. The literature suggests that although both landscape-scale planning and adaptive management are widely prescribed, they rarely occur in practice, so most research focuses on barriers to implementation rather than on outcomes (Allan and Curtis 2005; Halbert 1993; B. L. Johnson 1999; Stankey et al. 2003; Tonn, English, and Turner 2006; C. Walters 1997).

Chapter 2

1. As Botkin (1990) explains, preservationists historically regarded nature as self-regulating and human activity as unnatural disturbance; hence, nature was best left alone. Conservationists, by contrast, tried to figure out how to use nature in a way that would allow it to recover its balance. The former perspective prevailed in the more preservation-oriented agencies, such as the National Park Service. The latter dominated the utilitarian natural resource management agencies, such as the U.S. Forest Service and Bureau of Land Management.

2. Historian Donald Worster (1994) dates the transition from the classical to the "flux-of-nature" paradigm to a 1973 article by William Drury and Ian Nesbitt that characterized the forest as an erratic, shifting mosaic of trees and plants with no emerging order. According to ecologist C. S. Holling (1995), the revision in ecology arose out of extensive comparative studies, critical experimental manipulation of watersheds, paleontological reconstruction, and studies that linked ecosystem models and field research. It is important to note that the flux-of-nature paradigm did not replace, but rather subsumed, the equilibrium view, as most ecologists came to regard stable equilibria as limiting cases within a more general state of variability.

3. More specifically, conservation biology rests on a set of normative standards against which to measure action: first, diversity of organisms is good, and extinction is correspondingly bad; second, ecological complexity is good; third, evolution is good; and fourth, biotic diversity has intrinsic value, irrespective of its instrumental or utilitarian value (Soulé 1985).

4. The term "sprawl" refers to relatively low-density, noncontiguous, auto-dependent residential and nonresidential development that consumes large amounts of farmland and open space. University of Chicago planning professor Robert Bruegmann (2005) charges that commonly cited statistics linking population and land area consumed are inaccurate and misleading, and that land use should be compared not to population growth but to the increase in number of households. (According to Robert Burchell and his coauthors (2005), land in the United States is being consumed at three times the rate of household formation.) In any case, Bruegmann's critique elides the fundamental concern of sprawl detractors that more land is being developed to accommodate a smaller number of people. Bruegmann also rejects normative assessments of the costs of sprawl, a phenomenon that he says simply reflects people's growing ability to satisfy their desire for space, privacy, and mobility. His argument is consistent with libertarian values: by definition, what is good for individuals is good for society; according to this reasoning, there are no collective action problems.

5. There are 75,187 dams in the United States. All watersheds of greater than 750 square miles have some dams. Even the handful of rivers commonly cited as free flowing—the Upper Yellowstone, Colorado's Yampa River, the Virgin River of Utah, and the Middle Fork of the Salmon—have scores of dams in their tributaries (Graf 1999).

6. This does not mean they dismiss the importance of protecting individual species. The position espoused by most conservation biologists is that both "coarse-filter" (vegetation community-scale) and "fine-filter" (individual species-scale) approaches are needed to adequately protect the full range of organisms and processes.

7. The U.S. Public Interest Group (PIRG) has pointed out, however, that U.S. sewer systems are aging, and without significant investments in wastewater treatment infrastructure, sewage pollution levels are likely to rise.

8. Environmentalists had filed legal actions in 38 states, and the EPA was under court order to ensure that TMDL allocations were established (Scheberle 2004).

9. By the 1970s the lower 48 states had lost about half of their wetland endowment (from the time of European settlement), primarily as a result of agricultural conversion. Since the 1970s, however, wetland losses have declined steadily as agricultural conversion slowed. In 2006 the FWS reported that between 1998 and 2004 the rate of wetland acreage gained through restoration and creation exceeded losses for the first time (Dahl 2006). The study's author noted, however, that acreage figures say little about the quality or condition of wetlands; for example, open water ponds (including storm water retention ponds and decorative ponds) are replacing ecologically valuable estuarine and freshwater emergent wetlands, both of which continue to disappear.

10. State water rights rules vary by region. In the East, the "riparian" doctrine allows parties adjacent to rivers and streams to make reasonable use of those waters as long as such uses do not cause unreasonable harm to others. In the West, the governing doctrine is "prior appropriation," according to which those who made the earliest claim on a river have the highest priority rights to its water. In general, water rights are "usufructory," meaning the right must be exercised or is subject to revocation (Postel and Richter 2003).

11. Ideally, participants in a collaborative process jointly establish the rules of engagement, define the issues, design the collection and analysis of scientific data, help develop solutions, and aid in implementing decisions. Some practitioners add that collaborative processes are likely to generate the hypothesized benefits only if the facilitator is a well-trained professional (Susskind, McKearnan, and Thomas-Larmer 1999; Susskind 2005). Leach and Sabatier (2003), however, find that facilitators and coordinators need not be professionals in order to be effective and, moreover, that facilitators/coordinators are not essential to watershed collaboratives' efforts to implement restoration projects.

12. Scholars have spilled a great deal of ink defining consensus, debating its merits, and distinguishing between it and deliberation, and I will not rehearse those arguments in detail here. In brief, as political theorists Joshua Cohen and Joel Rogers (2003, 241) explain, to deliberate is "to debate the alternatives on the basis of considerations that all take to be relevant; it is a matter of offering reasons for alternatives, rather than merely stating a preference for one over another, with such preferences then subject to some rule of aggregation or submitted to bargaining." Seeking consensus is just one way of aggregating preferences.

13. There has been a lively debate among scholars and practitioners about the extent to which collaborative decision making can ameliorate structural differences in power. In the planning field, proponents of "communicative action" acknowledge deep structural inequalities but prescribe collaborative dialogue as a way to make incremental changes that may, over time, lead to more substantial change (Healey 2006; Innes 1996). These authors are hopeful about the transformative power of ideas, or policy discourses, that are generated collectively; they expect that in well-designed dialogues the force of reasoned argument, rather than power or status in a preexisting hierarchy, can be the deciding factor (Fischer 2003). By contrast, "critical realists," such as Susan Fainstein (2000, 2005), do not believe dialogue can lead to a "restructuration" of interests, and contend that although negotiation can ameliorate conflict, the resulting benefits for weaker groups are often meager. Political theorists Cohen and Rogers (2003) echo this concern when they worry that by failing to consider background conditions, proponents may overstate the ability of deliberative processes to neutralize power. A parallel debate rages in the field of environmental ethics, where environmental pragmatists, such as Ben Minteer (2002), advocate collaborative planning and criticize ecocentrist ethicists, such as J. Baird Callicott and Laura Westra, for seeking to impose their values on others.

14. Although social scientists often make claims for the superiority of either quantitative or qualitative methods, the two are best regarded as complementary. Quantitative analyses can provide information about the frequency of a phenomenon and reveal the presence or absence of relationships among variables; qualitative approaches elucidate the mechanisms by which attributes of a decision's structure and process translate into outcomes (Hedström and Swedberg 1998).

15. Growth pressures can cut both ways, of course; they can overwhelm growth management efforts, but they are also associated with strong resistance to new development as its negative side effects become more evident.

16. More technically, I sought variation on the dependent variable to ascertain which independent variables were causally significant.

17. Psychological research dating back to the 1920s has documented the halo effect, which may be related to cognitive dissonance. More recently, in their quantitative analysis of watershed groups, Leach, Pelkey, and Sabatier (2002) use stakeholders' perceptions of groups' impact on watersheds' environmental problems as a surrogate for environmentally protective outcomes; however, in a later work, they confirm that a "halo effect" makes participants likely to overestimate the group's impact on watershed conditions, and hence to affirm the need for objective measures (Leach and Sabatier 2005).

18. In a stressed or declining ecosystem, the relationship between adaptive management and precaution is particularly complicated. Stankey et al. (2003, 41) observe, "Acting in a risk-averse manner can suppress the experimental policies and actions needed to produce understanding that will reduce risk and uncertainty." On the other hand, Noss and Scott (1997) caution against using ecosys-

tem management experiments as an excuse to substitute one way of attempting to master nature for another. My point is simply that managers must have both the leeway and the resources to add more land or water to a system if information gleaned from monitoring suggests the amount set aside is insufficient.

19. That said, restoration may require active intervention, including reestablishing keystone species, controlling exotic plants and animals, restoring natural processes, etc. Ultimately, the question is whether human intervention aims to benefit natural systems or economic interests.

20. Karr (1993, 85–86) notes that "The existence of [ecological] integrity suggests that 'ecological health' is being protected. Ecological health is the condition when a system's inherent potential is realized, its condition is stable, its capacity for self-repair, when perturbed, is preserved, and minimal external support for management is needed."

Chapter 3

1. The 1971 Municipal Utility District Act created a mechanism for financing water, wastewater, and drainage improvements in suburban developments in Texas. The law authorizes municipal utility districts (MUDs) to provide for water and sewer systems and treatment plants, drainage improvements, and other services. A city must either consent to the formation of a MUD or provide the utility services itself (Butler and Myers 1984).

2. The HCP concept originated with an innovative plan to save habitat for endangered butterflies on San Bruno Mountain in northern California. In 1982 Congress codified the HCP option with an amendment to the ESA, section 10(a), which allows the FWS to grant an "incidental take permit"—that is, to allow the destruction of some endangered species or their habitat—in exchange for a plan to conserve sufficient habitat to ensure the species' survival.

3. The government entities represented were the city of Austin, Travis County, the Lower Colorado River Authority, the Texas Department of Highways and Public Transportation, the Texas General Land Office, and the Texas Parks & Wildlife Department.

4. The BAT members were Doug Slack, Texas A&M; Helen Ballew, Texas Nature Conservancy; David Steed, DLS Associates; John Cornelius, Fort Hood; William Elliott, Texas Department of Health; Joe Grzybowski, Central Oklahoma State University; Jim O'Donnell, Wild Basin Preserve; Jackie Poole, Texas Parks & Wildlife Department; James Reddell, Texas Memorial Museum; Chuck Sexton, city of Austin; and Rex Wahl, Texas Parks & Wildlife Department. Craig Pease of the University of Texas at Austin and Denice Shaw of North Texas State University consulted to the BAT. In addition, Kent Butler (Kent S. Butler & Associates), Clif Ladd (Espey, Huston & Associates), Joe Johnson (FWS), and David Tilton (FWS) served as advisers. (Ladd was originally a member of the BAT but resigned after his firm was chosen to write the HCP.)

5. In 1992, the World Resources Institute ranked Austin second among the 64 largest U.S. cities on its Green Cities Index, which is based on 14 environmental criteria (Pendleton 1992).

6. The draft plan called for Williamson County to raise about $13 million to buy 9,000 acres of habitat.

7. In 1985 developers had established the Southwest Travis County Road District and issued bonds on the assumption that property values would continue to rise. Shortly after the district issued the bonds, however, the real estate market collapsed, drying up funds to complete the project and leaving the county saddled with a growing debt.

8. City officials believed they had the authority to impose development fees in Austin's extraterritorial jurisdiction (ETJ) without legislation, although they anticipated court battles over this. But most of the valuable habitat was beyond the city's ETJ, and the county plainly lacked the authority to levy such fees.

9. Congress created the RTC in 1989 to replace the Federal Savings and Loan Insurance Corporation and respond to the insolvencies of about 750 savings and loan associations. The RTC owned thousands of acres of land in Texas that had been among the assets of the failed thrifts.

10. Oles claimed the plan's reliance on development fees would shift a significant share of its cost to developers—a move considered essential to garnering Judge Aleshire's support. Environmentalist Bill Bunch argued, however, that the new approach did not actually increase developers' share of the cost.

11. The FWS officially listed the Barton Springs salamander in May 2007.

12. Earth First! charged that the BCCP would safeguard only 29 percent to 37 percent of the Travis County habitat for the warbler and just 15 percent to 20 percent of the habitat for karst invertebrates. Sam Hamilton of FWS said these figures were incorrect and asserted that nearly 50 percent of warblers' current habitat would be protected, as would 93 percent to 95 percent of the caves occupied by rare invertebrates (Haurwitz 1993h). The difference arose out of whether they were counting occupied or potential habitat.

13. A survey commissioned in the early fall by TNC found a slight majority of county voters supported the bond issue, but most voters didn't go to the polls (Haurwitz 1993j).

14. Tax increment financing involves using the tax on the added value of property as a result of development made possible by participation in the HCP.

15. In the context of the ESA, "take" means to harass, harm, pursue, hunt, shoot, wound, kill, trap, capture, or collect any threatened or endangered species. Harm may include significant habitat modification that kills or injures a species by impairing its essential behavior.

16. The FWS and permit holders had little recourse in such cases because once the land was graded, it was impossible to prove the cleared land had been habitat or that clearing habitat had actually "taken" birds.

17. Usually, such disagreements have involved the county proposing to count land acquired for other purposes, such as water management, toward its target, and the city has resisted. In the latest twist, however, in 2006 the city of Austin proposed building a water treatment plant on part of the preserve and substituting another parcel that, although larger, was not actually songbird habitat. After a heated controversy, the city backed down and postponed a decision pending further study (Coppola 2007a). In December 2007 the city concluded a deal to purchase a privately owned parcel that was less environmentally sensitive than the previously proposed locations (Coppola 2007b).

Chapter 4

1. A growth management plan approved in February 1979 had separated the city into tiers to encourage infill development and discourage development on the outskirts. But a series of proposals that chipped away at the "future urbanizing area"—among them the massive 5,100-acre La Jolla Valley project, which included a Christian university, an industrial park, and housing—galvanized advocates of orderly growth (Calavita 1992).

2. According to the FWS (1993a), an aerial photograph of San Diego County taken in 1931 would have revealed 72 distinct patches of coastal sage scrub. By 1990, there were three times as many patches and each patch was, on average, one-tenth its 1931 size.

3. In 1988 the Environmental Protection Agency had sued the city for improperly treating its sewage. In response, the city had proposed an extensive upgrade whose biological impacts, according to the FWS, would have to be mitigated. (The EPA eventually issued a waiver for the sewer project, but by this time, the habitat conservation planning process was well under way.)

4. The San Diego Biodiversity Project and Palomar Audubon Society already had petitioned the FWS in September 1990; Atwood and the NRDC petitioned the FWS in December 1990 and the state in February 1991.

5. As noted in chapter 3, section 10 of the Endangered Species Act allows the FWS to issue "incidental take permits" for endangered species in exchange for the preparation of plans to conserve those species' habitat. The MSCP is only one of three HCP/NCCPs that has been prepared for San Diego County. It covers the South County; there is also an East County plan and a North County plan.

6. The Navy originally participated, but withdrew once biological analyses revealed the resource richness of several military complexes within the planning area. Twenty-two Native American tribes declined to participate from the outset.

7. Developers challenged the population viability analyses, which were used in comparing alternative preserve designs, because they were based on incomplete information.

8. Core areas are defined as areas that generally support a high concentration of sensitive biological resources that, if lost or fragmented, could not be replaced or mitigated elsewhere.

9. Ironically, although the CSS option included only 85,000 acres, most of those were privately owned, so the CSS was almost as expensive as the Multiple Habitats option but much less biologically robust.

10. Reflecting its continuing dual (and often conflicting) focus, the Working Group advised the jurisdictions to (1) preserve as much of the core biological resource areas and linkages as possible; (2) maximize the inclusion of public lands within the preserve; (3) maximize the inclusion of lands already conserved as open space; and (4) make the preserve affordable and share the costs equitably among all beneficiaries.

11. Developers had asked the city to allow development of 50 percent, not 25 percent, of property within the preserve. They lost this battle, which they waged only halfheartedly, but they did succeed in getting planners to deem large-lot residential development "conditionally compatible" with preservation of core areas.

12. In fire-prone southern California, brush management is an essential tool for preventing property damage.

13. Biologists Patrick Kelly and John Rotenberry (1993, 85) argue that "Reserve establishment in urbanized California is a pointless exercise in crisis management if those reserves are going to be gradually eroded away by external forces."

14. By this time, six other cities, as well as the Otay Water District, had also prepared subarea plans and were awaiting their approval.

15. The 171,917-acre figure represented an increase of 7,591 acres over the 1995 draft. The additional acreage reflected changes in the planning area boundaries that reduced the potential preserve acreage, as well as the deletion of 2,400 acres of private land, plus the addition of 10,000 acres of public land, as well as 4,250 acres of disturbed agricultural lands (City of San Diego 1998).

16. In many cases, clearing occurred despite requests by FWS officials to hold off until its biologists had reviewed project proposals and mitigation measures ordered by local officials (Silvern 1991b).

17. According to the plan, ten species were added based on wildlife policy clarifications (the agencies thought they were unlikely to occur within the study area, or the study area was not a significant portion of their range); three species were added based on improved preserve design; five species were added as a result of "additional evaluations"; 13 species were added as a result of new information and development of additional conservation measures (including protection standards for endemic species and vernal pools); and three species were deleted based on reevaluation of the data and conservation measures in the plan (City of San Diego 1998).

18. Subsequent lawsuits have challenged the Interior Department's "no surprises" policy, and the department has revised it, but the approved MSCP subarea plans' provisions remain in force.

19. The MSCP is the only plan for which the agency permitted separate subarea plans, rather than the umbrella plan (Wynn 2006). This approach, though it gave jurisdictions additional flexibility, has exacerbated the fragmentation

of management and monitoring. According to Jerre Stallcup (2006), the Implementation Committee has morphed into a five-county group known as the Southern California NCCP Partnership, coordinated by the Nature Conservancy. The group's main purpose is to lobby Washington, D.C., for money—at which it has been effective. The group meets rarely, and its annual meetings are not well attended. The Habitat Management Technical Committee began meeting regularly only because there was so little money or staff for management and monitoring.

Chapter 5

1. By the 1980s Everglades National Park was so dry in November and December that peripheral marshes had no fish. Storks adapted by postponing nesting, but that meant chicks arrived late, when rains had dispersed many of the fish, and birds often abandoned their nests, leaving chicks to starve (Boucher 1991).

2. Muck is the name given to fine soil that contains both organic remains of dead plants and sediments. Saltwater intrusion occurs when sufficient freshwater is pumped from coastal aquifers that more dense seawater flows into the freshwater column. All of these problems occurred before the C&SF project, but they were more severe and widespread after its completion.

3. Specifically, the act required the state's water districts to determine (1) the water supply needs of lakes and wetlands (duration, timing, and distribution of water); (2) minimum water levels and amount of time these levels need to be sustained to protect groundwater from saltwater intrusion; and (3) the minimum flows and levels of rivers and estuaries that will maintain stream flow characteristics and biological communities.

4. In 1994, after a series of workshops to synthesize contributors' work, Davis and Ogden published an edited volume that summarized the state of knowledge about Everglades ecology and laid out a set of principles that the region's scientists agreed should guide ecosystem restoration.

5. The report was inconclusive on the question of whether or not to remove structures: it noted that eliminating structures would reestablish natural patterns of wetland continuity, sheet flow, and animal movement, and would reduce the conduits for invasions by introduced species and pollutants. On the other hand, the report's authors made clear that it might be impossible to restore predrainage water flow rates, timing, and spatial patterns in the contemporary system of reduced water storage capacity and diminished wetland and recharge area. They acknowledged that adding structures would give managers more flexibility in operating the system. They emphasized, however, that the goal should be "an ecosystem that is resilient to both chronic stresses and catastrophic events with as little human intervention as possible" (Weaver and Brown 1993, 19).

6. Terry Rice, the hydrologist then at the helm of the Corps's Jacksonville district office, asked the commission to design its own conceptual plan for restoration and promised that if the commission's approach made sense, the Corps would incorporate it into the Restudy (Rice 2001).

7. Although the Reconnaissance Study was a Corps–led process, the Feasibility Study was a joint Corps/SFWMD process.

8. CERP planners responded to the park scientists' point by arguing that there had been too much soil loss for the system to handle the same volume of water it held a century earlier (Levin 2001). They rebuffed the criticisms of outside scientists by noting that they did not understand either the plan's details or the political context in which the plan had been forged.

9. The ten projects were (1) the C-44 basin storage reservoir, a 10,000-acre reservoir in Martin County; (2) the 50,000-acre EAA storage reservoir; (3) the Hillsboro (Site 1) impoundment, a 2,460-acre reservoir in Palm Beach County; (4) WCA 3A and 3B seepage management; (5) the C-11 impoundment and canal, a 1,600-acre STA and approximately eight miles of canal; (6) the 2,500-acre C-9 impoundment and STA; (7) the Taylor Creek/Nubbin Slough storage, a 5,000-acre reservoir and 5,000-acre STA; (8) a project to raise and bridge the east portion of the Tamiami Trail and fill the Miami Canal; (9) the North New River Canal improvements, to replace the function of the Miami Canal; and (10) the C-111 spreader canal. The four pilot projects were (1) the Caloosahatchee River (C-43) Basin ASR, (2) Lake Belt in-ground reservoir technology, (3) L-31 seepage management, and (4) wastewater reuse technology. WRDA 1999 had authorized two additional ASR pilot projects.

10. As noted earlier, many CERP designers support this goal in theory but do not believe it can be achieved, given the political and natural constraints posed by a highly altered system (Ogden 2006).

11. The objectives of Decomp include improving sheet flow, hydropatterns, and hydroperiods within WCA3 and Everglades National Park; promoting more natural hydrologic recession rates throughout ridge and slough, marl prairie, and rocky glade landscapes; reducing the pathways for the occurrence and dispersal of invasive exotic species; restoring, maintaining, and sustaining ridge-and-slough topography; maintaining the spatial extent and function of wetland resources in WCA3A, WCA3B, and Everglades National Park; restoring and recovering existing populations of migratory birds and their habitat; increasing fish and wildlife connectivity, including terrestrial species; increasing spatial extent and restoring vegetative composition, habitat function, and productivity of tree islands, and helping compensate for past losses; and restoring peat soils' depth and microtopography (NRC 2006).

12. These rules were required by a provision of the 1972 Florida Water Resources Act that had largely been ignored until lawsuits and prodding by Governor Lawton Chiles prompted the district to move forward on them.

13. The SFWMD has initiated two experimental projects—the Decomp Physical Model and the Loxahatchee Impoundment Landscape Assessment—aimed at reducing the uncertainty associated with (and hence resistance to) removing barriers to flow (NRC 2006). The Decomp project itself is delayed, however, by the legal requirement to finish Mod Waters first.

14. For example, according to biologist Andy Eller, the FWS fired him in 2004 for challenging the science used by the agency to approve road and housing

construction through panther territory in southwestern Florida. Marine biologist David Boyd said the Florida Park Service reassigned him after he pointed out environmental damages associated with a plan to widen U.S.1 into the Florida Keys. Herb Zebuth says Department of Environmental Protection workers were warned about negative consequences for opposing efforts to build the Scripps biotechnology park in the Everglades Agricultural Area. The SFWMD demoted ecologist Lou Toth after he publicly criticized progress on the Kissimmee River restoration, and fired ecologist Nick Aumen after he made critical remarks to a journalist (Santaniello 2005).

15. Whereas the state intends minimum flows and levels to protect water resources and ecological features that are currently experiencing or are threatened with significant harm, as part of CERP implementation the SFWMD must set initial water reservations to provide additional protection for fish and wildlife.

16. Recognizing the many challenges CERP faces, the Committee on Independent Scientific Review of Everglades Restoration Progress (NRC 2006) proposed, and CERP planners agreed to adopt, an incremental adaptive management approach that they hope will reduce public opposition to environmental restoration projects.

17. The projects to be expedited were originally scheduled for completion by 2015, with most done by 2010, but because of lack of federal funds would not be finished until 2022 in the absence of Acceler8 funds. To generate funds, the state issued "certificates of participation" that would be repaid with the SFWMD's tax revenue—an approach that enabled the district to avoid consulting voters in its 16 counties.

18. One exception was the 4,584-acre Harmony Ranch Project west of Hobe Sound, half of which occupied land that had been proposed for the Indian River Lagoon restoration. Although the SFWMD agreed to yield the land to development in 2004, saying it could substitute other land for the lost acreage, the Corps rejected the project, saying it would be "contrary to the public interest" (R. King 2006).

19. In 2003 the National Research Council affirmed an SCT white paper that hypothesized the movement of water was central to the formation and maintenance of the Everglades' rapidly disappearing ridge-and-slough landscapes, which in turn provide essential habitat for aquatic life and wading birds. Although they still do not fully understand the mechanisms by which the ridge-and-slough landscape have been degraded, many scientists believe that barriers to flow—including levees and canals—contribute significantly to its conversion to dense sawgrass stands (SCT 2003). CERP scientists are hopeful these insights will enhance plans for future restoration projects.

Chapter 6

1. Although commentators often speak of averages, it is misleading to do so because the amount of water flowing through the system fluctuates from as little

as 6 million acre-feet per year to as much as 60 million acre-feet (G. Martin 1999b; Zakin 2002). An acre-foot is 326,000 gallons.

2. Pelagic species, such as the delta smelt, are open-water fish. Anadromous fish, such as salmon, migrate up rivers from the sea to breed in freshwater.

3. The SWRCB, created in 1967, is responsible for ensuring the state's water quality, as well as for allocating and adjudicating water rights. The board issued its original decision on water quality for the Delta and Suisun Marsh in August 1978 (Water Right Decision 1485). The appellate court rejected the water quality standards and measures contained in that decision as insufficient to protect the Bay–Delta ecosystem, and charged the board with assessing the relationship between freshwater flows and Bay–Delta water quality. That process resulted in the revised 1988 draft decision.

4. The board justified its decision to separate decisions about water quality and water flow on the legal grounds that flow issues involved water rights, a state concern, whereas it set water quality standards in response to a federal mandate. In ecological terms, however, the decision made no sense: in an estuarine system, the mix of freshwater and salt water is a key determinant of water quality.

5. The EPA said that California would not be allowed to build new projects to divert more water, three of which had been on the drawing board since 1984, until it adopted standards that would protect the Bay–Delta ecosystem (Diringer 1991c).

6. Federal responsibilities affecting the Bay-Delta estuary and watershed include listing species and consulting under the Endangered Species Act; implementing the CVPIA; operating the CVP; reviewing and, when the state fails to do so under its delegation of federal Clean Water Act authority, promulgating water quality standards; and reviewing water development proposals under the Fish and Wildlife Coordination Act, the National Environmental Policy Act, section 404 of the Clean Water Act, and the Rivers and Harbors Act.

7. As part of the San Francisco Estuary Project, scientists aimed to devise a standard that could be used to measure freshwater flows out of the Delta, which they could not measure directly because of large variations in the tide. They recognized that the salinity gradient in the estuary was a good indicator of the amount of freshwater flowing through the system, so they devised X2, a measure of salt penetration into the estuary (Kimmerer 2005).

8. In 1995 the SWRCB formally adopted these standards as the centerpiece of its Water Quality Control Program (WQCP) for the Bay–Delta. The WQCP limited state and federal export pumping to 35 percent of Delta inflow from February to June, when estuarine fish breed, and 65 percent of inflow during the rest of the year; it also required the maintenance of low-salinity habitat (the X2 standard) in the estuary during the spring (Rosenkrans and Hayden 2005).

9. Environmentalists were well aware of the increasingly hostile climate in the U.S. Congress, especially after the Republican takeover in 1994. Nevertheless, some environmentalists and many fishing groups were skeptical about CALFED from the outset.

10. Patrick Wright, CALFED executive director from 2001 to 2005, argues that the goal was not consensus but broad-based support. Although he "abolished the word [consensus] after becoming director" (McClurg 2002, 6), it was nevertheless widely believed to be the program's goal, both internally and by observers.

11. Although planners were avoiding the term, canal proposals evoked memories of the "peripheral canal" that had mobilized the environmental community in the 1980s. Governor Jerry Brown had pushed a peripheral canal through the legislature, but a coalition of environmentalists and San Joaquin Valley farmers sponsored a referendum to stop the canal, and in 1982, after a bitter campaign, California voters decisively rejected the project.

12. Although generally received positively, the haste with which CALFED undertook ecosystem restoration was not always rewarded. For example, in 2003 some scientists charged that in its rush to acquire land, CALFED's $50 million wetland restoration plan had overlooked potential mercury problems (Leavenworth 2003). (Marshes and other wetlands can intensify problems with mercury, which is pervasive in California as a result of the state's mining history.)

13. The Pacific Decadal Oscillation is a natural fluctuation associated with warmer ocean temperatures in the northern Pacific. Variations in ocean temperatures play a huge role in salmon population levels, and during an up cycle in the ocean, there can be negative things going on in the watershed but still a boom in salmon; by contrast, during a down cycle in the ocean, improvements in the watershed may have a negligible impact (Luoma 2005). Many scientists believe that *both* improved ocean conditions and habitat restoration contributed to increases in salmon, particularly the winter run on the Sacramento River and the spring run on Butte Creek (Kier 2006; Swanson 2006).

14. At best the EWA was semiadaptive, in that it was never formally evaluated, so it is not clear that it actually yielded superior environmental outcomes.

15. Thus, the EWA baseline was actually eroded in two ways: the Interior Department changed its accounting in response to the court's ruling; it also began using CVPIA water to meet Clean Water Act objectives first, and using only the leftover water for fish.

16. Several times each year Department of Fish and Game staffers haul trawl nets through the Delta, Suisun Bay, and San Pablo Bay to survey fish species. Based on those surveys, they calculate abundance indices for the delta smelt and striped bass that indicate the abundance of those species in a given volume of water.

17. Additional research has shed light on other ways exports are affecting the smelt. One study suggested that the fittest fish hatch early in the year, and they are precisely the ones that are destroyed by winter pumping (Taugher 2006e). Research by scientists at the U.S. Geological Survey detected a significant correlation between high incidental take at the pumps and hydrodynamic conditions in the central and southern Delta caused by low San Joaquin River inflows and high water export rates (Bay Institute et al. 2007).

18. What does seem to have changed the MWD's position is the dual threat of climate change and massive levee failure.

19. Finally, in the fall of 2004 Congress passed a modest $389 million, six-year CALFED reauthorization bill. Although the bill no longer contained the "preauthorization" language for storage projects, according to Rep. Richard Pombo (R, Tracy), it made "storage the linchpin for implementation of all CALFED elements. The bill ensures that the program will be carried out in balance with new water storage or else the program will simply not exist" (Werner 2004).

20. After NMFS added winter-run salmon to the federal endangered species list in 1989, it ordered the Bureau of Reclamation to carry over 1.9 million acre-feet of water in the Shasta Reservoir to ensure the salmon would have sufficient cold water to survive a drought. Facing pressure to ship more water through the Delta, the bureau proposed to end the carryover storage requirement and reduce the stretch of river where it had to maintain cold water (Leavenworth 2004).

21. By February 2007 the EWA was already depleted for the year, after water officials used it to compensate for reduced pumping in January. Yet 2007 promised to be a dry year in which further pumping reductions were likely to be needed for fish (Taugher 2007c).

Chapter 7

1. See chapter 3, note 2, for a brief description of habitat conservation plans.

2. Both the Nature Conservancy and the Sonoran Institute declined to join the coalition, preferring to remain neutral. On the other hand, the area's most aggressive groups, the Southwest Center for Biological Diversity (later the Center for Biological Diversity) and Defenders of Wildlife, did join the coalition.

3. Immediately after the board's vote, Huckelberry—miffed that the proposal he had devised in collaboration with the FWS and Arizona Fish and Game had been rejected—relegated development of the Sonoran Desert plan to the Parks and Recreation Department. Within weeks, however, his staff had begun working closely with the coalition on the plan.

4. The board felt it needed to get new ordinances in place before August 21, when a new state law requiring compensation for downzoning was going to take effect.

5. The proposed Native Plant Protection ordinance required landowners who wanted rezonings to inventory and protect native plants. The Buffer Overlay Zone ordinance, which applied to parcels within one mile of public preserves, prohibited placing buildings or roads within 150 feet of a public preserve, or 300 feet if the county approved rezonings or specific plans; it also required landowners in the buffer to set aside 30 percent of their land as natural open space, or 50 percent in exchange for a rezoning or specific plan. The Hillside Development Zone Ordinance prohibited development within 300 feet of a protected peak or ridge if there was a rezoning or specific plan approval, or within 150 feet in other cases.

6. Some large developers initially supported the SDCP concept, hoping it would establish a set of predictable rules. Most quickly became skeptical or downright antagonistic, however, as the county's pro-environmental stance became more pronounced.

7. A year after the Design Review Committee began enforcing the ordinance, which had been tightened in 1998 to include driveways as part of the allowable square feet of grading on a hillside, the Board of Supervisors loosened it again in response to public complaints.

8. Within the planning area, Native American reservations account for 42 percent of the land; state lands comprise another 15 percent of the county's acreage; and the federal government—including the Bureau of Land Management, Forest Service, and FWS—controls 17 percent. There are also several cities and towns—Tucson, South Tucson, Oro Valley, Marana, and Sahuarita—that control land-use decision making inside their municipal boundaries, which cover about 12 percent of the county's area.

9. For instance, although some environmentalists argued that the county should include the jaguar on its list of priority vulnerable species, the STAT pointed out that the majority of the jaguar's range is in Mexico and it visits the county only sporadically, so the county's efforts would be unlikely to have much impact on its survival (Schulman 2007).

10. Arizona counties rely almost exclusively on the property tax for revenues.

11. The law, which amended the 1998 Growing Smarter Act, required counties and municipalities to update their comprehensive plans by the end of 2001 to include elements of land use, water resources, open space, environmental impacts, growth areas, and cost of growth (Duerksen and Snyder 2005).

12. Supervisor Ray Carroll did not attend the hearing because he had a conflict of interest with one of the properties that would be affected by the board's decisions.

13. In fact, the person who seemed to experience the greatest transformation during the SDCP process was Bill Arnold, who initially was "right in line with most people in the real estate/homebuilding industry" in thinking the SDCP was "completely bogus." Arnold's epiphany was not a result of stakeholder collaboration, however; instead, he became convinced early on—before the Steering Committee was formed—that at least some of the proponents of desert conservation were sincere. As a passionate outdoorsman he found himself persuaded by their arguments (B. Arnold 2006).

14. According to property rights activists Zimet and DuHamel, Huckelberry had promised them in an April 2002 letter that the county would apply its CLS only to properties that required a change in their legal land use, such as rezoning or an amendment to the comprehensive plan (T. Davis 2003c).

15. If the developer exceeds the developable percentage, (s)he must purchase land off site to offset the impacts on the CLS. The off-site mitigation ratio is 4:1 for biological core and special management areas. Originally the ratio was 3:1 for multiple-use management areas but was reduced to 2:1 in 2005.

16. For example, Huckelberry had proposed condemnation of Canoa Ranch and had threatened condemnation of private land near Davidson Canyon and Cienega Creek.

17. In January 2007 journalist Tony Davis pointed out that of the $91 million in bond money spent since 1997, only $2 million had gone to buy land on the Northwest Side.

18. The county had its missteps as well. For example, in the fall of 2000 the county's road crews cleared hundreds of ironwood and other trees for a Northwest Side road-widening project, in violation of its own native plant preservation rules. In 2002, the county endured another embarrassing incident when the think tank it had hired to conduct an economic study of the SDCP, Arizona State University's Morrison Institute for Public Policy, terminated its contract with the county in a dispute over data.

Chapter 8

1. The U.S. Fish and Wildlife Service and the Florida Game and Fresh Water Fish Commission warned that channelization would seriously damage the region's ecology. They proposed alternative flood-control plans that did not require converting the Kissimmee River into a canal, such as building levees between the uplands and the floodplain, but the Corps found the canal option to be the most cost-effective means of achieving flood control (Berger 1992; Blake 1980).

2. One consequence of the lack of flow was the accumulation of thick deposits of decomposing organic matter, which generate a high biological oxygen demand, in remnant river channels.

3. Benthic invertebrates are organisms that live on the bottom of a water body and have no backbone. Though often underappreciated, benthic invertebrates strongly influence energy flows in aquatic ecosystems; "The integrity of the freshwater supply depends on how various species make their living and contribute to complex food webs" (Covich et al. 1999, 125). Therefore, scientists recommend using the abundance, diversity, biomass, production, and species composition of benthic (and other) invertebrates as indicators of changing environmental conditions.

4. The water entering Lake Tohopekeliga contained more than 300 parts per billion (ppb) of phosphorus, but the phosphorus content of water entering C-38 below S-65 at the end of the Chain of Lakes was less than 5 ppb (USACE 1985).

5. The non-dechannelization alternative actually combined four options the Corps had been considering: flow-through marshes, pool-stage manipulation, impounded wetlands, and Paradise Run (USACE 1985).

6. When engineers dropped the high level of Lake Kissimmee from 53.75 feet above sea level to 52.50 feet, they exposed thousands of acres of lake bottom, which property owners soon claimed. If the state set the ordinary high-water line at 53.75 feet, however, that property would revert to being sovereign land, and no compensation would be required.

7. Florida's Environmental Land and Water Management Act allows the governor and his cabinet to designate a region as an Area of Critical State Concern.

Once an area is so designated, local governments must develop comprehensive plans and implement development regulations that are consistent with guiding principles devised by the state's Department of Community Affairs. To avert the controversy that accompanied the first few designations, the legislature created an alternative process in which resource planning and management committees are charged with coming up with a "voluntary, cooperative resource management program to resolve existing and prevent future problems, which may endanger those resources, facilities, and areas..." (Nicholas 1999, 1080).

8. Governor Graham approved this general plan and agreed to extend the life of the committee so that it could help with implementation. This was an unusual step: previous committees established under the statute had turned over implementation to the state Department of Community Affairs. But committee chair Timer Powers pointed out that the Kissimmee Valley was populated by determined individuals who would fiercely resist a state-imposed mandate (Anon. 1985).

9. A 1986 Florida Supreme Court decision had ruled that some submerged land was, indeed, state property.

10. The backfilling was complete in the sense that it extended across an uninterrupted stretch of river—as opposed to one option planners had considered, which would have backfilled only intermittently in order to reduce costs.

11. According to many observers, the shift in emphasis reflected not only changing public sentiment but also an entrepreneurial effort to capture federal money at a time when funding for traditional construction projects, such as dams and canals, was declining.

12. The Corps required the SFWMD to buy all land within the 100-year floodline. In February 1992 the SFWMD informed the Corps that board policy would preclude acquisition of residences south of U.S. Highway 98—a decision that satisfied most members of ROAR.

13. Although the restoration program's primary focus was re-creating the Kissimmee River's structure and function, some components reflect accommodations for landowners or infrastructure in the watershed. Among these projects are floodproofing selected agricultural and residential areas, converting potentially impacted septic tank systems to sewers, building a railroad bridge over the original river channel, and elevating a major highway across the historic floodplain.

14. The SFWMD actually bought somewhat more land than projected because it used a blocking approach: the floodplain boundary was a squiggly line, so the District accommodated landowners who did not want to be left with remnant parcels by purchasing out to the quarter-section boundary (Loftin 2007).

15. In November 2006 the SFWMD halted all flows from the upper basin to the Kissimmee River because of a severe drought in the region. Engineers reestablished flows on July 18, 2007. After 252 days without water, the river and floodplain had experienced substantial biological decline. Project managers expect, however,

that eventually restored biological communities will have similar species richness and diversity as communities in the prechannelization system.

16. In fact, the Headwaters Revitalization Project is so essential to the overall restoration that in 1992 the Department of the Army's Board of Engineers for Rivers and Harbors recommended, in its letter to the Chief of Engineers, that the headwaters component be completed prior to the initiation of construction in the lower basin.

Chapter 9

1. The lime that forms the basis of the tufa towers precipitates when freshwater springs under the lake come into contact with the lake water, which is full of carbonate ions (Hart 1996).

2. By the time land acquisition was complete, the city owned about 30,000 acres and held easements on thousands more.

3. The decline in waterfowl cannot be attributed solely to the shrinkage of Mono Lake; waterfowl populations have plummeted along the entire Pacific Flyway because of cumulative habitat loss.

4. Ironically, the DWP increased its diversions in response to repeated warnings from the state Water Rights Board that it could lose its claim on Mono Basin (and Owens Valley) water if it did not use it. The board recognized at the time that increased water withdrawals would harm the basin, but concluded it was required to approve them because the Water Commission Act states that domestic use of water is the highest use (SWRCB 1994b). The board therefore granted L.A. the right to appropriate 167,000 acre-feet, the entire flow of four of Mono Lake's five feeder streams (Bass 1979).

5. Because the DWP filed cross-complaints against 117 other Mono Basin landowners, some of whom were federal agencies, environmentalists' lawyers succeeded in moving the case to federal court, where it ended up in the hands of Judge Lawrence Karlton. In 1981, however, the case moved from Judge Karlton's courtroom to the California Supreme Court, after Karlton determined that the public trust doctrine was subsumed within the California water rights system.

6. Los Angeles appealed the case to the U.S. Supreme Court, which declined to review it, apparently convinced by the argument that the Mono Lake ruling arose entirely out of California law, not the U.S. Constitution (Mann 1983).

7. As Todd Kunioka and Lawrence Rothenberg (1993) explain, the Scenic Area designation—a provision in the California Wilderness Act of 1984—was an artfully crafted compromise: by stopping short of National Monument designation, including language explicitly protecting California's water rights, and keeping the land out of National Park Service hands, the bill satisfied L.A. interests; by providing protection for lands surrounding Mono Lake and including some of the craters, the bill gave the Mono Lake Committee and its allies some of the benefits they wanted as well.

8. Section 5437 of the California Fish and Game Code requires that every dam have two features: it must have a fishway, and it must let enough water pass to maintain "in good condition" the fish in the stream below the dam. To expedite the Mono Basin permits, the California Fish and Game Commission had agreed to let the city substitute a hatchery on Hot Creek for a fishery below Grant Lake Dam and had simply waived the prohibition on dewatering creeks below dams.

9. Although the members of the RTC agreed on the ultimate goal of restoring natural processes, the DWP's scientist disagreed with the Mono Lake Committee and its allies that some repairs to damaged portions would be necessary to allow natural hydrologic conditions to reestablish themselves.

10. In 1987 the Mono Lake Group expanded to include representatives of the Forest Service and Mono County.

11. First, water from the San Joaquin Valley would have had to come down the State Water Project aqueduct by way of the Metropolitan Water District of Southern California, potentially creating tension among the DWP and the Met's other customers. Second, many San Joaquin Valley agricultural officials vigorously opposed the notion of L.A. buying water from farmers, concerned about the impact of such deals on the long-term viability of agriculture in the region.

12. In fact, during the drought years after it was enjoined from using Mono Basin water, L.A. ramped up its purchases from the MWD substantially, going from about 78,600 acre-feet per year to 385,000 acre-feet (SWRCB 1994a).

13. The Mono Lake Committee supported arguments for a lake level of 6,405 or even 6,407 feet because of the benefits to waterfowl. The Delta lagoons, which provide important waterfowl habitat, disappeared when the lake fell below 6,400 feet. The steeper gradient of the shoreline now limits the formation of lagoons at levels below 6,400 feet.

14. Buttressing arguments for raising the lake level, in July 1993 the EPA had proposed redesignating the Mono Lake area as in violation of federal air quality standards because of its alkali dust storms (Forstenzer 1993).

15. The thalweg is a line drawn to join the lowest points along the length of a streambed as it slopes downward.

16. The Rush Creek return ditch, located below Grant Lake Reservoir, is the only way—other than spilling water over the dam—that water gets to the lower portion of Rush Creek.

17. The County Ponds emerged when the lake dropped below 6,405 feet. They filled up with irrigation water from the DeChambeau Ranch but dried up when irrigation stopped.

18. FERC tried to elide the controversy by neither accepting nor rejecting Southern California Edison's commitment to upgrade the Mill Creek return ditch. Recognizing that the peace that accompanied the settlement would only hold if the parties adhered to it in total, Edison agreed to do the construction even without the mandate from FERC (McQuilkin 2008).

19. All lakes have a seasonal mixing regime, and Mono Lake is ordinarily mon-omictic (mixing once a year). During meromixis, however, no mixing occurs: a freshwater layer floats on top of the more saline layer. The effect of meromixis on the lake's long-term productivity is uncertain. In the early years of meromixis (1996–1999), gull reproductive success was very low. Starting around 2000, as meromixis began to weaken, reproductive success rose, and in 2004 the average clutch size jumped from around 1.8 eggs per nest to 2.4 eggs per nest. (Meromixis is less likely at higher lake levels because the volume of freshwater inflow is a smaller percentage of the volume of the lake, so it was probably much less common prior to diversions.)

References

Adams, Craig. 1993. San Diego Sierra Club comments on Issue Paper #6. April 21. (On file with author.)

———. 2005. Personal communication.

Agrawal, Arun. 2000. "Small Is Beautiful, but Is Larger Better? Forest-Management Institutions in the Kumaon Himalaya, India." In Clark C. Gibson, Margaret A. McKean, and Elinor Ostrom, eds., *People and Forests: Communities, Institutions, and Governance* (Cambridge, MA: MIT Press), 57–85.

Alford, Andy. 2000. "Preserves Welcoming Careful Visitors." *Austin American–Statesman*, December 18.

Allan, Catherine, and Allan Curtis. 2005. "Nipped in the Bud: Why Regional Scale Adaptive Management Is Not Blooming." *Environmental Management* 36(3): 414–425.

Allen, T. F. H., and Thomas W. Hoekstra. 1992. *Toward a Unified Ecology* (New York: Columbia University Press).

Alliance for Habitat Conservation (AHC). 1994. "The Public Lands Alternative: A Solution to the MSCP Dilemma." January. (On file with author.)

Amy, Douglas J. 1990. "Environmental Dispute Resolution: The Promise and the Pitfalls." In Norman J. Vig and Michael E. Kraft, eds., *Environmental Policy in the 1990s* (Washington, DC: CQ Press), 211–234.

Anderson, Paul, and Heather Dewar. 1990. "Squabbles Delay Kissimmee Project." *Miami Herald*, March 11.

Andrews, Clinton J. 2002. *Humble Analysis: The Practice of Joint Fact-Finding* (Westport, CT: Praeger).

Angermeier, Paul L., and James R. Karr. 1994. "Biological Integrity Versus Biological Diversity as Policy Directives." *BioScience* 44(10): 690–697.

Anon.———. 1985. "Group's Plans on Target for River Basin's Future," *Fort Lauderdale Sun-Sentinel*, November 23.

———. 1986. "U.S. Owns 12,000 Acres Near Mono Lake, Appeals Court Says." *Los Angeles Times*, December 3.

———. 1989. "Everglades: Render Back to Nature." *Economist*, December 9.

———. 1991. "Session's Agenda Should Include Canyonlands Bill." *Austin American–Statesman*, July 15.

———. 1993. Bio-Guidelines Meeting Summary. January 27. (On file with author.)

———. 1997. "Restoring the Delta." *California Journal*, May 1.

———. 1998a. "Mono Lake Snapshot: Spring 1998." *Mono Lake Newsletter*, Spring/Summer.

———. 1998b. "Land Vision at Last." *Arizona Daily Star*, October 23.

———. 2000. "A Sneak Attack on the Everglades." *Tampa Tribune*, August 26.

———. 2001. "80% of Open Land Deemed Worthless by Pima County." *SAHBA Blue Print*, June.

———. 2002. "Caltrans' Mono Mistake." *Los Angeles Times*, July 8.

———. 2004. "From Saving Everglades to Subsidizing Growth." *Palm Beach Post*, November 26.

———. 2005. "NAS Report Calls for More Land Purchases, Above Ground Storage." *Greenwire*, January 25.

Anthony, Jerry. 2004. "Do State Growth Management Regulations Reduce Sprawl?" *Urban Affairs Review* 39(3): 376–397.

Applebaum, Stu. 2002. Everglades Oral History, South Florida Oral History Consortium. February 22.

Appleyard, Donald, and Kevin Lynch. 1974. *Temporary Paradise? A Look at the Special Landscape of the San Diego Region*. Report to the City of San Diego. (On file with author.)

Arnold, Bill. 2006. Personal communication.

Arnold, Craig Anthony. 2004. "Working Out an Environmental Ethic: Anniversary Lessons from Mono Lake." *Wyoming Law Review* 4(1): 1–55.

Arrow, Kenneth, Bert Bolin, Robert Costanza, Partha Dasgupta, Carl Folke, C. S. Holling, Bengt Owe-Jansson, Simon Levin, Karl-Goran Maler, Charles Perings, and David Pimentel. 1995. "Economic Growth, Carrying Capacity, and the Environment." *Science* 268 (April 25): 520–521.

Asher, Robert. 2003. Personal communication.

Associated Press (AP). 1993. "Deal is Struck in River Restoration," *Miami Herald*, February 24.

Atwood, Jonathan L. 1992. Letter to Dr. Dan Silver re draft biological standards and guidelines of Multiple Species Reserve Design. November 18. (On file with author.)

Avery, George. 1993. "Plan Would Help Environment, Economy." *Austin American–Statesman*, October 16.

Axelrod, Robert M. 1984. *The Evolution of Cooperation* (New York: Basic Books).

Bair, Bill. 1994. "Huge Project Begun to Restore Kissimmee River," *St. Petersburg Times*, April 24.

Balz, John. 2000. "Everglades Plan Gets Boost." *St. Petersburg Times*, September 7.

Bancroft, Ann. 1994a. "Big Step for Saving Mono Lake." *San Francisco Chronicle*, September 21.

———. 1994b. "L.A. Gives Up—Mono Lake Saved." *San Francisco Chronicle*, September 29.

Banta, Bob. 1990. "Wildlife Panel Hears Plans for Habitat Areas." *Austin American–Statesman*, June 16.

Barber, Benjamin. 1984. *Strong Democracy: Participatory Politics for a New Age* (Berkeley: University of California Press).

Barbour, Michael. 1996. "Ecological Fragmentation in the Fifties." In William Cronon, ed., *Uncommon Ground: Rethinking the Human Place in Nature* (New York: W. W. Norton), 233–255.

Bardach, Eugene, and Robert Kagan. 1982. *Going by the Book: The Problem of Regulatory Unreasonableness* (Philadelphia: Temple University Press).

Barnett, Cynthia. 2007. *Mirage: Florida and the Vanishing Water of the Eastern U.S.* (Ann Arbor: University of Michigan Press).

Barnett, Ernie. 2002. Everglades Oral History, South Florida Oral History Consortium. February 1.

Barnum, Alex. 1996. "Bay Area's Wetland Renaissance." *San Francisco Chronicle*, October 25.

———. 1998. "Squabbles Among Farmers, Environmentalists Slowed Progress to a Dribble." *San Francisco Chronicle*, December 19.

Barringer, Felicity. 2004. "Thriving Bald Eagle Finds Its Way off Endangered List." *New York Times*, May 19.

Bass, Ron. 1979. "The Troubled Waters of Mono Lake." *California Journal* 9: 349–351.

Baumgartner, Frank R., and Bryan D. Jones. 1993. *Agendas and Instability in American Politics* (Chicago: University of Chicago Press).

Bay Institute, Center for Biological Diversity, and Natural Resources Defense Council. 2007. Petition to the State of California Fish and Game Commission and Supporting Information for Listing the Delta Smelt (*Hypomesus transpacificus*) as an Endangered Species Under the California Endangered Species Act. (On file with author.)

Beatley, Timothy. 1994. *Habitat Conservation Planning: Endangered Species and Urban Growth* (Austin: University of Texas Press).

———. 1998. "The Vision of Sustainable Communities." In Raymond J. Burby, ed., *Cooperating with Nature: Confronting Natural Hazards with Land-Use Planning for Sustainable Communities* (Washington, DC: Joseph Henry Press), 233–262.

————. 2000. "Preserving Biodiversity: Challenges for Planners." *Journal of the American Planning Association* 66(1): 5–20.

Beatley, Timothy, and Kristy Manning. 1997. *The Ecology of Place: Planning for Environment, Economy, and Community* (Washington, DC: Island Press).

Beattie, Mollie. 1996. "An Ecosystem Approach to Fish and Wildlife Conservation." *Ecological Applications* 6(3): 696–699.

Beck, Michael. 1995. Letter to the City of San Diego Natural Resources, Culture & Arts Committee. September 27. (On file with author.)

Beierle, Thomas C., and Jerry Cayford. 2002. *Democracy in Practice: Public Participation in Environmental Decisions* (Washington, DC: Resources for the Future).

Berger, John J. 1992. "The Kissimmee Riverine–Floodplain System." In National Research Council, *Restoration of Aquatic Ecosystems* (Washington, DC: National Academies Press), 477–496.

Berkes, Fikret. 1999. *Sacred Ecology* (Philadephia: Taylor & Francis).

————. 2004. "Rethinking Community-Based Conservation." *Conservation Biology* 18(3): 621–630.

Best, Ronnie. 2006. Personal communication.

Bingham, Lisa B., David Fairman, Daniel J. Fiorino, and Rosemary O'Leary. 2003. "Fulfilling the Promise of Environmental Conflict Resolution." In Rosemary O'Leary and Lisa B. Bingham, eds., *The Promise and Performance of Environmental Conflict Resolution* (Washington, DC: Resources for the Future), 329–351.

Biological Assessment Team (BAT). 1990. *Comprehensive Report of the Biological Assessment Team.* (On file with author.)

Biological Task Force for Preserve Design, San Diego County, California (Biological Task Force). 1992. Draft Biological Standards and Guidelines for Multiple Species Preserve Design. October 21. (On file with author.)

Birke, Charles L. 1993. Letter to Janis Sammartino, Esq., re MSCP. (On file with author.)

Blake, Nelson Manfred. 1980. *Land into Water—Water into Land: A History of Water Management in Florida* (Tallahassee: University Presses of Florida).

Blue Ribbon Task Force, Delta Vision (BRTF). 2007. "A Vision for Durable Management of a Sustainable Delta." First, embryonic draft, September. Available at http://deltavision.ca.gov.

Bobker, Gary. 2005. Personal communication.

Booth, William. 1992. "A River's Life Hangs in Balance as Congress Considers a Remedy." *Washington Post*, February 16.

Born, Stephen M., and Kenneth D. Genskow. 1999. *Exploring the "Watershed Approach": Critical Dimensions of State–Local Partnerships.* Report 99-1, Final report of the Four Corners Watershed Innovators Initiative, University of Wisconsin, Madison Extension.

Botkin, Daniel B. 1990. *Discordant Harmonies: A New Ecology for the Twenty-first Century* (New York: Oxford University Press).

Botkin, Daniel B., Wallace S. Broecker, Lorne G. Everett, Joseph Shapiro, and John A. Wiens. 1988. *The Future of Mono Lake: Report of the Community and Organization Research Institute "Blue Ribbon Panel" for the Legislature of the State of California* (Riverside: University of California Water Resources Center).

Boucher, Norman. 1991. "Smart as Gods." *Wilderness*, Winter, 11–21.

———. 1995. "Back to the Everglades." *Technology Review*, August–September, 24–35.

Boxall, Bettina. 2005. "Mono Lake Parcel is Purchased," *Los Angeles Times*, March 17.

———. 2006a. "The Delicate Act of Juggling Water." *Los Angeles Times*, April 16.

———. 2006b. "Delta Smelt's Fate Worries Scientists." *Los Angeles Times*, April 17.

———. 2006c. "Governor Seeks Restructuring of Water Program." *Los Angeles Times*, April 21.

Boyarsky, Bill. 1986a. "Perilous Political Waters." *Los Angeles Times*, March 23.

———. 1986b. "Bradley Supports Increased Water Flow into Mono Lake." *Los Angeles Times*, August 28.

———. 1987. "Scientists See a Stark Future for Mono Lake," *Los Angeles Times*, August 5.

Bradshaw, Ben. 2003. "Questioning the Credibility and Capacity of Community-based Resource Management." *Canadian Geographer* 47(2): 137–150.

Brandt, Alf W. 2002. "An Environmental Water Account: The California Experience." *University of Denver Water Law Review* 51: 426–456.

Brandt, Laura. 2002. Personal communication.

Braun, David. 2003. Personal communication.

Brazil, Eric. 2001. "State Fails Test on Endangered Fish." *San Francisco Chronicle*, March 24.

Brick, Philip, and Edward P. Weber. 2001. "Will the Rain Follow the Plow? Unearthing a New Environmental Movement." In Philip Brick, Donald Snow, and Sara Van de Wetering, eds., *Across the Great Divide: Explorations in Collaborative Conservation and the American West* (Washington, DC: Island Press), 15–24.

Brody, Samuel D. 2003. "Implementing the Principles of Ecosystem Management in Local Land Use Planning." *Population and Environment* 24(6): 511–540.

Bruegmann, Robert. 2005. *Sprawl: A Compact History* (Chicago: University of Chicago Press).

Brunner, Ronald D. 2002. "Problems of Governance." In Ronald D. Brunner, Christine H. Colburn, Christina M. Cromley, Roberta Klein, and Elizabeth A. Olson, eds., *Finding Common Ground* (New Haven, CT: Yale University Press), 1–47.

Brunner, Ronald D., Toddi A. Steelman, Lindy Co-Juell, Christina M. Cromley, Christine M. Edwards, and Donna W. Tucker. 2005. *Adaptive Governance: Integrating Science, Policy, and Decision Making* (New York: Columbia University Press).

Burby, Raymond J., and Peter J. May. 1997. *Making Governments Plan: State Experiments in Managing Land Use* (Baltimore: Johns Hopkins University Press).

Burchell, Robert W., Anthony Downs, Barbara McCann, and Sahan Mukherji. 2005. *Sprawl Costs: Economic Impacts of Unchecked Development* (Washington, DC: Island Press).

Burnham, Michael. 2007. "The Next Frontier: Vanishing Soil, Shifting Economics Have Developers Eyeing Canefields." *Greenwire*, September 28.

Butler, Kent, and Dowell Myers. 1984. "Boomtime in Austin, Texas." *Journal of the American Planning Association* 50(4): 447–458.

Calavita, Nico. 1992. "Growth Machines and Ballot Box Planning: The San Diego Case." *Journal of Urban Affairs* 14(1): 1–24.

CALFED Bay–Delta Program (CALFED). 2000a. *Final Programmatic Environmental Impact Statement/Environmental Impact Report*. July.

———. 2000b. *Programmatic Record of Decision*, August 28.

———. 2005. *Annual Report 2004* (Sacramento: California Bay–Delta Program).

Campbell, Carolyn. 2006. Personal communication.

Canaday, Jim. 2007. Personal communication.

Caputo, Marc. 2004. "Developers Accused of Making Water Grab." *Miami Herald*, March 31.

Carruthers, John I. 2002. "The Impacts of State Growth Management Programmes: A Comparative Analysis." *Urban Studies* 39(11): 1959–1982.

Cestero, Barbara. 1999. "Beyond the Hundredth Meeting: A Field Guide to Collaborative Conservation on the West's Public Lands" (Tucson, AZ: Sonoran Institute).

Chase, Alston. 1986. *Playing God in Yellowstone: The Destruction of America's First National Park* (New York: Harcourt Brace).

Chase, Carolyn. 1997. "The Devil Is in the Details." *Earth Times*, February, 8–10.

Chesnick, Joyesha, and Blake Morlock. 1999a. "Balancing Act, Special Report." *Tucson Citizen*, April 19.

———. 1999b. "Balancing Act, Special Report," *Tucson Citizen*, April 20.

Christensen, Norman L., Ann M. Bartuska, James H. Brown, Stephen Carpenter, Carla D'Antonio, Rober Francis, Jerry F. Franklin, James A. MacMahon, Reed F. Noss, David J. Parsons, Charles H. Peterson, Monica G. Turner, and Robert G. Woodmansee. 1996. "The Report of the Ecological Society of America Committee on the Scientific Basis for Ecosystem Management." *Ecological Applications* 6(3): 665–691.

City of Austin, Environmental and Conservation Services Department. 1991. *Balcones Canyonlands Conservation Plan*, Executive Summary, 2nd, rev. draft. April. (On file with author.)

City of Austin and Travis County, Texas. 1996. *Habitat Conservation Plan and Final Environmental Impact Statement*. March. (On file with author.)

City of San Diego. 1998. *Final Multiple Species Conservation Program MSCP Plan*. August. (On file with author.)

——. 2007. *MSCP 2006 Annual Report.*

Clark, Jamie Rappaport. 1999. "The Ecosystem Approach from a Practical Point of View." *Conservation Biology* 13(3): 679–681.

Clifford, Frank. 1998. "Delta Dam Proposal Gathering Momentum." *Los Angeles Times*, November 25.

Coddon, David L. 1991. "What Price Progress?" *San Diego Union–Tribune*, September 6.

Coggins, George Cameron. 2001. "Of Californicators, Quislings, and Crazies: Some Perils of Devolved Collaboration." In Philip Brick, Donald Snow, and Sarah Van de Wetering, eds., *Across the Great Divide: Explorations in Collaborative Conservation and the American West* (Washington, DC: Island Press), 163–171.

Coglianese, Cary. 2001. "Is Consensus an Appropriate Basis for Regulatory Policy?" In Eric W. Orts and Kurt Deketelaere, eds., *Environmental Contracts* (Boston: Kluwer Law International), 93–113.

Coglianese, Cary, and Jennifer Nash. 2002. "Policy Options for Improving Environmental Management in the Private Sector." *Environment* 44(9): 11–23.

Cohen, Joshua, and Joel Rogers. 2003. "Power and Reason." In Archon Fung and Erik Olin Wright, eds., *Deepening Democracy: Institutional Innovations in Empowered Participatory Governance* (New York: Verso), 237–255.

Cohn, Jeffrey P. 1998. "Negotiating Nature." *Government Executive*, February, 50–53.

Collier, Bill. 1990a. "123,500-Acre Preserve Urged for Vireo," *Austin American–Statesman*, February 7.

——. 1990b. "A Little Bird, a Lot of Real Estate." *Austin American–Statesman*, February 8.

——. 1990c. "Endangered Species Habitat Charted." *Austin American–Statesman*, March 31.

——. 1990d. "$86 Million Proposed to Buy Land for Habitat," *Austin American-Statesman*, June 30.

——. 1990e. "Endangered Species Plan Enters Critical Phase." *Austin American–Statesman*, November 19.

————. 1990f. "Preserve Plan Unveiled; 64,202 Acre Habitat for Endangered Species Proposed," *Austin American-Statesman*, December 5.

————. 1991a. "Hoped-for Refuge Still Hard Sell." *Austin American–Statesman*, April 21.

————. 1991b. "Key Changes Recommended in Habitat Plan." *Austin American–Statesman*, November 28.

————. 1991c. "Landowners, Groups Unite to Challenge Conservation Plan." *Austin American-Statesman*, December 15.

————. 1992a. "2 Appointments to Conservation Committee Criticized." *Austin American–Statesman*, January 9.

————. 1992b. "Consultants Give Panel a New Habitat Plan Draft." *Austin American–Statesman*, February 8.

————. 1992c. "Conservation Plan Opposed by Aleshire." *Austin American–Statesman*, February 27.

————. 1992d. "Environmental Battle Being Fought on Many Fronts." *Austin American–Statesman*, June 25.

————. 1992e. "Business, Nature Groups Endorse Proposition 10 to Fund Preserve." *Austin American-Statesman*, July 30.

Collins, Michael. 2002. Everglades Oral History, South Florida Oral History Consortium. June 11.

Cone, Marla. 1993. "DWP Agrees to Take Less Mono Lake Water." *Los Angeles Times*, December 14.

————. 1994. "Mono Lake Plan Could Slash L.A. Water Supply." *Los Angeles Times*, September 8.

————. 1997. "San Diego OKs Broadest Conservation Plan in U.S." *Los Angeles Times*, March 19.

Connally, Kevin. 2006. Personal communication.

————. 2007. Personal communication.

Coppola, Sarah. 2007a. "Austin Delays Water Plant—Again," *Austin American-Statesman*, August 10.

————. 2007b. "Austin at Last Picks Site for Water Treatment Plant," *Austin American-Statesman*, December 14.

Cortner, Hanna J., and Margaret A. Moote. 1999. *The Politics of Ecosystem Management* (Washington, DC: Island Press).

County of San Diego. 1997. *Multiple Species Conservation Program: County of San Diego Subarea Plan*. October 22.

————. 2007. *MSCP 2006 Annual Report*.

Covich, Alan P., Margaret A. Palmer, and Todd A Crowl. 1999. "The Role of Benthic Invertebrate Species in Freshwater Ecosystems." *Bioscience* 49(2): 119–127.

Crandall, Robert W. 1983. *Controlling Industrial Pollution: The Economics and Politics of Clean Air* (Washington, DC: Brookings Institution).

Crook, Michael. 1989. "Kissimmee Plan Opens Cost Rift," *Miami Herald*, November 5.

Cummins, Ken. 1990. "Corps of Engineers Wants New Conservationist Image." *Fort Lauderdale Sun–Sentinel*, June 3.

Curtius, Mary. 1998. "S.F. Bay: Cleaner but Still a Ways to Go." *Los Angeles Times*, September 10.

Cusick, Daniel. 2005. "Stakeholders Worry Everglades Is in Jeopardy." *Greenwire*, March 25.

———. 2006a. "Leadership Failures, Disputes Crippled Key Project—Interior Report." *Greenwire*, April 20.

———. 2006b. "Fla. River's Rebirth Lifts Spirit of Beleaguered Agency." *Land Letter*, November 2.

Cutting, Lisa. 2006. "12 Years and Counting: Mono Basin Restoration Progress Report." *Mono Lake Newsletter*, Spring, 4–8.

Dahl, T. E. 2006. *Status and Trends of Wetlands in the Coterminous United States 1998 to 2004* (Washington, DC: U.S. Department of the Interior, Fish and Wildlife Service).

Daly, Herman E. 1996. *Beyond Growth: The Economics of Sustainable Development* (Boston: Beacon Press).

Davies, J. Clarence, and Jan Mazurek. 1997. *Regulating Pollution: Does the U.S. System Work?* (Washington, DC: Resources for the Future).

Davis, Martha. 2007. Personal communication.

Davis, Michael. 2002. Everglades Oral History, South Florida Oral History Consortium. March 6.

Davis, Steven M., and John C. Ogden, eds. 1994. *Everglades: The Ecosystem and Its Restoration* (Boca Raton, FL: St. Lucie Press).

Davis, Tony. 1998a. "Challenging Sprawl." *Arizona Daily Star*, February 8.

———. 1998b. "Conservation Plan Gets County Nudge." *Arizona Daily Star*, October 28.

———. 1999a. "Canoa 'No' Vote Changes Rules on Rezonings." *Arizona Daily Star*, January 17.

———. 1999b. "Desert Sprawl." *High Country News*, January 18.

———. 1999c. "Newly Tightened Slope Law Assailed by Builders, Owners." *Arizona Daily Star*, June 13.

———. 2000a. "Species List Ups Price of Desert Plan." *Arizona Daily Star*, January 3.

———. 2000b. "Critics Say Protection Plan for Desert Lacks Public Input." *Arizona Daily Star*, November 12.

———. 2001a. "Sonoran Plan Disputes Outlined." *Arizona Daily Star*, February 8.

———. 2001b. "Canoa Could Set Growth Precedent." *Arizona Daily Star*, March 19.

———. 2001c. "Sonoran Desert Salvation Unveiled." *Arizona Daily Star*, March 22.

———. 2001d. "Business Panel Wants Smaller Group Involved." *Arizona Daily Star*, April 3.

———. 2001e. "State, County at Odds over Desert Plan." *Arizona Daily Star*, April 22.

———. 2001f. "Hull Asks Feds to Stop Pima County Desert Conservation Plan." *Arizona Daily Star*, May 21.

———. 2001g. "Supervisor, Builder Trade Blame for Housing Costs." *Arizona Daily Star*, October 14.

———. 2001h. " 'Show of Force' Sought on Pima Land Use Plan." *Arizona Daily Star*, December 12.

———. 2001i. "New Comprehensive Plan OK'd." *Arizona Daily Star*, December 19.

———. 2002. "Still Time to Save the Best Space." *Arizona Daily Star*, December 15.

———. 2003a. "55 Desert Species Get Protection." *Arizona Daily Star*, February 9.

———. 2003b. "Desert Panel OKs Platform." *Arizona Daily Star*, March 8.

———. 2003c. "Desert Plan Panel's Mixed Advice." *Arizona Daily Star*, May 27.

———. 2003d. "San Diego's Habitat Triage." *High Country News*, November 10.

———. 2004a. "Doubts Cloud Pima County, Ariz., Administrator's Promise on Open Space." *Arizona Daily Star*, May 24.

———. 2004b. "New Home-Building Permits Shoot Through the Roof in Tucson, Ariz." *Arizona Daily Star*, August 28.

———. 2005. "Open-Space Rule Could Be Scaled Back." *Arizona Daily Star*, June 20.

———. 2006. "Price Tag for Open Space Put at $2.6B." *Arizona Daily Star*, August 3.

———. 2007. "Open-Space Lags on Northwest Side." *Arizona Daily Star*, January 14.

Dewar, Heather. 1994. "Prescription Proposed for Ailing Glades." *Miami Herald*, January 13.

Diamond, Jared. 2005. *Collapse: How Societies Choose to Succeed or Fail* (New York: Viking).

Diringer, Elliot. 1991a. "SF Bay Makes Waves in Water Politics." *San Francisco Chronicle*, April 18.

———. 1991b. "State Panel Skirts Key Water Issue." *San Francisco Chronicle*, May 2.

———. 1992a. "Wilson to Propose Long-Term Plan for State's Water." *San Francisco Chronicle*, April 6.

———. 1991d. "EPA Says State Must Give Priority to Protecting Bay," *San Francisco Chronicle*, October 2.

———. 1992b. "Ultimatum to U.S. on Bay, Delta." *San Francisco Chronicle*, July 31.

———. 1992c. "EPA Pressures State to Protect Bay, Delta." *San Francisco Chronicle*, August 25.

———. 1993a. "New Plan for Bay Eases Impact on Cities, Farms." *San Francisco Chronicle*, March 9.

———. 1993b. "U.S. to Push New Rules on State Water." *San Francisco Chronicle*, December 15.

Dombeck, Michael P. 1996. "Thinking like a Mountain: BLM's Approach to Ecosystem Management." *Ecological Applications* 6(3): 699–702.

Downs, Anthony. 1994. *New Visions for Metropolitan America* (Washington, DC: Brookings Institution Press).

———. 2005. "Smart Growth: Why We Discuss It More Than We Do It." *Journal of the American Planning Association* 71(4): 367–380.

Drier, Peter, John Mollenkopf, and Todd Swanstrom. 2001. *Place Matters: Metropolitics for the Twenty-first Century* (Lawrence: University Press of Kansas).

Dryzek, John. 1990. *Discursive Democracy: Politics, Policy, and Political Science* (New York: Cambridge University Press).

Duane, Timothy P. 1997. "Community Participation in Ecosystem Management." *Ecology Law Quarterly* 24: 771–797.

Duerksen, Christopher, and Cara Snyder. 2005. *Nature-Friendly Communities: Habitat Protection and Land Use Planning* (Washington, DC: Island Press).

DuHamel, Jonathan. 2006. Personal communication.

Durbin, Kathie, and Paul Larmer. 1997. "The Feds Won't Enforce the ESA." *High Country News*, August 4.

Ebbin, Marc. 1997. "Is the Southern California Approach to Conservation Succeeding?" *Ecology Law Quarterly* 24: 695–706.

Echeverria, John D. 2001. "No Success like Failure: The Platte River Collaborative Watershed Planning Process." *William and Mary Environmental Law and Policy Review* 25: 559–604.

Eckersley, Robyn. 2002. "Environmental Pragmatism, Ecocentrism, and Deliberative Democracy: Between Problem-Solving and Fundamental Critique."

In Ben Minteer and Bob Pepperman Taylor, eds., *Democracy and the Claims of Nature: Critical Perspectives for a New Century* (Lanham, MD: Rowman & Littlefield), 49–69.

Ehrmann, John R., and Barbara L. Stinson. 1999. "Joint Fact-Finding and the Use of Technical Experts." In Lawrence Susskind, Sarah McKearnan, and Jennifer Thomas-Larner, eds., *The Consensus Building Handbook* (Thousand Oaks, CA: Sage), 375–399.

Ellis, Virginia. 1989. "Judge Halts L.A. Diversion of Water from Mono Basin." *Los Angeles Times*, August 23.

———. 1990. "State Backs Environmentalists on Mono Lake Water." *Los Angeles Times*, March 29.

———. 1991a. "Lake Must Rise Before Streams May Be Diverted." *Los Angeles Times*, April 19.

———. 1991b. "L.A. Backing State Plan for Mono Lake Drought." *Los Angeles Times*, May 8.

Eng, Larry, and Gail Kobetich. 1994a. Memo to David Flesh re comments on the Multiple Species Conservation Program (MSCP) Working Group's MSCP Draft Framework Plan. March 17. (On file with author.)

Eng, Larry, and Gail Kobetich. 1994b. Memo to David Flesh re comments on the Multiple Species Conservation Program (MSCP) Working Group's Draft Framework Plan. June 7. (On file with author.)

Ernst, Howard R. 2003. *Chesapeake Bay Blues: Science, Politics, and the Struggle to Save the Bay* (Lanham, MD: Rowman & Littlefield).

Esty, Daniel C., and Marian R. Chertow. 1997. "Thinking Ecologically: An Introduction." In Marion R. Chertow and Daniel C. Esty, eds., *Thinking Ecologically* (New Haven, CT: Yale University Press), 1–16.

Everglades Coalition. 1993. *The Greater Everglades Ecosystem Restoration Plan*. July. (On file with author.)

Fainstein, Susan S. 2000. "New Directions in Planning Theory." *Urban Affairs Review* 35(4): 451–478.

———. 2005. "Planning Theory and the City." *Journal of Planning Education and Research* 25(2): 121–130.

Fairbanks, Janet. 2003. Personal communication.

Finnigan, Daryl, Thomas I. Gunton, and Peter W. Williams. 2003. "Planning in the Public Interest: An Evaluation of Civil Society Participation in Collaborative Land Use Planning in British Columbia." *Environments* 31(3): 12–29.

Fiorino, Daniel J. 2004. "Flexibility." In Robert F. Durant, Daniel J. Fiorino, and Rosemary O'Leary, eds., *Environmental Governance Reconsidered: Challenges, Choices, and Opportunities* (Cambridge, MA: MIT Press), 393–425.

Fischer, Frank. 2000. *Citizens, Experts, and the Environment* (Durham, NC: Duke University Press).

————. 2003. *Reframing Public Policy: Discursive Politics and Deliberative Practices* (New York: Oxford University Press).

Florin, Hector. 2007. "Sector Plan Out," *Palm Beach Post*, November 27.

Forester, John. 1989. *Planning in the Face of Power* (Berkeley: University of California Press).

Forstenzer, Martin. 1993. "EPA Cites Bad Air at Mono." *Los Angeles Times*, July 11.

Foster, Sheila. 2002. "Environmental Justice in an Era of Devolved Collaboration." *Harvard Environmental Law Review* 26: 459–498.

Francis, George. 1993. "Ecosystem Management." *Natural Resources Journal* 33: 315–345.

Frank, Leonard S. 1993. Letter to Mayor Susan Golding re Council docket Item no. 127, Clean Water Program Resolution of Intention (ROI). June 14. (On file with author.)

Franklin, Jerry. 1997. "Ecosystem Management: An Overview." In A. Haney and M. S. Boyce, eds., *Ecosystem Management: Applications for Sustainable Forest and Wildlife Resources* (New Haven, CT: Yale University Press), 21–53.

Franz, Damon. 2002. "State, Federal Officials Say WRDA Will Not Harm Restoration." *Land Letter*, September 26.

Freeman, Jody. 1997. "Collaborative Governance in the Administrative State." *UCLA Law Review* 45: 1–98.

Freeman, Jody, and Daniel A. Farber. 2005. "Modular Environmental Regulation." *Duke Law Journal* 54(4): 795–912.

Fung, Archon, and Erik Olin Wright, eds. 2003. *Deepening Democracy: Institutional Innovations in Empowered Participatory Governance* (New York: Verso).

Geddes, Barbara. 2003. *Paradigms and Sand Castles: Theory Building and Research Design in Comparative Politics* (Ann Arbor: University of Michigan Press).

George, Alexander L., and Andrew Bennett. 2005. *Case Studies and Theory Development in the Social Sciences* (Cambridge, MA: MIT Press).

Gilbert, Nancy. 1992. Memo to Jerre Stallcup re Multiple Species Conservation Program Preserve design. March 3. (On file with author.)

Gilpin, Michael. 1992. "Comments on Population Viability Analysis for the California Gnatcatcher Within the MSCP Study Area." June 7. (On file with author.)

Gledhill, Lynda. 1998. "Davis Picks Water Advisory Board." *San Francisco Chronicle*, November 20.

Glen, Alan. 2003. Personal communication.

Gordon, John, and Jane Coppock. 1997. "Ecosystem Management and Economic Development." In Marion R. Chertow and Daniel C. Esty, eds., *Thinking Ecologically* (New Haven, CT: Yale University Press), 37–48.

Gottlieb, Robert, and Margaret Fitzsimmons. 1991. *Thirst for Growth: Water Agencies as Hidden Government in California* (Tucson: University of Arizona Press).

Governor's Commission for a Sustainable South Florida (GCSSF). 1995. "The Initial Report of the Governor's Commission for a Sustainable South Florida."(On file with author.)

———. 1996. "The Conceptual Plan of the Governor's Commission for a Sustainable South Florida." (On file with author.)

———. 1999. "Restudy Plan Report." (On file with author.)

Graf, William. 1999. "Dam Nation: A Geographic Census of Large American Dams and Their Hydrologic Impacts." *Water Resources Research* 35(4): 1305–1311.

Graham, Mary. 1999. *The Morning After Earth Day* (Washington, DC: Brookings Institution Press).

Greene, Juanita. 1983a. "Water-Management Fight Simmers Along Kissimmee." *Miami Herald*, March 6.

———. 1983b. "Restoring River Poses Tough Challenge." *Miami Herald*, September 19.

Greene, Juanita, and Randy Loftis. 1983. "Quiet Kissimmee: Stage Set for Storm." *Miami Herald*, August 8.

Greer, Keith. 2003. Personal communication.

———. 2004. "Habitat Conservation Planning in San Diego County, California: Lessons Learned After Five Years of Implementation." *Environmental Practice* 6(3): 230–239.

Grijalva, Raul. 2006. Personal communication.

Grossi, Mark, and E. J. Schultz. 2006. "Farmers, Environmentalists Agree on $800 m Restoration." *Fresno Bee*, September 14.

Grossman, Elizabeth. 2002. *Watershed: The Undamming of American Rivers* (New York: Counterpoint).

Groth, Darren, Ian Gill, Mark McMillin, Denis Moser, Jim Whalen, Keith Johnson, Len Frank, Mike Madigan, and Kim Kilkenny. 1995. Letter to Mayor Susan Golding re joint BIA–Alliance position on the Multiple Species Conservation Program (MSCP). August 31. (On file with author.)

Grumbine, R. Edward. 1994. "What Is Ecosystem Management?" *Conservation Biology* 8(1): 27–38.

———. 1997. "Reflections on 'What Is Ecosystem Management?'" *Conservation Biology* 11(1): 41–47.

Grunewald, Christopher. 1998. "San Diego Multiple Species Conservation Program (MSCP) Plan." In Peter Aengst, Jeremy Anderson, Jay Chamberlin, Christopher Grunewald, Susan Loucks, and Elizabeth Wheatley, *Balancing Public Trust & Private Interest: Public Participation in Habitat Conservation*

Planning (Ann Arbor: University of Michigan School of Natural Resources and Environment), Appendix A.

Grunwald, Michael. 2002a. "Growing Pains in Southwest Florida." *Washington Post*, June 25.

———. 2002b. "An Environmental Reversal of Fortune." *Washington Post*, June 26.

———. 2006. *The Swamp: The Everglades, Florida, and the Politics of Paradise* (New York: Simon & Schuster).

Gruson, Kerry. 1983a. "Flooding Poses Threat to Everglades' Ecology." *New York Times*, July 25.

———. 1983b. "Saving the Everglades, with a Passion." *New York Times*, September 25.

Gunderson, Lance H., C. S. Holling, and Stephen S. Light. 1995. "Barriers Broken and Bridges Built: A Synthesis." In Lance H. Gunderson, C. S. Holling, and Stephen S. Light, eds., *Barriers and Bridges to the Renewal of Ecosystems and Institutions* (New York: Columbia University Press), 489–532.

Gunningham, Neil. 1995. "Environment, Self-Regulation, and the Chemical Agency: Assessing Responsible Care." *Law & Policy* 17(1): 57–108.

Gunton, Thomas I., J. C. Day, and Peter W. Williams. 2003. "Evaluating Collaborative Planning: The British Columbia Experience." *Environments* 31(3): 1–11.

Hafenbrack, Josh. 2006. "Land, Cost Key to North–South Bidding for Biotech Prize." Fort Lauderdale *Sun–Sentinel*, January 14.

Hain, J. Christopher. 2004. "Water Projects Get Rush Money." *Palm Beach Post*, October 15.

Haitch, Richard. 1987. "Costly Reversal of a U.S. Project." *New York Times*, February 15.

Halbert, C. L. 1993. "How Adaptive Is Adaptive Management? Implementing Adaptive Management in Washington State and British Columbia." *Reviews in Fisheries Science* 1: 261–263.

Hanna, Mark. 2007. Personal communication.

Hansen, Kevin. 1984. "South Florida's Water Dilemma: A Trickle of Hope for the Everglades." *Environment* 26: 14–20, 40–42.

Harrison, Kathryn. 1999. "Cooperative Approaches to Environmental Protection." *Journal of Industrial Ecology* 2(3): 51–72.

Hart, John. 1996. *Storm over Mono: The Mono Lake Battle and the California Water Future* (Berkeley: University of California Press).

Hartig, John H., Michael A. Zarull, Thomas M Heidtke, and Hemang Shah. 1998. "Implementing Ecosystem-Based Management: Lessons from the Great Lakes." *Journal of Environmental Planning and Management* 41(1): 45–75.

Haurwitz, Ralph K. M. 1993a. "Austin Ecological Effort Praised." *Austin American–Statesman*, March 14.

———. 1993b. "Environmentalists Objecting to Key Parts of Balcones Plan." *Austin American–Statesman*, April 25.

———. 1993c. "Developers, Tired of Balcones Wait, File Flurry of Subdivision." *Austin American–Statesman*, June 27.

———. 1993d. "Behind Austin's Diversity of Wildlife Lies an Environment Just as Complex." *Austin American–Statesman*, July 18.

———. 1993e. "Amphibian Finds Itself in Middle of Austin Debate." *Austin American–Statesman*, July 19.

———. 1993f. "Babbitt Gives Key Support to BCCP." *Austin American–Statesman*, August 8.

———. 1993g. "Committee to Lobby for $48.9 Million BCCP Bond Issue." *Austin American–Statesman*, September 11.

———. 1993h. "Babbitt Urges Voters to OK $48.9 Million Bond Proposition." *Austin American–Statesman*, October 25.

———. 1993i. "Voters Reject BCCP Bonds." *Austin American–Statesman*, November 3.

———. 1993j. "Low Vote Count Doomed BCCP." *Austin American–Statesman*, November 4.

———. 1996. "Urban Habitat." *Austin American–Statesman*, April 21.

———. 1997a. "Bill Could Undo Balcones Plan." *Austin American–Statesman*, April 3.

———. 1997b. "Building Nibbles Best Habitat for Warblers." *Austin American–Statesman*, September 24. (Hard copy only. On file with author.)

Healey, Patsy. 1993. "Planning Through Debate: The Communicative Turn in Planning Theory." In Frank Fischer and John Forester, eds., *The Argumentative Turn in Policy Analysis and Planning* (Durham, NC: Duke University Press), 233–253.

———. 2006. *Collaborative Planning: Shaping Places in Fragmented Societies*, 2nd ed. (New York: Palgrave Macmillan).

Hedström, Peter, and Richard Swedberg. 1998. *Social Mechanisms: An Analytical Approach to Social Theory* (New York: Cambridge University Press).

Hennessey, Ann. 2003. "Crestridge Ecological Reserve Survives Vandalism with the Help of Volunteers." *Outdoor California*, September–October.

Hillman, Mick, Graeme Aplin, and Gary Brierley. 2003. "The Importance of Process in Ecosystem Management: Lessons from the Lachlan Catchment, New South Wales, Australia." *Journal of Environmental Planning and Management* 46(2): 219–237.

Hite, Justin. 2005. "2004 Best Year for Mono's Gulls in 22-Year Study." *Mono Lake Newsletter*, Summer.

Hockenstein, Jeremy B., Robert N. Stavins, and Bradley W. Whitehead. 1997. "Crafting the Next Generation of Market-Based Environmental Tools." *Environment* 39(4): 12–20, 30–33.

Holling, C. S. 1995. "What Barriers? What Bridges?" In Lance H. Gunderson, C. S. Holling, and Stephen S. Light, eds., *Barriers and Bridges to the Renewal of Ecosystems and Institutions* (New York: Columbia University Press), 1–34.

———. 1996. "Surprise for Science, Resilience for Ecosystems, and Incentives for People." *Ecological Applications* 6(3): 733–735.

Holling, C. S., ed. 1978. *Adaptive Environmental Assessment and Management* (New York: Wiley).

Holling, C. S., and Gary K. Meffe. 1996. "Command and Control and the Pathology of Natural Resource Management." *Conservation Biology* 10(2): 328–337.

Hood, Laura C. 1998. *Frayed Safety Nets: Habitat Conservation Planning Under the Endangered Species Act* (Washington, DC: Defenders of Wildlife).

Hopkins, Heidi. 1997a. "Mill Creek: Sharing Scarce Water." *Mono Lake Newsletter*, Winter.

———. 1997b. "Settlement Reached on Mono Basin Restoration." *Mono Lake Newsletter*, Fall.

———. 1997c. "Community Energizes Public Process." *Mono Lake Newsletter*, Fall.

Horton, Tom. 2003. *Turning the Tide: Saving the Chesapeake Bay* (Washington, DC: Island Press).

Houck, Oliver. 1999. "TMDLs IV: The Final Frontier." *Environmental Law Reporter* 29(10): 469–479.

Huckelberry, Chuck. 2001. Board of Supervisors memorandum: Sonoran Desert Conservation Plan progress report and update. October 9.

———. 2006. Personal communication.

Hundley, Norris, Jr. 2001. *The Great Thirst*, rev. ed. (Berkeley: University of California Press).

Huntington, Charles, and Sari Sommarstrom. 2000. *An Evaluation of Selected Watershed Councils in the Pacific Northwest and Northern California*. Report prepared for Pacific Rivers Council and Trout Unlimited, Eugene, OR.

Hurchalla, Maggy. 2003. Personal communication.

Immen, Wallace. 1984. "Nature Proves Better at Cleansing Water, Fostering Wildlife." *Toronto Globe and Mail*, November 5.

Imperial, Mark T. 1999. "Institutional Analysis and Ecosystem-Based Management: The Institutional Analysis and Development Framework." *Environmental Management* 24(4): 449–465.

Imperial, Mark T., and Timothy Hennessey. 2000. *Environmental Governance in Watersheds* (Washington, DC: National Academy of Public Administration).

Ingram, Erik. 1990. "Stormy Marin Meeting on Bay's Water Quality." *San Francisco Chronicle*, August 21.

Innes, Judith E. 1996. "Planning Through Consensus Building." *Journal of the American Planning Association* 62(4): 460–472.

————. 1999. "Evaluating Consensus Building." In Lawrence Susskind, Sarah McKearnan, and Jennifer Thomas-Larner, eds., *The Consensus Building Handbook* (Thousand Oaks, CA: Sage), 631–675.

Innes, Judith E., and David Booher. 1999. "Consensus Building and Complex Adaptive Systems: A Framework for Evaluating Collaborative Planning." *Journal of the American Planning Association* 65(4): 412–423.

Innes, Judith E., Sarah Connick, and David Booher. 2007. "Informality as a Planning Strategy: Collaborative Water Management in the CALFED Bay–Delta Program." *Journal of the American Planning Association* 73(2): 195–210.

Innes, Judith E., Sarah Connick, Laura Kaplan, and David E. Booher. 2006. "Collaborative Governance in the CALFED Program: Adaptive Policy Making for California Water." Working paper 2006-1 (Institute of Urban and Regional Development, University of California, Berkeley, and the Center for Collaborative Policy, California State University, Sacramento).

Innes, Judith E., Judith Gruber, Michael Neuman, and Robert Thompson. 1994. *Coordinating Growth and Environmental Management Through Consensus Building* (Berkeley: California Policy Seminar, University of California).

Interagency Ecosystem Management Task Force. 1995. *The Ecosystem Approach: Healthy Ecosystems and Sustainable Economies*, vol. 1. June.

Interagency Task Force on Mono Lake. 1979. *Report of Interagency Task Force on Mono Lake* (Sacramento, CA: State Water Resources Control Board).

Jackson, Robert B., Stephen R. Carpenter; Clifford N. Dahm, Diane M. McKnight, Robert J. Naiman, Sandra L. Postel, and Steven W. Running. 2001. "Water in a Changing World." *Ecological Applications* 11(4): 1027–1045.

Jacobs, Paul. 1997. "State Wildlife Agency Struggles to Manage New Holdings." *Los Angeles Times*, April 28.

Jaffe, Matthew. 2001. "The Glorious Sonoran." *Sunset*, January 1.

Jensen, Mari. 2001. "Science, Politics and Preservation." *Tucson Citizen*, October 25.

John, DeWitt. 1994. *Civic Environmentalism: Alternatives to Regulation in States and Communities* (Washington, DC: CQ Press).

————. 2004. "Civic Environmentalism." In Robert F. Durant, Daniel J. Fiorino, and Rosemary O'Leary, eds., *Environmental Governance Reconsidered: Challenges, Choices, and Opportunities* (Cambridge, MA: MIT Press), 219–254.

Johnson, Barry L. 1999. "Introduction to the Special Feature: Adaptive Management—Scientifically Sound, Socially Challenged?" *Conservation Ecology* 3(1): 10. Available at http://www.ecologyandsociety.org/vol3/iss1/art10/.

Johnson, Bart R., and Ronald Campbell. 1999. "Ecology and Participation in Landscape-Based Planning Within the Pacific Northwest." *Policy Studies Journal* 27(3): 502–529.

Johnson, Jenny. 2000. "Land Mine." *Austin Chronicle*, March 3.

Johnson, Robert. 2003. Personal communication.

Johnson, Steve. 2005. Personal communication.

Jones, Robert A. 1991. "Mono Lake: A Test for a New Regime." *Los Angeles Times*, May 8.

Juarez, Marcario, Jr. 2001. "Group to Contest Conservation Plan." *Arizona Daily Star*, August 30.

Kahn, Matthew. 2000. "The Environmental Impact of Suburbanization." *Journal of Policy Analysis and Management* 19(4): 569–586.

Kareiva, Peter. 1993. Letter to Dr. R. McMillen re Population Viability Analysis for the California Gnatcatcher Within the MSCP Study Area," June 20. (On file with author.)

Karkkainen, Bradley. 2001/2002. "Collaborative Ecosystem Governance: Scale, Complexity, and Dynamism." *Virginia Environmental Law Journal* 21: 189–243.

———. 2002. "Environmental Lawyering in the Age of Collaboration." *Wisconsin Law Review*: 555–574.

Karr, James R. 1993. "Measuring Biological Integrity: Lessons from Streams." In Steven Woodley, James Kay, and George Francis, eds., *Ecological Integrity and the Management of Ecosystems* (Delray Beach, FL: St. Lucie Press), 83–115.

Katz, Leah Alyssa. 2000. "The Art of Collaboration: Creating the County of San Diego's Multiple Species Conservation Program." Senior research project, Urban Studies and Planning Department, University of California, San Diego.

Kay, Jane. 2001. "Turning the Tide," *San Francisco Chronicle*, March 11.

———. 2006. "It's Rising and Healthy." *San Francisco Chronicle*, July 29.

Kay, Michele. 1991a. "Growth Depends on Habitat Plan's Success." *Austin American–Statesman*, July 17.

———. 1991b. "Preserve Proposal Forks Path for Future." *Austin American–Statesman*, August 25.

Keiter, Robert B. 1998. "Ecosystems and the Law: Toward an Integrated Approach." *Ecological Applications* 8(2): 332–341.

———. 2003. *Keeping Faith with Nature: Ecosystems, Democracy, and America's Public Lands* (New Haven, CT: Yale University Press).

Kelly, Patrick A., and John T. Rotenberry. 1993. "Buffer Zones for Ecological Reserves in California: Replacing Guesswork with Science." In *Interface Between Ecology and Land Development in California* (Los Angeles: Southern California Academy of Sciences), 85–92.

Kemmis, Daniel. 1990. *Community and the Politics of Place* (Norman: University of Oklahoma Press).

Kempton, Willett, James S. Boster, and Jennifer A. Hartley. 1995. *Environmental Values in American Culture* (Cambridge, MA: MIT Press).

Kenney, Douglas S. 2000. *Arguing About Consensus: Examining the Case Against Western Watershed Initiatives and Other Collaborative Groups Active in Natural Resources Management* (University of Colorado School of Law, Natural Resources Law Center).

Kier, Bill. 2006. Personal communication.

Kimmerer, Wim. 2005. Personal communication.

King, Pat. 2006. Personal communication.

King, Robert P. 1999. "Scientists: No Easy Way to Restore Everglades, Biscayne Bay," Cox News Service, January 20.

———. 2004. "Future of Water for Everglades in Doubt." *Palm Beach Post,* April 26.

———. 2006. "Development Called Threat to Everglades Restoration." *Palm Beach Post,* January 29.

Knopman, Debra S., Megan M. Susman, and Marc K. Landy. 1999. "Civic Environmentalism: Tackling Tough Land-Use Problems with Innovative Governance." *Environment* 41(10): 24–33.

Koebel, Joseph W., Jr. 1995. "An Historical Perspective on the Kissimmee River Restoration Project." *Restoration Ecology* 3(3): 149–159.

———. 2007. Personal communication.

Koehler, Don. 2003. Personal communication.

Kondolf, G. Mathias, and Elisabeth R. Micheli. 1995. "Evaluating Stream Restoration Projects." *Environmental Management* 19(1): 1–15.

Koontz, Tomas M. 2005. "We Finished the Plan, So Now What? Impacts of Collaborative Stakeholder Participation on Land Use Policy." *Policy Studies Journal* 33(3): 459–481.

Koontz, Tomas M., Toddi A. Steelman, JoAnn Carmin, Katrina Smith Korfmacher, Cassandra Moseley, and Craig W. Thomas. 2004. *Collaborative Environmental Management: What Roles for Government?* (Washington, DC: Resources for the Future).

Koontz, Tomas M., and Craig W. Thomas. 2006. "What Do We Know and Need to Know About the Environmental Outcomes of Collaborative Management?" *Public Administration Review* 66: 111–121.

Kraus, Mark. 2006. Personal communication.

Kunioka, Todd, and Lawrence S. Rothenberg. 1993. "The Politics of Bureaucratic Competition: The Case of Natural Resource Policy." *Journal of Policy Analysis and Management* 12(4): 700–725.

Landy, Marc K., Marc J. Roberts, and Stephen R. Thomas. 1994. *The Environmental Protection Agency: Asking the Wrong Questions,* expanded ed. (New York: Oxford University Press).

Lange, Jonathan I. 2001. "Exploring Paradox in Environmental Collaborations." In Philip Brick, Donald Snow, and Sarah Van de Wetering, eds., *Across the Great*

Divide: Explorations in Collaborative Conservation and the American West (Washington, DC: Island Press), 200–209.

LaRue, Steve. 1994. "Golding Seeks to Hasten Creation of 165,000-Acre Wildlife Preserve." *San Diego Union–Tribune*, May 1.

Larmer, Paul. 1996. "A Colorado County Tries a Novel Approach: Work the System." *High Country News*, May 13.

Lawrence, Steve. 1985. "Residents Fight Officialdom to Keep Mono Lake Wet." *Los Angeles Times*, October 27.

Layzer, Judith A. 2002. "Citizen Participation and Government Choice in Local Environmental Controversies." *Policy Studies Journal* 30(2): 193–207.

Leach, William D., and Neil W. Pelkey. 2001. "Making Watershed Partnerships Work: A Review of the Empirical Literature." *Journal of Water Resources Planning and Management* 127(6): 378–385.

Leach, William D., Neil W. Pelkey, and Paul A. Sabatier. 2002. "Stakeholder Partnerships as Collaborative Management: Evaluation Criteria Applied to Watershed Management in California and Washington." *Journal of Policy Analysis and Management* 21(4): 645–670.

Leach, William D., and Paul Sabatier. 2003. "Facilitators, Coordinators, and Outcomes." In Rosemary O'Leary and Lisa B. Bingham, eds., *The Promise and Performance of Environmental Conflict Resolution* (Washington, DC: Resources for the Future), 148–171.

Leach, William D., and Paul A. Sabatier. 2005. "Are Trust and Social Capital the Keys to Success? Watershed Partnerships in California and Washington." In Paul A. Sabatier, Will Focht, Mark Lubell, Zev Trachtenberg, Arnold Vedlitz, and Marty Matlock, eds., *Swimming Upstream: Collaborative Approaches to Watershed Management* (Cambridge, MA: MIT Press), 233–258.

Leavenworth, Stuart. 2003. "Restoration Plan Stirs Concern." *Sacramento Bee*, October 22.

———. 2004. "Major Shift Mapped for Delta Water." *Sacramento Bee*, September 26.

Leavitt, Christy. 2006. *Troubled Waters: An Analysis of Clean Water Act Compliance, July 2003–December 2004* (Washington, DC: U.S. PIRG Education Fund).

Lee, Charles. 2007. Personal communication.

Lee, Kai N. 1993. *Compass and Gyroscope: Integrating Science and Politics for the Environment* (Washington, DC: Island Press).

Lee, Mike. 2005. "Jury Still Out on Whether Landmark Habitat Plan Is Living Up to Promise." *San Diego Union–Tribune*, June 18.

———. 2006. "Judge Assails San Diego's Landmark Habitat Plan." *San Diego Union–Tribune*, October 14.

Leslie, Jacques. 2005. *Deep Water: The Epic Struggle over Dams, Displaced People, and the Environment* (New York: Farrar, Straus and Giroux).

Levin, Ted. 2001. "Reviving the River of Grass." *Audubon*, July–August. Available at http://magazine.audubon.org/features0107/ecology/ecology0107. html.

Librach, Austan. 1995. "Endangered Species Protection: The Austin, Texas, Experience." June. (On file with author.)

———. 2003. Personal communication.

Little Hoover Commission (LHC). 2005. *Still Imperiled, Still Important* (Sacramento: State of California, November 17). (On file with author.)

Little, Jane Braxton. 1997. "Mono Lake: Victory over Los Angeles Turns into Local Controversy." *High Country News*, December 8.

Lodge, Thomas E. 1998. *The Everglades Handbook: Understanding the Ecosystem* (Boca Raton, FL: St. Lucie Press).

Loftin, M. Kent. 2007. Personal communication.

Loftin, M. Kent, Louis A. Toth, and Jayantha T. B. Obeysekera. 1990. *Kissimmee River Restoration, Alternative Plan Evaluation & Preliminary Design Report* (West Palm Beach: South Florida Water Management District).

Loftis, Randy. 1984. "Sentiment Runs Deep Against River Project." *Miami Herald*, March 18.

———. 1987. "U.S. Aid Urged for Kissimmee Restoration." *Miami Herald*, January 18.

Logan, John R., and Harvey L. Molotch. 1987. *Urban Fortunes: The Political Economy of Place* (Berkeley: University of California Press).

Logan, John R., and Min Zhou. 1989. "Do Suburban Growth Controls Control Growth?" *American Sociological Review* 54 (June): 461–471.

Los Angeles Department of Water and Power (LADWP). 2006. "Compliance with State Water Resources Control Board Order nos. 98-05 and 98-07." May. (On file with author.)

Lowi, Theodore J. 1999. "Frontyard Propaganda." *Boston Review* 24(5): 17–18.

Lowry, William R. 2003. *Dam Politics: Restoring America's Rivers* (Washington, DC: Georgetown University Press).

Lubchenco, Jane. 2002. "State of the Planet 2002: Science and Sustainability." Paper presented at the Earth Institute, Columbia University, May 13. Available at www.earth.Columbia.edu/sop2002/sopagenda.html.

Lubell, Mark. 2004. "Collaborative Environmental Institutions: All Talk and No Action?" *Journal of Policy Analysis and Management* 23(3): 549–573.

———. 2005. "Do Watershed Partnerships Enhance Beliefs Conducive to Collective Action?" In Paul A. Sabatier, Will Focht, Mark Lubell, Zev Trachtenberg, Arnold Vedlitz, and Marty Matlock, eds., *Swimming Upstream: Collaborative Approaches to Watershed Management* (Cambridge, MA: MIT Press), 201–232.

Ludwig, Donald, Ray Hilborn, and Carl Walters. 1993. "Uncertainty, Resource Exploitation, and Conservation: Lessons from History." *Science* 260 (April 2): 17, 36.

Lund, Jay, Ellen Hanak, William Fleenor, Richard Howitt, Jeffrey Mount, and Peter Moyle. 2007. *Envisioning Futures for the Sacramento–San Joaquin Delta* (San Franscisco: Public Policy Institute of California).

Luoma, Sam. 2005. Personal communication.

Lurie, Alan. 2001. "Affordable Housing a Nightmare for Tucson's Minorities." *Southern Arizona Home Builder*, September. (On file with author.)

MacVicar, Tom. 2001. Everglades Oral History, South Florida Oral History Consortium. May 20.

Mann, Jim. 1983. "Supreme Court Declines Mono Lake Hearing." *Los Angeles Times*, November 8.

Mansbridge, Jane. 1980. *Beyond Adversary Democracy* (New York: Basic Books).

Marshall, Arthur R. 1971/1972. "Statement to Governor Reuben O'D. Askew from the Governor's Conference on Water Management in South Florida." *Water Management Bulletin* 5(3).

———. 1980. "A Critique of Water Management in South Florida," *Water Management Bulletin*, November 20–21 (West Palm Beach, FL: Central and Southern Florida Flood Control District).

Marshall, Rob. 2006. Personal communication.

Martin, Glen. 1998. "Study Proposed on Idea of Raising Height of Shasta Dam by 6 Feet." *San Francisco Chronicle*, October 13.

———. 1999a. "Divvying Up Our Water." *San Francisco Chronicle*, June 25, A1.

———. 1999b. "Farmers, Fish Share in Ballyhooed Water Plan." *San Francisco Chronicle*, June 26.

———. 1999c. "Dams Making Way for Salmon." *San Francisco Chronicle*, November 9, A3.

———. 2002. "California's Water Pact Threatened," *San Francisco Chronicle*, April 14.

———. 2006a. "No Simple Answer Seen for Drastic Decline of Delta Fish." *San Francisco Chronicle*, February 28.

———. 2006b. "Sick San Joaquin River on Brink of a New Life." *San Francisco Chronicle*, September 14.

Martin, Ken. 1990. "Panel: Development Strictures Are Real." *Austin Business Journal*, March 12–18, 3.

Martinez, Sylvia. 1991. "Endangered Species Flap Takes Its Toll on Tax Rolls." *Austin American–Statesman*, July 24.

Mazmanian, Daniel A., and Michael E. Kraft. 1999. "The Three Epochs of the Environmental Movement." In Daniel A. Mazmanian and Michael E. Kraft, eds., *Toward Sustainable Communities: Transition and Transformations in Environmental Policy* (Cambridge, MA: MIT Press), 3–41.

Mazzotti, Frank. 2006. Personal communication.

McCloskey, Michael. 1996. "The Skeptic: Collaboration Has Its Limits." *High Country News*, May 13.

McClure, Robert. 1994. "Corps Shows Off Everglades Plans." *Fort Lauderdale Sun-Sentinel*, December 1.

McClurg, Sue. 2002. "CALFED Today: A Roundtable Discussion." *Western Water*, September/October, 4–17.

McCully, Patrick. 2001. *Silenced Rivers: The Ecology and Politics of Large Dams* (New York: Zed Books).

McKinley, Laurie. 1993. "Comments on Proposed Council Policy on Mitigation of Impacts to Biological Resources." May 27. (On file with author.)

McLeod, K. L., J. Lubchenco, S. R. Palumbi, and A. A. Rosenberg. 2005. "Scientific Consensus Statement on Marine Ecosystem-Based Management." Available at http://compassonline.org/marinescience/solutions_ecosystem.asp.

McQuilkin, Geoffrey. 2005. "High Flows Benefit Mill Creek This Year," *Mono Lake Newsletter*, Fall, 4, 10.

———. 2007. Personal communication.

Meffe, Gary, and C. Ronald Carroll. 1994. *Principles of Conservation Biology* (Sunderland, MA: Sinauer Associates).

Meffe, Gary K., Larry A. Nielsen, Richard L. Knight, and Dennis A. Schenborn. 2002. *Ecosystem Management: Adaptive, Community-Based Conservation* (Washington, DC: Island Press).

Melnick, R. Shep. 1983. *Regulation and the Courts: The Case of the Clean Air Act* (Washington, DC: Brookings Institution).

Meltzer, Erica. 2007. "Grant to Gauge Conservation Plan's Success." *Arizona Daily Star*, May 21.

Millennium Ecosystem Assessment. 2003. *Ecosystems & Human Well-Being* (Washington, DC: Island Press).

Miller, Bartshe. 2003. "Ten Years Later: An Education Program Grows Up with the Trees." *Mono Lake Newsletter*, Winter, 6–8.

Minteer, Ben. 2002. "Deweyan Democracy and Environmental Ethics." In Ben Minteer and Bob Pepperman Taylor, eds., *Democracy and the Claims of Nature: Critical Perspectives for a New Century* (Lanham, MD: Rowman & Littlefield), 33–48.

Mooney, Michael G. 2006. "Growth, Climate Change Could Spell Disaster for Valley, State Waterways." *Modesto Bee*, October 8.

Moote, Margaret A., and Mitchell P. McClaran. 1997. "Viewpoint: Implications of Participatory Democracy for Public Land Planning." *Journal of Range Management* 50: 473–481.

Morgan, Curtis. 2001a. "Water Plan Goes Underground." *Miami Herald*, February 2.

———. 2001b. "Utilities Try to Stop Glades Water Rules." *Miami Herald*, March 8.

———. 2001c. "Elevated Road Urged for Glades Restoration." *Miami Herald*, June 23.

———. 2002. "Florida's Kissimmee River Begins to Recover from Man-made Damage." *Miami Herald*, December 15.

———. 2003. "Florida Releases Final Blueprint of Everglades Restoration Plan." *Miami Herald*, November 5.

———. 2004a. "Judge for Florida Everglades Case Removed from Earlier, Similar Case." *South Florida Sun–Sentinel*, February 21.

———. 2004b. "South Florida Anglers Fear Loss of Canal Access." *Miami Herald*, May 3.

———. 2005. "New Glades Plan: 2 Bridges." *Miami Herald*, August 23.

———. 2006. "Mining Debate Centers on Risk, Benefits." *Miami Herald*, May 8.

Morgenstern, Richard D., and William A. Pizer, eds. 2006. *Reality Check: The Nature and Performance of Voluntary Environmental Programs in the United States, Europe, and Japan* (Washington, DC: Resources for the Future).

Mottola, Daniel. 2005. "Birds vs. People at BCP." *Austin Chronicle*, February 25.

Mulliken, John. 1985a. "Alternatives to Kissimmee Report Urged." *Fort Lauderdale Sun–Sentinel*, February 26.

———. 1985b. "Group's Plans on Target for River Basin's Future." *Fort Lauderdale Sun–Sentinel*, November 23.

———. 1986. "Graham Seeks U.S. Funds to Help Restore Kissimmee." *Fort Lauderdale Sun-Sentinel*, June 1.

———. 1987. "Kissimmee Washout Forces New Dam Study." *Fort Lauderdale Sun–Sentinel*, June 30.

———. 1988. "To Repair a River." *Fort Lauderdale Sun-Sentinel*, August 17.

Murphy, Dean E. 1993. "Study Urges Halving L.A.'s Mono Lake Ration." *Los Angeles Times*, May 29.

Murphy, Dennis. 1999. "Case Study." In K. Norman Johnson, Frederick Swanson, Margaret Herring, and Sarah Greene, eds., *Bioregional Assessments* (Washington, DC: Island Press), 230–247.

National Academy of Public Administration (NAPA). 1995. *Setting Priorities, Getting Results: A New Direction for the U.S. Environmental Protection Agency* (Washington, DC: National Academy of Public Administration).

National Research Council (NRC), Mono Basin Ecosystem Study Committee. 1987. *The Mono Basin Ecosystem: Effects of a Changing Lake Level* (Washington, DC: National Academy Press).

National Research Council (NRC). 2002. *Regional Issues in Aquifer Storage and Recovery for Everglades Restoration: A Review of the ASR Regional Study Project Management Plan of the Comprehensive Everglades Restoration Plan* (Washington, DC: National Academies Press).

National Research Council (NRC), Committee on Restoration of Greater Everglades Ecosystem. 2005. *Re-engineering Water Storage in the Everglades: Risks and Opportunities* (Washington, DC: National Academies Press).

National Research Council (NRC), Committee on Independent Scientific Review of Everglades Restoration Progress. 2006. *Progress Toward Restoring the Everglades: The First Biennial Review* (Washington, DC: National Academies Press).

Nelson, Barry, Christina Swanson, Spreck Rosenkranz, Zeke Grader, and Bill Jennings. 2006. Memo re recommendations for additional actions to protect Delta fisheries. June 5. (On file with author.)

Newborn, Steve. 1999. "Kissimmee Begins the Rebending." *Tampa Tribune*, June 11.

———. 2000. "Kissimmee Restoration." *Tampa Tribune*, July 16.

Nicholas, James C. 1999. "State and Regional Land Use Planning: The Evolving Role of the State." *St. John's Law Review* 73(4): 1069–1089.

Nickelsburg, Stephen M. 1998. "Mere Volunteers? The Promise and Limits of Community-Based Environmental Protection." *Virginia Law Review* 84(7): 1371–1409.

Nintzel, Jim. 1998. "Arresting Development." *Tucson Weekly*, May 28–June 3.

Nolte, Carl. 1990. "Delta Bass Population at Record Low." *San Francisco Chronicle*, August 4.

Nordheimer, Jon. 1987. "Lake's Rescue Threatens Everglades." *New York Times*, July 23.

Norton, Richard K. 2005. "More and Better Local Planning." *Journal of the American Planning Association* 71(1): 55–71.

Noss, Reed F. 1983. "A Regional Landscape Approach to Maintain Diversity." *Bioscience* 33: 700–706.

Noss, Reed F., and Allen Y. Cooperrider. 1994. *Saving Nature's Legacy* (Washington, DC: Island Press).

Noss, Reed F., Eric Dinerstein, Barrie Gilpin, Brian J. Miller, John Terborgh, and Steve Trombulak. 1999. "Regional and Continental Restoration." In Michael E. Soulé and John Terborgh, eds., *Continental Conservation: Scientific Foundations of Regional Reserve Networks* (Washington, DC: Island Press), 99–128.

Noss, Reed F., and J. Michael Scott. 1997. "Ecosystem Protection and Restoration: The Core of Ecosystem Management." In M. Boyce and A. Hanley, eds., *Ecosystem Management: Applications to Sustainable Forest and Wildlife Resources* (New Haven, CT: Yale University Press), 239–264.

Noss, Reed, and Laura Hood Watchman. 2001. "Report of Independent Peer Reviewers: Sonoran Desert Conservation Plan." October. (On file with author.)

Oberbauer, Thomas. 2003. Personal communication.

Odum, Eugene P. 1972. "Ecosystem Theory in Relation to Man." In John A. Wiens, ed., *Ecosystem Structure and Function* (Corvallis: Oregon State University Press), 11–24.

Office Supervisor, Fish and Wildlife Enhancement Region 1, Ventura Office, Ventura, California (Office Supervisor). 1992. E-mail memo to Nancy Gilbert re draft Population Viability Analysis (PVA) for the California gnatcatcher prepared by Ogden, June 8. (On file with author.)

Ogden Environmental and Energy Services Co. (Ogden). 1992. "Biological Objectives and Criteria for Identifying Preserve Planning Areas for the Multiple Species Conservation Program," June 1992. (On file with author.)

———. 1996. *Biological Monitoring Plan for the Multiple Species Conservation Program*. April 23. (On file with author.)

Ogden, John. 1999. "Case Study." In K. Norman Johnson, Frederick F. Swanson, Margaret Herring, and Sarah Greene, eds., *Bioregional Assessments* (Washington, DC: Island Press), 168–186.

———. 2003. Personal communication.

O'Leary, Rosemary. 2003. "Environmental Policy in the Courts." In Norman J. Vig and Michael E. Kraft, eds., *Environmental Policy: New Directions for the 21st Century*, 5th ed. (Washington, DC: CQ Press), 151–173.

O'Leary, Rosemary, T. Nabatchi, and L. B. Bingham. 2004. "Environmental Conflict Resolution." In Robert F. Durant, Daniel J. Fiorino, and Rosemary O'Leary, eds., *Environmental Governance Reconsidered: Challenges, Choices, and Opportunities* (Cambridge, MA: MIT Press), 323–354.

Olszewski, Lori, and Vlae Kershner. 1993. "Gov. Wilson Backs Away on Water Policy." *San Francisco Chronicle*, April 2.

Orfield, Myron. 1997. *Metropolitics: A Regional Agenda for Community Stability* (Washington, DC: Brookings Institution Press).

Orians, Gordon H. 1995. "Thought for the Morrow: Cumulative Threats to the Environment." *Environment* 37(7): 6–14, 33–36.

Ostrom, Elinor. 1990. *Governing the Commons: The Evolution of Institutions for Collective Action* (New York: Cambridge University Press).

———. 2001. "Reformulating the Commons." In Joanna Burger, Elinor Ostrom, and Richard Norgaard, eds., *Protecting the Commons: A Framework for Resource Management in the Americas* (Washington, DC: Island Press), 17–41.

Oyola-Yemaiel, Arthur. 1999. "Towards the Formation of a Sustainable South Florida: An Analysis of Conflict Resolution and Consensus Building in the South Florida Ecosystem Restoration Initiative." Unpublished dissertation (Miami: Florida International University).

Ozawa, Connie. 1991. *Recasting Science: Consensual Procedures in Public Policy Making* (Boulder, CO: Westview Press).

Paehlke, Robert. 2001. "Spatial Proportionality: Right-Sizing Environmental Decision-Making." In Edward A. Parson, ed., *Governing the Environment: Persistent Challenges, Uncertain Innovations* (Toronto: University of Toronto Press), 73–123.

Pease, Craig. 2006. Personal communication.

Pellow, David. 1999. "Negotiation and Confrontation: Environmental Policymaking Through Consensus." *Society & Natural Resources* 12(3): 189–203.

Pendleton, Scott. 1992. "Austin, Texas: A City That's Trying to Stay Unspoiled." *Christian Science Monitor*, January 29.

Perry, David A., and Michael P. Amaranthus. 1997. "Disturbance, Recovery, and Stability." In Kathryn A. Kohm and Jerry Franklin, eds., *Creating a Forestry for the 21st Century* (Washington, DC: Island Press), 31–56.

Perry, Tony. 1999. "Plan Designed to Bring Delta Foes Together Inflames Debate Instead." *Los Angeles Times*, September 16.

———. 2000. "Joint Effort to Save Sacramento–San Joaquin Delta Strikes Snags." *Los Angeles Times*, October 1.

Perry, Tony, and Frank Clifford. 1998. "Water War Nearly Over, Babbitt Says." *Los Angeles Times*, December 18.

Peterson, M. Nils, Markus J. Peterson, and Tarla Rai Peterson. 2005. "Conservation and the Myth of Consensus." *Conservation Biology* 19(3): 762–767.

Peterson, Paul. 1981. *City Limits* (Chicago: University of Chicago Press).

Petit, Charles. 1991. "EPA Vetoes Key Part of State Water Plan." *San Francisco Chronicle*, September 4.

———. 1995. "Just Add Water." *San Francisco Chronicle*, November 26.

Pettigrew, Richard. 1995. Cover letter to the GCSSF's *Initial Report 1995*. October 1. (On file with author.)

———. 2003. Personal communication.

Pew Oceans Commission. 2003. *America's Living Oceans: Charting a Course for Sea Change* (Arlington, VA: Pew Oceans Commission). Available at www.pewoceans.org.

Pickett, S. T. A., and Richard S. Ostfeld. 1995. "The Shifting Paradigm in Ecology." In Richard L. Knight and Sarah F. Bates, eds., *A New Century for Natural Resources Management* (Washington, DC: Island Press), 261–278.

Pickett, S. T. A., V. T. Parker, and P. L. Fiedler. 1992. "The New Paradigm in Ecology: Implications for Conservation Biology Above the Species Level." In P. L. Fiedler and S. K. Jain, eds., *Conservation Biology* (New York: Chapman and Hall).

Pickett, S. T. A., and P. S. White, eds. 1985. *The Ecology of Natural Disturbance and Patch Dynamics* (New York: Academic Press).

Pittman, Craig. 1999a. "Bush Pledges Support for Everglades Restoration." *St. Petersburg Times*, January 22.

———. 1999b. "Everglades Water Storage Plan Full of Unknowns." *St. Petersburg Times*, June 30.

Pollak, Daniel. 2001. *The Future of Habitat Conservation? The NCCP Experience in Southern California* (Sacramento: California Research Bureau).

Poole, B. 2007. "Drought, Increasing Costs Threaten Older Ranchers' Traditions." *Tucson Citizen*, June 2.

Poole, Deana, and Hector Florin. 2006. "Officials Begin to Ponder Plans for Mecca Land." *Palm Beach Post*, May 25.

Popper, Frank J. 1981. *The Politics of Land-Use Reform* (Madison: University of Wisconsin Press).

Postel, Sandra, and Brian Richter. 2003. *Rivers for Life: Managing Water for People and Nature* (Washington, DC: Island Press).

Pralle, Sarah B. 2006. *Branching Out, Digging In: Environmental Advocacy and Agenda Setting* (Washington, DC: Georgetown University Press).

President's Council on Sustainable Development (PCSD). 1996. *Sustainable America: New Consensus for Prosperity, Opportunity, and a Healthy Environment for the Future* (Washington, DC: U.S. Government Printing Office).

Rabe, Barry G. 1999. "Sustainability in a Regional Context: The Case of the Great Lakes Basin." In Daniel A. Mazmanian and Michael E. Kraft, eds., *Toward Sustainable Communities: Transition and Transformations in Environmental Policy* (Cambridge, MA: MIT Press), 247–281.

Raymond, Leigh. 2006. "Cooperation Without Trust: Overcoming Collective Action Barriers to Endangered Species Protection." *Policy Studies Journal* 34(1): 37–57.

Rees, William E. 2000. "Patch Disturbance, Ecofootprints, and Biological Integrity: Revisiting the Limits to Growth (or Why Industrial Society Is Inherently Unstable)." In David Pimentel, Laura Westra, and Reed F. Noss, eds., *Ecological Integrity: Integrating Environment, Conservation, and Health* (Washington, DC: Island Press), 139–156.

Reese, April. 2005. "Tucson Area Nears Completion of Ambitious Conservation Plan." *Arizona Daily Star*, April 14.

Reis, Greg. 1998. "Lakewatch." *Mono Lake Newsletter*, Fall.

———. 1999. "Lakewatch." *Mono Lake Newsletter*, Winter.

———. 2000. "Lakewatch: Average Runoff Years Haven't Led to Lake Rise." *Mono Lake Newsletter*, Fall.

———. 2007. "The Streams Wait Patiently as We Learn," *Mono Lake Newsletter*, Winter.

Rheem, Donald L. 1987. "Putting Bends Back in a River to Help the Everglades." *Christian Science Monitor*, January 26.

Rice, Terry. 2001. Everglades Oral History, South Florida Oral History Consortium. March 8.

Rieke, Elizabeth Ann. 1996. "The Bay–Delta Accord: A Stride Toward Sustainability." *University of Colorado Law Review* 67: 341–369.

Ring, Dick. 2002. Everglades Oral History, South Florida Oral History Consortium. May 17.

Ritter, John. 1998. "A Plan for Cease-fire in California Water Wars." *USA Today*, December 21.

Rivera, Jorge, Peter de Leon, and Charles Koerber. 2006. "Is Greener Whiter Yet? The Sustainable Slopes Program After Five Years." *Policy Studies Journal* 34(2): 195–221.

Robitaille, Stephen. 2003. "Water War Talk." *California Journal*, August 1.

Roderick, Kevin. 1989. "Selling a Lake." *Los Angeles Times*, September 24.

———. 1991. "Rationing, Slow Growth Favored to Offset Drought." *Los Angeles Times*, January 31.

Rodgers, Terry. 1991. "Catching Gnats—and Flak." *San Diego Union–Tribune*, May 4.

Rolfe, Allison. 2000. "Mapping the MSCP Process: Habitat Conservation Planning in the San Diego Region." Unpublished master's thesis (San Diego State University).

———. 2003. Personal communication.

Romney, Lee. 2002. "The State Highway Proposal Hits Mono Lake Roadblock." *Los Angeles Times*, August 4.

Rosenkrans, Spreck, and Ann H. Hayden. 2005. *Finding the Water: New Water Supply Opportunities to Revive the San Francisco Bay–Delta Ecosystem* (New York: Environmental Defense).

Ruhl, J. B. 1995. "Biodiversity Conservation and the Ever-Expanding Web of Federal Laws Regulating Nonfederal Lands: Time for Something Completely Different?" *University of Colorado Law Review* 66: 555–673.

Sabel, Charles, Archon Fung, and Bradley Karkkainen. 2000. *Beyond Backyard Environmentalism* (Boston: Beacon Press).

Salt, Terrence "Rock." 2003. Personal communication.

Santaniello, Neil. 2000. "Bringing a River Back to Life." *Fort Lauderdale Sun–Sentinel*, June 25.

———. 2005. "Some South Florida Scientists Claim They Were Fired for Being Outspoken." *South Florida Sun–Sentinel*, June 19.

Savitz, Jacqueline. 1999. "Compensating Citizens." *Boston Review* 24(5): 20–21.

Sax, Joseph L. 2000. "Environmental Law at the Turn of the Century: A Repertorial Fragment of Contemporary History." *California Law Review* 88(6): 2375, 2377–2402.

Scheberle, Denise. 2004. "Devolution." In Robert F. Durant, Daniel J. Fiorino, and Rosemary O'Leary, eds., *Environmental Governance Reconsidered: Challenges, Choices, and Opportunities* (Cambridge, MA: MIT Press), 361–392.

Schneiders, Robert Kelley. 1999. *Unruly River: Two Centuries of Change Along the Missouri* (Lawrence: University Press of Kansas).

Schrader-Freschette, Kristin. 1996. "Throwing Out the Bathwater of Positivism, Keeping the Baby of Objectivity: Relativism and Advocacy in Conservation Biology." *Conservation Biology* 10(3): 912–914.

Schulman, Alexis. 2007. "Bridging the Divide: Increasing Local Ecological Knowledge in U.S. Natural Resource Management." Unpublished master's thesis (Cambridge, MA: Department of Urban Studies and Planning, MIT).

Science and Technical Advisory Team (STAT). n.d. "Goal and Objectives for the Biological Element of the Sonoran Desert Conservation Plan." Available at www.pima.gov/cmo/sdcp/STAT/HabitatGoal.htm.

Science Coordination Team (SCT). 2003. "The Role of Flow in the Everglades Ridge and Slough Landscape." South Florida Ecosystem Restoration Working Group. January 14.

Scientific Review Panel (SRP). 1993. *Southern California Coastal Sage Scrub NCCP Conservation Guidelines* (Sacramento: California Department of Fish & Game and California Resources Agency).

Scott, James C. 1998. *Seeing like a State: How Certain Schemes to Improve the Human Condition Have Failed* (New Haven, CT: Yale University Press).

Scott, J. Michael, Elliott A. Norse, Hector Arita, Andy Dobson, James A. Estes, Mercedes Foster, Barrie Gilbert, Deborah B. Jensen, Richard L. Knight, David Mattson, and Michael E. Soulé. 1999. "The Issue of Scale in Selecting and Designing Biological Reserves." In Michael E. Soulé and John Terborgh, eds., *Continental Conservation: Scientific Foundations of Regional Reserve Networks* (Washington, DC: Island Press), 19–37.

Sexton, Chuck. 2003. Personal communication.

Shabecoff, Philip. 1986. "Program Aims to Rescue Everglades from 100 Years of the Hand of Man." *New York Times*, January 20.

Shen, Hsieh Wen, Guillermo Tabios III, and James A. Harder. 1994. "Kissimmee River Restoration Study." *Journal of Water Resources Planning and Management* 120(3): 330–349.

Sheridan, Tom. 2006. Personal communication.

Showley, Roger M. 2005. "End of the Line?" *San Diego Union–Tribune*, January 9.

Shutkin, William A. 2000. *The Land That Could Be: Environmentalism and Democracy in the Twenty-first Century* (Cambridge, MA: MIT Press).

Silva, Mark, and Heather Dewar. 1990. "Martinez Picks Ambitious Plan to Restore Kissimmee River." *Miami Herald*, January 17.

Silver, Dan. 1993a. Comments on working draft MSCP Plan. January 10. (On file with author.)

———. 1993b. Letter to Karen Scarborough. August 2. (On file with author.)

———. 1995. Letter to Keith Greer re draft biology guidelines. November 8. (On file with author.)

———. 2003. Personal communication.

Silvern, Drew. 1991a. "Protected Status for Gnatcatcher Turned Down." *San Diego Union–Tribune*, August 31.

———. 1991b. "Bird Loses More Land While Fate Is Debated." *San Diego Union–Tribune*, October 13.

Sloan, Gene. 1994. "Bringing Back Kissimmee Marshes and Wildlife." *USA Today*, April 22.

Slocum, Ken. 1987. "Return of the River." *Wall Street Journal*, September 8.

Sneider, Daniel. 1996. "Mono Lake's Resurrection Is a Model for Watershed Battles." *Christian Science Monitor*, August 1.

Snow, Donald. 2001. "Coming Home: An Introduction to Collaborative Conservation." In Philip Brick, Donald Snow, and Sarah Van de Wetering, eds., *Across the Great Divide: Explorations in Collaborative Conservation and the American West* (Washington, DC: Island Press), 1–11.

Snow, Lester. 2005. Personal communication.

Society of American Foresters. 1993. *Sustaining Long-term Forest Health and Productivity* (Bethesda, MD: Society of American Foresters).

Soulé, Michael E. 1985. "What Is Conservation Biology." *Bioscience* 35(11): 727–734.

South Florida Ecosystem Restoration Task Force (SFERTF). 2000. "Appendix E: Integrated Science Plan." In *Coordinating Success: Strategy for Restoration of the South Florida Ecosystem*, vol. 2. July 31. (On file with author.)

South Florida Water Management District (SFWMD). 2004. *2004 South Florida Environmental Report*, Executive Summary. August 19.

———. 2006. *2006 South Florida Environmental Report*, Executive Summary.

———. 2007. *2007 South Florida Environmental Report*, Executive Summary.

Southern California Natural Communities Conservation Planning Scientific Review Panel (SRP). 1993. "Scientific Review Panel Conservation Guidelines and Documentation" (California Resources Agency, California Department of Fish and Game, August). (On file with author.)

Spencer, Wayne. 2003. Personal communication.

Spivy-Weber, Frances. 1998. "Mono Basin Air Quality Still an Issue." *Mono Lake Newsletter*, Fall.

Stahl, Andy. 2001. "Ownership, Accountability, and Collaboration." In Philip Brick, Donald Snow, and Sarah Van de Wetering, eds., *Across the Great Divide* (Washington, DC: Island Press), 194–199.

Stallcup, Jerre. 2004. Personal communication.

————. 2006. Personal communication.

Stankey, George H., Bernard T. Bormann, Clare Ryan, Bruce Shindler, Victoria Sturtevant, Roger N. Clark, and Charles Philpot. 2003. "Adaptive Management and the Northwest Forest Plan." *Journal of Forestry* 101(1): 40–46.

Stanley, Thomas R., Jr. 1995. "Ecosystem Management and the Arrogance of Humanism." *Conservation Biology* 9(3): 255–262.

Stanush, Michele. 1990a. "Developers Fear Impact of Species Protection." *Austin American–Statesman*, March 8.

————. 1990b. "Environment Protection Plan Called Boon to Austin." *Austin American–Statesman*, March 9.

State Water Resources Control Board (SWRCB). 1993. *Draft Environmental Impact Report for the Review of the Mono Basin Water Rights of the City of Los Angeles* (Sacramento, CA: State Water Resources Control Board).

————. 1994a. *Final Environmental Impact Report for the Review of the Mono Basin Water Rights of the City of Los Angeles* (Sacramento, CA: State Water Resources Control Board).

————. 1994b. Mono Lake Basin Water Rights Decision 1631, *Decision and Order Amending Water Right Licenses to Establish Fishery Protection Flows in Streams Tributary to Mono Lake and to Protect Public Trust Resources at Mono Lake and in the Mono Lake Basin*. September 28 (Sacramento, CA: State Water Resources Control Board).

————. 1998. State Water Rights Order 98-05 (Sacramento, CA: State Water Resources Control Board).

State of Florida, Office of the Governor. 1994. Executive Order 94-54, Governor's Commission for a sustainable South Florida. (On file with author.)

Stegner, Page. 1981. "Water and Power." *Harper's*, March, 61–70.

Steinhart, Peter. 1980. "The City and the Inland Sea." *Audubon* 82(5): 98–125.

Steinzor, Rena I. 2000. "The Corruption of Civic Environmentalism." *Environmental Law Reporter* 30: 10909–10921.

Stewart, Robert W. 1988. "75% Cut in Water Diversion by L.A. Urged to Protect Mono Lake Basin." *Los Angeles Times*, September 21.

Stiffler, Lisa. 2005. " 'Political Realities' Helped Shape Urban Preserve." *Seattle Post–Intelligencer*, May 5.

Stone, Deborah. 2003. *Policy Paradox: The Art of Political Decision Making*, rev. ed. (New York: W.W. Norton).

Strahl, Stuart. 2002. Personal communication.

Susskind, Lawrence. 2005. "Resource Planning, Dispute Resolution, and Adaptive Governance." In John T. Scholz and Bruce Stiftel, eds., *Adaptive Governance and Water Conflict: New Institutions for Collaborative Planning* (Washington, DC: Resources for the Future), 141–149.

Susskind, Lawrence, and Jeffrey Cruikshank. 1987. *Breaking the Impasse: Consensual Approaches to Resolving Public Disputes* (New York: Basic Books).

Susskind, Lawrence, Sarah McKearnan, and Jennifer Thomas-Larmer, eds. 1999. *The Consensus Building Handbook* (Thousand Oaks, CA: Sage).

Swanson, Tina. 2006. Personal communication.

Switzer, Jacqueline Vaughn. 1997. *Green Backlash: The History and Politics of Environmental Opposition in the United States* (Boulder, CO: Lynne Rienner).

Szaro, Robert C., William T. Sexton, and Charles R. Malone. 1998. "The Emergence of Ecosystem Management as a Tool for Meeting People's Needs and Sustaining Ecosystems." *Landscape and Urban Planning* 40(1–3): 1–7.

Tarlock, A. Dan. 1996. "Environmental Law: Ethics or Science." *Duke Environmental Law & Policy Forum* 7: 193–223.

Taugher, Mike. 2005a. "Water Planning Thrown into Turmoil." *Contra Costa Times*, October 15.

———. 2005b. "Pumps, Clams Eyed as Culprits in Delta Species Decline." *Contra Costa Times*, November 15.

———. 2006a. "Federal Guidelines for Delta Plan Have Scientific Flaws, Report Says." *Contra Costa Times*, January 7.

———. 2006b. "Policy Threatens to Eclipse Science on Delta, Miller Says." *Contra Costa Times*, February 28.

———. 2006c. "Salinity May Be Greatest Threat to Dwindling Delta Smelt." *Contra Costa Times*, April 23.

———. 2006d. "Permit on Delta Draws Dispute." *Contra Costa Times*, May 23.

———. 2006e. "Scientists Say Timing Affects Fish," *Contra Costa Times*, October 25.

———. 2007a. "California's Changing Climate." *Contra Costa Times*, January 22.

———. 2007b. "Plan to Get More out of Delta Denied." *Contra Costa Times*, January 31.

———. 2007c. "Delta Woes May Whet Water War." *Contra Costa Times*, February 26.

———. 2007d. "Water Agency Put on Notice." *Contra Costa Times*, March 24.

———. 2007e. "Lawsuit Targets Delta Pumps." *Contra Costa Times,* June 20.

———. 2007f. "Judge Rejects Plea to Halt Pumping." *Contra Costa Times*, June 23.

———. 2007g. "Delta Water Supply Slashed." *Contra Costa Times*, September 1.

Taylor, Ronald B. 1986. "Mono Lake Group Wins Round, Slows Diversion of Creek." *Los Angeles Times*, August 15.

Tewdwr-Jones, M., and P. Allmendinger. 1998. "Deconstructing Communicative Rationality: A Critique of Habermasian Collaborative Planning." *Environment and Planning A* 30: 1975–1989.

Thomas, Craig. 2003. *Bureaucratic Landscapes: Interagency Cooperation and the Preservation of Biodiversity* (Cambridge, MA: MIT Press).

Thomas, Jack Ward. 1996. "Forest Service Perspective on Ecosystem Management." *Ecological Applications* 6(3): 703–705.

Thompson, Don. 2004. "Mono Lake Scenic Area May Face Development." Associated Press, November 16.

———. 2006. "Pumps May Fuel Delta Decline," Associated Press, June 3.

Thompson, Paul. 1996. "Pragmatism and Policy: The Case of Water." In Andrew Light and Eric Katz, eds., *Environmental Pragmatism* (New York: Routledge).

Tobin, Mitch. 2001a. "Traffic Woes Top Issue, Poll Finds." *Arizona Daily Star*, June 24.

———. 2001b. "Pima Officials Seeking New Conservation Rules." *Arizona Daily Star*, October 31.

———. 2002. "Endangered Species—Or Land Grab?" *Arizona Daily Star*, March 24.

Tonn, Bruce, Mary English, and Robert Turner. 2006. "The Future of Bioregions and Bioregional Planning." *Futures* 38: 379–381.

Toth, Louis A. 1990. "Impacts of Channelization on the Kissimmee River Ecosystem." In M. Kent Loftin, Louis A. Toth, and Jayantha T. B. Obeysekera, eds., *Proceedings of the 1988 Kissimmee River Restoration Symposium* (West Palm Beach: South Florida Water Management District), 47–56.

———. 1995. "Principles and Guidelines for Restoration of River/Floodplain Ecosystems—Kissimmee River, Florida." In John Cairns, Jr., ed., *Rehabilitating Damaged Ecosystems* (Boca Raton, FL: Lewis Publishers), 49–73.

———. 2005. "A Process for Selecting Indicators for Estuarine Projects with Broad Ecological Goals." In Stephen A. Bortone, ed., *Estuarine Indicators* (New York: CRC Press), 437–449.

———. 2006. Personal communication.

Toth, Louis A., D. Albrey Arrington, and Glenn Begue. 1997. "Headwaters Restoration and Reestablishment of Natural Flow Regimes: Kissimmee River of Florida." In J. E. Williams, C. A. Wood, and M. P. Dombeck, eds., *Watershed Restoration Principles and Practices* (Bethesda, MD: American Fisheries Society), 425–442.

Toth, Louis A., and Nicholas G. Aumen. 1994. "Integrating Multiple Issues in Environmental Restoration and Resource Enhancement Projects in Southcentral Florida." In John Cairns, Jr., Todd V. Crawford, and Hal Salwasser, eds., *Implementing Integrated Environmental Management* (Blacksburg: Virginia Polytechnic Institute and State University), 61–78.

Turner, John, and Jason Rylander. 1997. "Land Use: The Forgotten Agenda." In Marian R. Chertow and Daniel C. Esty, eds., *Thinking Ecologically* (New Haven, CT: Yale University Press), 60–75.

Union of Concerned Scientists (UCS). 2006. "Global Warming: Mississippi Delta." Available at http://www.ucsusa.org/gulf/gcplacesmis.html.

University of California San Diego TV (UCSD TV). 1996. "Crosscurrents." Documentary. (On file with author.)

U.S. Army Corps of Engineers (USACE). 1985. *Central and Southern Florida, Kissimmee River, Florida: Final Feasibility Report and Environmental Impact Statement* (Jacksonville, FL: U.S. Army Corps of Engineers).

———. 1992. *Final Integrated Feasibility Report and Environmental Impact Statement: Environmental Restoration, Kissimmee River, Florida* (Jacksonville, FL: U.S. Army Corps of Engineers).

———. 1994. *Central and Southern Florida Project, Comprehensive Review Study: Reconnaissance Report* (Jacksonville, FL: U.S. Army Corps of Engineers).

U.S. Army Corps of Engineers and the South Florida Water Management District (USACE & SFWMD). 1999. *Central and Southern Florida Project Comprehensive Review Study: Final Integrated Feasibility Report and Programmatic Environmental Impact Statement* (Jacksonville, FL: Jacksonville District, USACE).

———. 2006. *Central and Southern Florida Project Comprehensive Everglades Restoration Plan: 2005 Report to Congress.* Available at http://www. evergladesplan.org/pm/program_docs/cerp_report_congress_2005.aspx.

U.S. Commission on Ocean Policy. 2004. *An Ocean Blueprint for the 21st Century* (Washington, DC: U.S. Commission on Ocean Policy). Available at http://www.oceancommission.gov.

U.S. Environmental Protection Agency (USEPA). 1992. "In Florida, Corps of Engineers' Kissimmee River Restoration Aims to Return to Pre-channelization Environmental Conditions." *Nonpoint Source News–Notes* 18 (January–February).

———. 1994. *The New Generation of Environmental Protection* (Washington, DC: U.S. EPA).

———. 1999. "EPA's Framework for Community-Based Environmental Protection." EPA 237-K-99-001. February.

U.S. Fish and Wildlife Service (USFWS). 1990. "Golden-Cheeked Warbler (*Dendroica chrysoparia*) Status Summary." January 11. (On file with author.)

———. 1991. "Scope of Work for the Multiple Species Conservation Program of the City of San Diego Clean Water Program." (On file with author.)

———. 1993a. "Determination of Threatened Status for the Coastal California Gnatcatcher." *Federal Register*, 50 CFR Part 17, March 30.

———. 1993b. "Special Rule Concerning Take of the Threatened Coastal California Gnatcatcher." *Federal Register*, 58 FR 16742, March 30.

U.S. Government Accountability Office (USGAO). 2007. *Restoration Is Moving Forward but Is Facing Significant Delays, Implementation Challenges, and Rising Costs.* GAO-07-520. May.

Vierebome, Peggy. 1990. "Habitat Plan Searching for a Home Among Wary Officials." *Austin American–Statesman*, June 14.

Vogel, Cathleen. n.d. "Central and Southern Florida Project Comprehensive Review Study: Road Map or Roadblock for the Future? A Case Study in Water Resource Planning in the Age of Ecosystem Management." (On file with author.)

Vogel, David. 2005. *The Market for Virtue: The Potential and Limits of Corporate Social Responsibility* (Washington, DC: Brookings Institution Press).

Vorster, Peter. 2006. Personal communication.

Vosler, Sybil. 2003. Personal communication.

Wackernagel, Mathis and William E. Rees. 1996. *Our Ecological Footprint: Reducing the Human Impact on the Earth* (British Columbia: New Society Publishers).

Wade, Malcolm (Bubba), Jr. 2001. Everglades Oral History, South Florida Oral History Consortium. April 3.

Wallace, Mary G., Hanna J. Cortner, Margaret A. Moote, and Sabrina Burke. 1996. "Moving Toward Ecosystem Management: Examining a Change in Philosophy for Resource Management." *Journal of Political Ecology* 3: 1–36.

Walters, Carl. 1997. "Challenges in Adaptive Management of Riparian and Coastal Ecosystems." *Conservation Ecology* 1(2): 1. Available at http://www.consecol.org/vol1/iss2/art1/.

Walters, Dan. 2007. "A New Jolt on Water Quandary." *Sacramento Bee*, September 9.

Warner, Kee, and Harvey Molotch. 2000. *Building Rules: How Local Controls Shape Community Environments and Economics* (Boulder, CO: Westview Press).

Wassmer, Robert W. 2006. "The Influence of Local Containment Policies and Statewide Growth Management on the Size of United States Urban Areas." *Journal of Regional Science* 46(1): 25–65.

Weaver, J., and B. Brown, eds. 1993. *Federal Objectives for the South Florida Restoration* (Miami: Science Subgroup of the South Florida Management and Coordination Working Group, November 5).

Weber, Edward P. 1998. *Pluralism by the Rules: Conflict and Cooperation in Environmental Regulation* (Washington, DC: Georgetown University Press).

———. 2000. "A New Vanguard for the Environment: Grass-Roots Ecosystem Management as a New Environmental Movement." *Society & Natural Resources* 13: 237–259.

———. 2003. *Bringing Society Back In: Grassroots Ecosystem Management, Accountability, and Sustainable Communities* (Cambridge, MA: MIT Press).

Weiser, Matt. 2005a. "A Massive Restoration Program May Have Nothing Left to Save," *High Country News*, May 30.

———. 2005b. "Anglers Unite on Delta Water," *Sacramento Bee*, July 2.

———. 2005c. "New Low for Tiny Delta Fish," *Sacramento Bee*, October 31, 2005.

Weiss, Jeffrey. 1986. "Club Steps Up Fight for River Plan." *Miami Herald*, February 18.

Werner, Erica. 2004. "House Passes $395 million CalFed Bill, Sends It to President." Associated Press, October 6.

Western Governors' Association (WGA). 1998. *Principles for Environmental Management in the West*. Resolution 98-001, February 24 (Denver).

Whalen, James E. 1992a. Letter to Jeffrey Okun re draft MSCP implementation equity issues (draft dated 9/29/92). October 14. (On file with author.)

————. 1992b. Letter to David Flesh, MSCP project manager. December 16. (On file with author.)

————. 1993a. Bio-guidelines meeting summary. January 28. (On file with author.)

————. 1993b. Letter to David Flesh. April 20. (On file with author.)

————. 1993c. "Comments regarding MSCP issue paper no. 9." June 9. (On file with author.)

————. 1995. Letter to Valerie Stallings, chair of the NRC&A Committee, re June 5 staff report/MSCP. June 13. (On file with author.)

————. 1999. "Management Review." In K. Norman Johnson et al., eds., *Bioregional Assessments* (Washington, DC: Island Press), 255–261.

Whalen, P. J., Louis A. Toth, Joseph W. Koebel, and Patricia L. Strayer. 2002. "Kissimmee River Restoration: A Case Study." *Water Science and Technology* 45: 55–62.

Wheeler, Douglas P. 1996. "An Ecosystem Approach to Species Protection." *NR&E*, Winter, 7–9.

Whitfield, Estus. 2007. Personal communication.

Wievel, Wim, Joseph Persky, and Mark Senzik. 1999. "Private Benefits and Public Costs: Policies to Address Suburban Sprawl." *Policy Studies Journal* 27(1): 96–114.

Wilkinson, Charles. 1992. *Crossing the Next Meridian* (Washington, DC: Island Press).

Williams, Gary E., David H. Anderson, Stephen G. Bousquin, Christine Carlson, David J. Colangelo, J. Lawrence Glenn, Bradley L. Jones, Joseph W. Koebel, Jr., and Jennifer Jorge. 2007. "Kissimmee River Restoration and Upper Basin Initiatives." In *2007 South Florida Environmental Report* (West Palm Beach: South Florida Water Management District).

Williams, Vince. 1990. "Management and Mis-management of the Upper Kissimmee River Basin Chain of Lakes." In M. Kent Loftin, Louis A. Toth, and Jayantha T. B. Obeysekera, eds., *Proceedings of the 1988 Kissimmee River Restoration Symposium* (West Palm Beach: South Florida Water Management District), 9–29.

Wilson, Janet. 1991. "Habitat Plan to Save Taxes, Man Says." *Austin American–Statesman*, April 11.

Wilson, Matt, and Eric Weltman. 2000. "Government's Job." *Boston Review* 24(5): 13–14.

Wondollek, Julia M., and Steven L. Yaffee. 2000. *Making Collaboration Work: Lessons from Innovation in Natural Resource Management* (Washington, DC: Island Press).

Wood, Christopher A. 1994. "Ecosystem Management: Achieving the New Land Ethic," *Renewable Resources Journal* 12(1): 6–12.

Woody, Theresa. 1993. "Grassroots in Action: The Sierra Club's Role in the Campaign to Restore the Kissimmee River." *Journal of the North American Benthological Society* 12(2): 201–205.

Worster, Donald. 1994. *Nature's Economy: A History of Ecological Ideas*, 2nd ed. (New York: Cambridge University Press).

Wright, Patrick. 2001. "Fixing the Delta: The CALFED Bay–Delta Program and Water Policy Under the Davis Administration." *Golden Gate University Law Review* 31: 331–350.

———. 2006. Personal communication.

Wullschleger, J. G., S. J. Miller, and L. J. Davis. 1990. "An Evaluation of the Effects of the Restoration Demonstration Project on the Kissimmee River Fishes." In M. Kent Loftin, Louis A. Toth, and Jayantha Obeysekera, eds., *Proceedings of the 1988 Kissimmee River Restoration Symposium* (West Palm Beach: South Florida Water Management District).

Wyman, Scott. 2006. "Broward Officials Consider Restrictions on Water as Growth Dries Up Supply." *South Florida Sun-Sentinel*, June 15.

Wynn, Susan. 2006. Personal communication.

Yaffee, Steven L., Ali F. Phillips, Irene C. Frentz, Paul Hardy, Susanne Maleki, and Barbara E. Thorpe. 1996. *Ecosystem Management in the United States: An Assessment of Current Experience* (Washington, DC: Island Press).

Yin, Ming, and Jian Sun. 2007. "The Impacts of State Growth Management Programs on Urban Sprawl in the 1990s." *Journal of Urban Affairs* 29(2): 149–179.

Young, Gordon. 1981. "The Troubled Waters of Mono Lake." *National Geographic*, October, 504–519.

Young, Samantha. 2006. "Federal Agency to Review Effects of Water Pumping on Delta Fish." Associated Press State & Local Wire, July 8.

Zakin, Susan. 2002. "Delta Blues." *High Country News*, September 30.

Zaneski, Cyril. 1999. "Army Corps Plans to Double Pace of Everglades Restoration." *Miami Herald*, March 29.

Zinn, Jeffrey A., and Claudia Copeland. 2001. "IB97014: Wetlands Issues." CRS issue brief for Congress, August 7. Available at www.ncseonline.org/NLE/CRSreports/wetlands/wet-5.cfm.

Index

American and Comparative Environmental Policy

Sheldon Kamieniecki and Michael E. Kraft, series editors